FINANCIAL LITERACY

Introduction to the Mathematics of Interest, Annuities, and Insurance

---

Second Edition

Kenneth Kaminsky

University Press of America,® Inc.
Lanham · Boulder · New York · Toronto · Plymouth, UK

University Press of America,® Inc.
4501 Forbes Boulevard
Suite 200
Lanham, Maryland 20706
UPA Acquisitions Department (301) 459-3366

Estover Road
Plymouth PL6 7PY
United Kingdom

Printed in the United States of America
British Library Cataloging in Publication Information Available

Library of Congress Control Number: 2010932652
ISBN: 978-0-7618-5309-1 (paperback : alk. paper)
eISBN: 978-0-7618-5310-7

™ The paper used in this publication meets the minimum requirements of American National Standard for Information Sciences—Permanence of Paper for Printed Library Materials, ANSI Z39.48-1992

To Amy, David and Jonathan

Table of Contents

Preface to the first edition

This book contains material sufficient for a one semester introductory course in the mathematics of compound interest and life contingencies. The topics covered include the elements of the compound interest, annuities-certain, mortality tables, life annuities, and life insurance. The mathematical background required for the book is knowledge of basic algebra, including logarithms. No previous knowledge of the mathematics of finance is assumed. However, I have included material containing somewhat more advanced topics. The sections containing this material are starred (*), and may be omitted without loss of continuity.

Most sections of the book contain exercise sets. In addition, there are several sets of miscellaneous exercises at the ends of chapters. My intention is that there be some exercises for which the student does not have a priori knowledge of what tools are going to be required. An answer section providing answers to most exercises is included. In addition, a Solutions Manual containing more detailed solutions to most exercises is available.

The first five chapters cover the main ideas of compound interest, from transactions involving a single payment, to more complex transactions with multiple payments, including payment sequences that may remain level, increase, or decrease. Chapter 6 introduces mortality tables and contains a brief discussion of the probabilities associated with such tables. Expectation of life and a brief look at the force of mortality are also included. The mathematics of compound interest from the first five chapters together with the notions of probability, expectation and mortality tables from Chapter 6 provide the background and motivation for the study of pure endowments, life annuities, and life insurance treated in Chapters 7 and 8. Chapter 9 covers reserves and gross premiums.

Following Chapter 9 is a collection of 146 review problems with solutions. A Chapter Key indicating the chapter on which each problem is based follows these problems.

I have included life tables and commutation tables that allow students to compare premiums and benefits at different interest rates, and also allow for the estimation of rates of change of these quantities.

I am deeply grateful to Amy Kaminsky, David Kaminsky, Matt Foss, Su Dorée, and Adam Roesch for reading the earlier versions of the manuscript and suggesting changes that have led to significant improvements in the presentation. I thank the many students who made suggestions and/or found errors. As the author, I claim all remaining errors as my own.

I also want to thank Pilar Crespo for generously giving the use of her apartment in Madrid where the greater part of this book was written.

Kenneth Kaminsky
August, 2003

Preface to the second edition

This is the second edition of a text we have used at Augsburg College since 2003 for our introductory Mathematics of Finance course.

I have made many changes, large and small. I have moved the section on Bonds from Chapter 4 to Chapter 3, where it is a better fit; added a brief section on Common and Preferred Stocks to Chapter 4; dropped the section on rates of change; added many exercises, and replaced the Answers section with a Solutions section. The Review Problems with Solutions that followed Chapter 9 in the first edition will now be available, with a few additional problems, as a free, downloadable pdf linked to

www.augsburg.edu/home/math/faculty/kaminsky/finlit.html.

An interactive program for projects and additional problems will also be available at the same web-site .

In the first edition, we used a trial-and-error approach to solving problems requiring numerical solutions. Since there are now many reasonably priced scientific calculators with solvers on the market, in the second edition I give equal time to these solvers, and illustrate their use on such problems. For students with calculators not having solvers, I continue to illustrate or give guidance to the trial-and-error.

I am grateful to Amy Kaminsky, Jonathan Kaminsky, and Matt Foss for reading the manuscript and suggesting many improvements. I also thank the many students who made suggestions and/or found errors in the first edition. As the author, I still claim all remaining the errors as my own.

Kenneth Kaminsky
July, 2010

Part I—Payments-Certain

Chapter 1—Variations on the Theme $A_n = P(1 + i)^n$: Some General Principles of Compound Interest

1.1 Preliminaries

When we borrow money from a bank, or from a friend, and pay a fee for the privilege, we are paying rent for the borrowed money. Similarly, when we deposit money in a bank and earn a fee, the bank is renting our money. Of course, the bank usually pays at a lower rate for our money than we do for theirs. The rent we pay, or the fee that gets paid to us, is called *interest.* In this book, the kind of interest we deal with is called *compound* interest, which means that interest is paid on the current amount owed or on deposit. If you deposit $100 and have $105 after a year, then the interest you earn in the second year will be based on $105, rather than on the original $100 amount deposited. This is called *compounding*. There will be more on this soon.

In this chapter, you will learn some of the general principles of compound interest, go through several examples illustrating the basic concepts, and review some of the mathematical tools you are going to need to know if you want to be able to figure things out for yourself.

Some of the problems you will learn how to solve require a few mathematical preliminaries. We will review a few rules from algebra that you are going to need very soon.

The meaning of a^n: If n is a positive integer, then the expression a^n is the product $\overbrace{a \cdot a \cdot \ldots \cdot a}^{n \text{ factors}}$. We call a the base, and n the *exponent* of the expression a^n. For example, $3^4 = 3 \cdot 3 \cdot 3 \cdot 3 = 81$, and $(1.03)^2 = (1.03)(1.03) = 1.0609$. In 3^4, 3 is the base, and 4 is the exponent. In $(1.03)^2$, 1.03 is the base, and 2 is the exponent.

If n is a positive integer, then a^{-n} means $1/a^n$. For example, $2^{-5} = 1/2^5 = 1/(2 \cdot 2 \cdot 2 \cdot 2 \cdot 2) = 1/32 = 0.03125$.

If n is not an integer, and a is positive, it will be enough for us to say that you can calculate a^n on your scientific calculator. However, you should note that if n is between two consecutive integers m and $m + 1$, then either $a^m < a^n < a^{m+1}$ or $a^{m+1} < a^n < a^m$. For example, suppose you want to find $(1.04)^{2.374}$. Since 2.374 is between 2 and 3, $(1.04)^{2.374}$ should be between $1.04^2 = 1.0816$ and $1.04^3 = 1.124864$. You should check on your calculator (yours should have an x^y, y^x, or a ^ key) to find that $(1.04)^{2.374} = 1.09758...$, which is certainly between 1.0816 and 1.124864. A special case of non-integer n is when n is the reciprocal of an integer. For example, $3^{1/5}$ is one way of writing the fifth *root* of 3. This means that $(3^{1/5})\cdot(3^{1/5})\cdot(3^{1/5})\cdot(3^{1/5})\cdot(3^{1/5}) = 3$. If you check your calculator, you will find $3^{1/5} = 1.2457309. . . .$ More generally, $a^{1/n}$ denotes the nth root of a. You may recall that this is also sometimes written $\sqrt[n]{a}$. The two expressions have exactly the same meaning.

Remark: When I put three dots (. . .) at the end of a number, it means that there is more of the number that I am not showing. My calculator actually shows 1.09758243309 for $(1.04)^{2.374}$, and I even got my computer to show 1.097582433092538535726258264050608067493, but there is usually not much point to writing down that much of the number. Still, it *is* important to *calculate* with as much of the number as you can. In other words, I advise you to use the full capacity of your calculator and not to round off or truncate numbers until after the final calculation. You should not need to write down the results of every step that your calculator shows you. You should try to learn to use the calculator's memory without writing down all intermediate steps.

To give another example, suppose you want to calculate $(.98)^{12.4}$. According to what I said earlier, this should be between $(.98)^{12}$ and $(.98)^{13}$. In fact, it is. You get $(.98)^{12} = .7847...$ and $(.98)^{13} = .7690...$ on your calculator. You will also find that $(.98)^{12.4} = .7784...$, which *is* between $(.98)^{12}$ and $(.98)^{13}$ as advertised: $(.98)^{13} < (.98)^{12.4} < (.98)^{12}$.

How we use the rule $(a^m)^n = a^{mn}$: Suppose you want to solve an equation like $(1 + i)^9 = 3$ for i. First, you need to reduce one side of the equation to $1 + i$. To do this, raise both sides of the equation to the power 1/9. This is the same as taking the 9th root of both sides. What you get is $\{(1 + i)^9\}^{1/9} = (1 + i)^{9(1/9)} = (1 + i)^1 = 1 + i = 3^{1/9}$. Subtracting 1 from both sides gives us $i = 3^{1/9} - 1 \approx 1.12983 - 1 = 0.12983. . . .$

Logarithms and the "Log Rule:" To what power do we need to raise 10 in order to get 100? The answer is 2, since $10^2 = 100$. To what power do we need to raise 10 in order to get 1,000? The answer is 3, since $10^3 = 1,000$. To what power do we need to raise 10 in order to get 500? It isn't 2.5, although that might be your first thought. In fact, $10^{2.5} = 316.227766...$, quite a bit short of 500. If you play around with the x^y (or y^x, or ^) key on your calculator for a while, you will find that

$10^{2.69897}$ will get you reasonably close to 500. As you will see in the next definition, 2.69897... is the *common logarithm* of 500. It is the power to which we must raise 10 to get 500.

Definition of Common Logarithm: The *common logarithm* of a positive number x, written $\log(x)$, is the power to which we must raise 10 in order to get x. That is, $\log(x)$ is defined by the equation

$$10^{\log(x)} = x. \qquad \blacksquare$$

Also from the definition, for any number y,

$$\log\left(10^{y}\right) = y,$$

because $\log\left(10^{y}\right)$ is the power we have to raise 10 to in order to get 10^{y}

So,

$$\log(100) = \log(10^{2}) = 2$$
$$\log(1{,}000) = \log(10^{3}) = 3$$
$$\log(500) = \log(10^{2.69897\ldots}) = 2.69897\ldots$$

Remark: The '$\blacksquare$' you see above is the symbol that means a definition, derivation, or example is finished.

Although logarithms obey several rules, we will be using only one of them in this book. We call it:

The Log Rule: If a is a positive number, and m is any number for which $a^{m} = b$, then

$$m = \frac{\log(b)}{\log(a)}$$

Derivation of The Log Rule: Before deriving the Log Rule, we first need to show that for $a > 0$, $\log(a^{m}) = m(\log(a))$. We know from the definition that $10^{\log(x)} = x$, so it follows that $10^{\log\left(a^{m}\right)} = a^{m}$. On the other hand, $10^{m(\log(a))} = 10^{(\log(a))m} = (10^{\log(a)})^{m} = a^{m}$. We have now shown that the two powers of 10, $10^{\log\left(a^{m}\right)}$ and $10^{m(\log(a))}$, are equal. If two powers of 10 are equal, the powers must be equal as well. Thus, $\log(a^{m}) = m\left(\log(a)\right)$. To derive the Log Rule, we start by assuming that $a^{m} = b$, and solve for m. If we take the logarithms of both sides of $a^{m} = b$, we get $\log(a^{m}) = \log(b)$. Substituting $m(\log(a))$ for $\log(a^{m})$, we get

$m(\log(a)) = \log b$. Divide both sides of this last equation by $\log(a)$ and we get the Log Rule, $m = \log(b)/\log(a)$. ▮

Any time you need to solve an equation of the form $a^m = b$ for the exponent, m, you will need to use the Log Rule. Because this is the only way that logarithms are going to arise in this book, I won't tell you any more about them, except to say that you will find a button for the logarithm (log) on your scientific calculator. You might use it now to verify that log(500) = 2.69897000434...

Remark: A few scientific calculators have an *ln* button (which indicates the *natural logarithm*), but not a *log* button, while most have both. Both *ln* and *log* are log buttons, but to different bases. Fortunately, this does not affect the Log Rule. You can check for yourself that even although $\log(a) \neq \ln(a)$ unless $a = 1$,

$$\frac{\log(b)}{\log(a)} = \frac{\ln(b)}{\ln(a)},$$

always, so your *ln* button is just as good as a *log* button throughout this book. Here are some examples illustrating The Log Rule.

Example 1.1: Solve the equation $(1.065)^n = 3$.

Solution: For $a = 1.065$, $b = 3$, and $m = n$, the Log Rule gives

$$n = \log(3)/\log(1.065) =$$

$$0.47712125472.../0.02734960777... = 17.4452686360...$$

As a check, you can verify that $(1.065)^{17.445268636} \approx 3$ on your calculator, where the symbol ≈ means 'approximately equal to'. ▮

Example 1.2: Solve the equation $(0.8013)^n = .4$.

Solution: The inputs for the Log Rule are $a = 0.8013$, $b = .4$, and $m = n$. The solution is $n = \log(.4)/\log(.8013) = -0.397940.../-0.096204... = 4.13638167562...$ As a check, you might verify that $(0.8013)^{4.13638167562} \approx 0.4$ on your calculator.

Exercises 1.1

1.1.1 Without using your calculator, between what two whole numbers must $3^{3.6}$ lie?

1.1.2 Evaluate $9^{1.5}$ without using your calculator. *Hint:* $1.5 = 3/2$.

1.1.3 Solve the equation $1.09^n = 2$ for n.

1.1.4 Solve the equation $(0.89)^n = .11$ for n.

1.2 Introduction to Compound Interest

Example 1.3: At birth, Florence got a \$100 gift that was deposited in an account paying interest at the rate of 5% compounded annually (i.e., once per year). Florence turned 93 and has just found out about her account.

a) How much did Florence's account have after one year? Two years? Three years?
b) How much does Florence have in her account now?

Solution:
a) What does it mean to say that Florence's money earns interest at the rate of 5% compounded annually? It means that at the end of one year, Florence had \$100 + \$100(.05) = \$100(1 + .05) = \$100(1.05) = \$105. The .05 and the 1.05 are very important, as you will see. 5% = .05 is the *interest rate*, and 1.05 is the *accumulation factor* for this interest rate.

original investment → \$100 + \$100(.05) ← *original investment × interest rate*
original investment → \$100·(1 + .05) ← *interest rate plus 1*
original investment → \$100·(1.05) ← *accumulation factor*
investment after one year → \$105

To find how much was in Florence's account after two years, we multiply her \$105 by 1.05 again. But, if you look at the derivation above, \$105 = \$100(1.05), so \$105(1.05) = {\$100(1.05)}(1.05) = \$100$(1.05)^2$ = \$110.25.

investment after one year → \$105·(1.05) ← *accumulation factor*
original investment → \$100·$(1.05)^2$ ← *accumulation factor squared*
investment after two years → \$110.25

The important thing to pay attention to is the $(1.05)^2$ because it points to a pattern that will emerge on how to compound Florence's money year by year. In fact, after three years, Florence had {\$100$(1.05)^2$}(1.05) = \$100$(1.05)^3$ = \$115.7625 which rounds to \$115.76.

b) Is the pattern clear? To find how much is in Florence's account today, we multiply \$100 by $(1.05)^{93}$ to get \$100(93.4554...) ≈ \$9,345.55. ▮

The more general result we can glean from this example can be described as follows: After n years, the *accumulated value*, or *future value*, A, of the *initial amount*, or *principal amount*, P, deposited for n years at *interest rate i* compounded annually, is

$$A = P(1+i)^n. \tag{1.1}$$

We call $(1+i)^n$ an *n-year accumulation factor* for the initial amount. We say that P is *accumulated* for n years. The quantity $(1+i)$ is the 1-year accumulation factor, or more simply, just the *accumulation factor.* We call i the *annual rate of interest.*

Remark: The (1.1) you see following the expression above, means that we will refer back to the expression, and similar ones, as *Formula 1.1* or *Equation 1.1*.

Remark: We will sometimes use the *subscript* or *function* notation A_n or $A(n)$ to emphasize that we are calculating a quantity A whose value depends on a variable, n. Just remember that $A_n = P(1+i)^n$ or $A(n) = P(1+i)^n$ means that A is evaluated at the value n. We will use subscript and function notation in several places later in the text. In some cases there will be more than one variable between the parentheses. For example, the notation $S(P, i, n, d)$ which appears first in Chapter 3 will mean that we need to substitute the values of P, i, n, and d into the formula given by S. This will become clear when the time comes.

In the example above, we were given P = \$100, i = 5% = .05, and various values of n, and were asked to find A_n for these various values. In the next problem, we will want to find P given A_n, i, and n.

Remark: Throughout Chapters 1 through 5 of this book, we will assume that all payments are certain to be made. That is to say, they will not depend on anyone's survival. In this sense, such payments are called *payments certain*. Things will change in Chapters 6—9 where in most cases, payments will depend on the survival of a life.

Example 1.4: Suppose that Martha can earn 7.5% per annum (i.e., annually) on her investment, and that she hopes to have \$8,000 in three years. How much does she need to start with now to accomplish this?

Solution: Here, $A = A_3 = \$8,000$ is to be the accumulated value, $i = 7.5\% = 0.075$ is the interest rate, $1 + i = 1.075$ is the accumulation factor, $n = 3$ years, and P, the initial value, is what we are looking for. Using Formula 1.1, we have $\$8,000 = P(1.075)^3$, and we need to find P. Dividing both sides of this expression by $(1.075)^3$, we get $P = \$8,000/(1.075)^3 = \$8,000(1.075)^{-3} = \$6,439.68$. In other words, if Martha were to invest \$6,439.68 now at 7.5% per year, in three years she would have \$8,000.

The result from this example can be described more generally as follows: The *present value*, or *discounted value*, P, of the future amount, A, deposited or invested for n years at interest rate i per year may be expressed

$$P = A(1+i)^{-n}. \tag{1.2}$$

We call $(1+i)^{-n}$ a *discount factor* for A, and we say that P is the *present value of A discounted n years.* We sometimes abbreviate *present value* by *PV* and *future value* by *FV*.

Example 1.5: Román has \$5,000 to invest. He wants this money to earn enough to become \$8,000 in four years. What annual interest rate does he need to get?

Solution: Copying Formula 1.1 with $P = \$5,000$, $n = 4$, and $A_4 = \$8,000$ we write

$$\$8,000 = \$5,000(1 + i)^4,$$

and we need to solve this for i. What we do is divide both sides by \$5,000 to get $(1 + i)^4 = 1.6$; next, take fourth roots of both sides to get $\{(1 + i)^4\}^{1/4} = 1 + i = (1.6)^{1/4} \approx 1.12468$; subtract 1 from both sides to get $i \approx .12468 = 12.468\%$. Román needs to get about 12.468% on his investment to turn \$5,000 into \$8,000 in four years. ∎

To generalize, what we just did was to solve Equation 1.1 for i to get

$$i = \left(A/P\right)^{1/n} - 1. \tag{1.3}$$

Derivation of Formula 1.3: Starting with Equation 1.1, $A = P(1+i)^n$, divide both sides by P to get $A/P = (1+i)^n$. Next, take the nth root of both sides to get $(A/P)^{1/n} = \{(1+i)^n\}^{1/n} = 1+i$. Finally, subtract 1 from both sides to arrive at Formula 1.3: $i = \left(A/P\right)^{1/n} - 1$.

Example 1.6: If Esther has an investment earning 9% annually, how long will it take her to double her money?

Solution: Here, the amount of Esther's investment is not specified, so we are free to call it what we will. Call it P. Actually, since each dollar is to become two dollars, we could just as well let $P = 1$. We want to find the number of years n for which

$$2P = P(1.09)^n.$$

Dividing both sides by P, we get

$$(1.09)^n = 2,$$

showing that the value of P did not matter here. This is an equation of the form $a^m = b$, with $a = 1.09$, $b = 2$ and $m = n$. The Log Rule gives $n = \log(2)/\log(1.09) \approx 8.043$. In a little over eight years at 9% annual interest, Esther's investment, whatever its size, will double in value ∎

Now we have solved Equation 1.1 for n. The generalization is

$$n = \frac{\log(A/P)}{\log(1+i)}. \tag{1.4}$$

Exercises 1.2

1.2.1 Suppose an ancestor of yours opened an account in the amount of \$4,000 on January 1, 1776 and that this account earned interest at the annual rate of 2.5%. If the account were left to grow, how much would have been in the account on January 1, 2002? *Note:* That's 226 years.

1.2.2 An acquaintance of Al's assures him that if he invests \$15,000 now in a certain new internet stock, he will see his money double in $2\frac{1}{2}$ years. What annual rate of interest is Al's acquaintance assuring him of?

1.2.3 Eddie bought a work of art whose value has increased by 50% while he has owned it. He figured out that his investment increased in value 12% annually. How long ago did Eddie buy his work of art?

1.2.4 Faramarz is wondering how he will accumulate the \$10,000 he thinks he will need as the down payment on a house four years from now. If his money will earn 6% per annum, how much would Faramarz need to start with now so that he will have his \$10,000 in four years?

1.2.5 José Luis invested \$500 in a fund that earned enough interest to accumulate to \$750 three years later.
a) What annual rate of interest was José Luis's fund earning?
b) How much will José Luis's investment be worth five years after making it?

1.2.6 Sukie has an investment that grows at a constant annual rate. Her investment was worth \$11,000 four years ago and was worth \$14,500 one year ago.
a) What is the annual rate of growth of the investment? (The annual rate of growth would be the interest rate if the investment were on deposit in a bank account.)
b) What will the investment be worth one year from now?

1.2.7 What interest rate does Sita need to get on her investment of \$6,000 in order for it to quadruple in 10 years?

1.2.8 Barbro has \$12,000 invested at 7.84% annually. She is going to need \$17,500 in 5 years. Is this investment going to do it for her?

1.2.9 Joanna does not remember the interest rate her bank is paying. She does remember making a single deposit of \$500 on July 3, 2004, and that on April 28, 2009, her balance was \$698.18. What annual rate of interest has Joanna been earning? *Hint:* Joanna's money has been in the bank for 1,760 days. Take 1 year = 365.25 days.

1.2.10 Shimon has promised to repay a loan with a single payment of \$4,130.35 in 3 years. If the interest rate for the loan was 8.5%, how much did Shimon originally borrow? That is, what is the *PV* of \$4,130.35 in 3 years at 8.5% annual interest?

1.3 Time Diagrams, Equations of Value, and Focal Dates

Equations 1.1—1.4 are equivalent to one another. This means that we can obtain any one of them from any of the others. In particular, Equations 1.1 and 1.2 show us something about how money has a *time value*. That is, the value of money, when exposed to interest, changes over time. What $A = P(1+i)^n$ tells us is the value that P becomes when exposed to interest at rate i over n years. What $P = A(1+i)^{-n}$ tells us is the value of A, *discounted* n years back, at interest rate i.

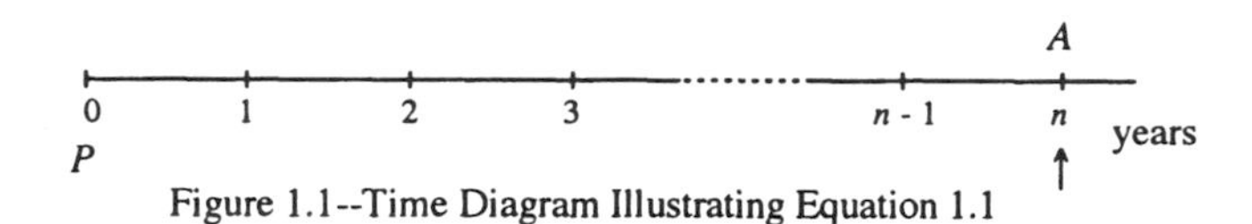

Figure 1.1--Time Diagram Illustrating Equation 1.1

Formulas 1.1 and 1.2 are examples of *equations of value*. Such equations can be illustrated with the use of *time diagrams*. The time diagram shown above in Figure 1.1, illustrates Equation 1.1. It shows the amount of money, P, deposited at time 0, its accumulated amount A after n years, and an upward arrow indicating the *focal date* (or *comparison date*): the time-point at which the *current value* of invested money is being calculated. In this case, the focal date is n years after the original deposit.

Why does an equation of value need a focal date? Consider the following three points: First, both sides of any equation are necessarily equal. Second, equations of value equate money values. Third, the value of money changes over time. Thus, in order for both sides of an equation of value to be equal, the sides must reflect the value of an investment at a single point in time—that time being the focal date. Both sides of any equation of value represent the current value of the investment at the time indicated by the focal date.

Take, as an example, a 2¢ mug of beer on July 4, 1776. That sounds cheap. But suppose the value of the dollar increased by 2.2% annually since 1776. Then, 2¢ on July 4, 1776 would be equivalent to $\$0.02(1.022)^{235} \approx \$3.33$ on July 4, 2011. Here, July 4, 2011 acts as the focal date for comparing 2¢ with $3.33 some 235 years earlier, with annual interest at 2.2%. We could also say that July 4, 1776 acts as the focal date for comparing the value of $3.33 some 235 years earlier, with interest at 2.2% annually. You can think of the mug of beer as not changing, but its current value changes as we move the focal date—it moves from 2¢ in the year 1776 to $3.33 in the year 2011, or it moves from $3.33 in the year 2011 to 2¢ in the year 1776.

So, the focal date is the one time at which we want all payments evaluated. Payments at different times will be 'brought' to the focal date by multiplying by powers of the accumulation factor, $1 + i$.

The equations of value $A = P(1+i)^n$ and $P = A(1+i)^{-n}$ each equate a single deposit to a single withdrawal from the perspective of a particular focal date. In $A = P(1+i)^n$, the focal date is n years after the initial deposit of P. The equation gives the accumulated value, or future value of P in n years. In $P = A(1+i)^{-n}$, the focal date is at the time of the initial deposit. The equation gives the present value or discounted value of A.

The time diagram below illustrates Equation 1.2. The only difference between it and the one of Figure 1.1 is that the upward arrow showing the focal date is now at time 0, signifying that we are evaluating both amounts at time 0.

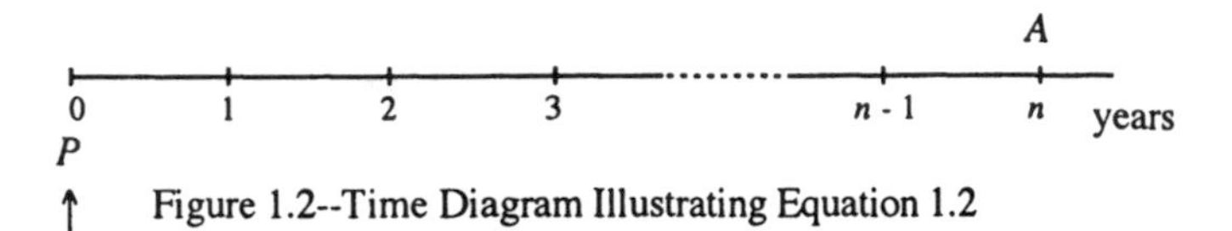

Figure 1.2--Time Diagram Illustrating Equation 1.2

Example 1.7: On July 4, 2004, Juan Carlos put \$5,000 in the bank. On July 4, 2009, the account had accumulated to \$6,534.80. How much was in Juan Carlos's account on July 4, 2007, assuming a constant rate of compound interest?

Solution: First, look at a time diagram for this problem. We have taken July 4, 2004 time 0, and added 1 for each year. Thus, July 4, 2007 corresponds to time 3, and July 4, 2009 corresponds to time 5.

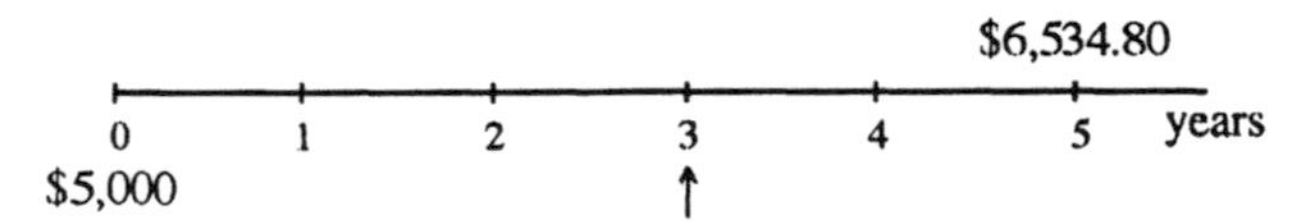

Figure 1.3--Time Diagram Illustrating Example 1.5

We have to do two things. We have to determine i from the information given, and then we have to 'bring' the value of the account forward 3 years from time 0 *or* back 2 years from time 5. From the time diagram, we see that $\$5{,}000(1 + i)^5 = \$6{,}534.80$. Solving for i, we apply Equation 1.3, to get:

$$i = (6{,}534.80/5{,}000)^{1/5} - 1 \approx .055 = 5.5\%$$

almost exactly. What we want is the amount 3 years after the deposit (or, equivalently, 2 years before July 4, 2009). This means that we can find the accumulated value of \$5,000 at 5.5% per annum 3 years after July 4, 2004, or we can find the discounted value of \$6,534.80 at 5.5% per annum 2 years before July 4, 2009. We will do it both ways. First, accumulating for 3 years, from July 4, 2004,

$$\$5{,}000(1.055)^3 \approx \$5{,}871.21.$$

Discounting for two years from July 4, 2009, we get

$$\$6{,}534.80(1.055)^{-2} \approx \$5{,}871.21.$$

Either way, this shows that Juan Carlos had \$5,871.21 in his account on July 4, 2007. ▮

Example 1.8: MaryPaul has an account earning 7.8% annually. She opened the account with \$19,149.16. By January 1, 2009 the amount in the account had doubled. When did MaryPaul open her account?

Solution: Let's call January 1, 2009 time n and work backwards to time 0, where MaryPaul opened the account. Then we will see what n turns out to be, because n will be the number of years after MaryPaul opened her account. Here is what a time diagram for this might look like (Note that n might not be a whole number.):

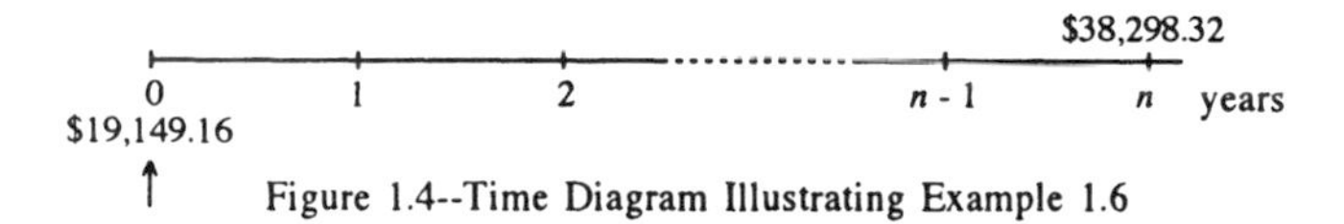

Figure 1.4--Time Diagram Illustrating Example 1.6

We are looking for n in the equation $A = P(1 + i)^n$, where $P =$ \$19,149.16, $A = 2(\$19{,}149.16) = \$38{,}298.32$, and $i = 0.078$. Thus, Equation 1.4 applies and we find

$$n = \frac{\log(38{,}298.32/19{,}149.16)}{\log(1+0.078)} = \frac{\log(2)}{\log(1.078)} \approx 9.229 \text{ years},$$

so MaryPaul opened the account 9.229 years before January 1, 2009. At an average of 365.25 days per year, this is $9.229 \cdot 365.25 \approx 3{,}371$ days. With a little work, we find that MaryPaul opened her account around October 11, 1999. ▮

Exercises 1.3

1.3.1 In your own words, explain the function of time diagrams.

1.3.2 In your own words, explain the function of equations of value.

1.3.3 In your own words, explain the function of focal dates.

1.3.4 Samantha borrowed \$250 from her sister and promised to pay her back one year later with a payment of \$265. They later agreed that because Samantha would have difficulty repaying the loan on time, she could pay the appropriate amount one year late. How much should Samantha pay two years after borrowing the money? Make a time diagram.

1.3.5 When Joel borrowed $3,000 from his in-laws, he promised to repay them with a single payment of $3,800 in three years. But Joel got a big raise and is ready to repay the loan in two years. What amount is appropriate for him to pay his in-laws? Make a time diagram.

1.3.6 Freddie put $10,000 in the bank 5 years ago today. Five years from today, his balance will reach $21,560.19. Assuming compound interest, how much does Freddie have in his account today? Make a time diagram.

1.3.7 After waiting for a certain period (call it n years) since depositing some money, Sarita has a balance of $12,094.47. If she waits for another n years, her balance will be $18,298.76. Assuming compound interest, how much did Sarita deposit? Make a time diagram.

1.3.8 Richa borrowed a sum from Steffan and was to repay the loan with a payment of $4,000 in 4 years. Instead, she will repay the loan early with a payment of $3,200 in 3 years. Assuming compound interest, how much did Richa borrow? Make a time diagram.

1.3.9 Doug made a deposit six years ago. If he waits another four years, his money will have doubled. Assume the rate of compound interest to be constant over the entire ten-year period indicated. Make time diagrams and write equations of value with:

a) the focal date at six years ago;
b) the focal date at today;
c) the focal date at four years from now.
d) Show that the three equations of value are equivalent, and solve any one of them for i.

1.3.10 Diana loaned $2,000 to her friend Edwina agreeing that Edwina would repay the loan with a single payment of $2,140 in two years. What annual rate of compound interest was Diana charging Edwina?

1.3.11 How long would it take for your money to double if it earned 6% annually?

1.4 Depreciation

Until now, we have looked at assets that *appreciate*, or increase in value over time. As you know, many types of assets *depreciate*, or lose value over time. You may have an asset that loses value every year. You probably do if you own a late model car. Unless the car is some kind of collector's item, it loses part of its value every year. Farm machinery and machines for industry typically lose part of their value every year. Perhaps the most common example of a depreciating asset, besides the car, is the home computer.

We will first take a look at a particular situation in which an asset loses a constant percentage of its value every year.

Example 1.9: Sloss just bought a fancy laptop for $3,499. Suppose that at each year's end, it will lose 40% of the value it had at the beginning of the year.

a) Make a table tracing the value of the computer to the end of five years.

b) Find out how many years it would take for Sloss's computer to be worth $100.

Solution:

a) We will begin with the table and then go through a few of the calculations.

Year	*Depreciated Value*
0	$3,499.00
1	$2,099.40
2	$1,259.64
3	$755.78
4	$453.47
5	$272.08

Table 1.1—Depreciation of Sloss's Laptop

At the end of the first year, Sloss's computer is worth $3,499 less 40% of $3,499, or $3,499(1 - .4) = $3,499(.6) = $2,099.40. At the end of two years, the computer is worth $\$3{,}499(1-.4)^2$ =

$3,499(.6)^2$= $1,259.64. In fact, after n years, Sloss's computer has a value of $\$3{,}499(1-.4)^n = \$3{,}499(.6)^n$.

b) To find out when Sloss's computer will be worth $100, we set

$$\$3{,}499\,(.6)^n = \$100.$$

Dividing both sides by $3,499, we see that we need to solve $(.6)^n = 100/3{,}499$. This is a Log Rule problem with $a = .6$, $b = 100/3{,}499 = .02857\ldots$, and $m = n$. We find

$$n = \frac{\log(.02857\ldots)}{\log(.6)} = 6.959\ldots \text{ years,}$$

so, after about 6 years, 11.5 months, Sloss's computer is worth $100. ∎

We call the percentage loss per year of an asset the *rate of depreciation*, and we denote it with a δ (Greek letter 'delta'). Thus, if an asset has an initial worth of B, then after n years, its worth, D, may be calculated by the formula

$$D = B(1 - \delta)^n. \tag{1.5}$$

Example 1.10: Ronna bought a tractor for $225,000. After 10 years, it was valued at $10,000. Find the annual rate of depreciation, assuming the rate to be constant.

Solution: We are looking for the rate of depreciation δ satisfying

$$\$225{,}000(1 - \delta)^{10} = \$10{,}000.$$

Dividing both sides by $225,000, we get

$$(1 - \delta)^{10} = 10{,}000/225{,}000 = 2/45.$$

This is similar to what we did in Equation 1.3. We raise both sides of the last equation to the power 1/10 (that is, we take 10th roots) to get

$$\{(1-\delta)^{10}\}^{1/10} = 1 - \delta = (2/45)^{1/10} = 0.732456\ldots.$$

Solving this for δ, we get

$$\delta = 1 - 0.732456\ldots = 0.267543\ldots$$

This means that Ronna's tractor loses about 26.75% of its value every year, so the annual rate of depreciation is about 26.75%. ∎

How do we handle accumulation and depreciation when the interest rates and percentages of depreciation are not constant? The next two examples present some possible scenarios.

Example 1.11: Lennart's computer depreciated by 50% the first year, by 40% of its remaining value the second year, by 32% of its remaining value the third year, by 26% of its remaining value the fourth year, and by 21% of its remaining value the fifth year. Show the calculations following the value of Lennart's computer over 5 years and put the results in a table. Assume that the computer's original cost was $3,500.

Solution: The calculations go as follows:

$3,500(.5) = $1,750 after one year,
$3,500(.5)(.6) = $1,050 after 2 years,
$3,500(.5)(.6)(.68) = $714 after 3 years,
$3,500(.5)(.6)(.68)(.74) = $528.36 after 4 years, and
$3,500(.5)(.6)(.68)(.74)(.79) ≈ $417.40 after 5 years.

In tabular form:

Year	*Depreciated Value*
0	$3,500.00
1	$1,750.00
2	$1,050.00
3	$714.00
4	$528.36
5	$417.40

Table 1.2—Lennart's Computer ∎

Example 1.12: Tony and Carol made an investment of $15,000 five years ago. The investment grew by 10% in the first year, by 12% in the second year, by 9% in the third year, by 6% in the fourth year, and by 15% in the fifth year. Show the calculations following the value of Tony and Carol's investment over 5 years and put the results in a table.

Solution: Here are the calculations:

$15,000(1.10) = $16,500 after one year,
$15,000(1.10)(1.12) = $18,480 after 2 years,
$15,000(1.10)(1.12)(1.09) = $20,143.20 after 3 years,
$15,000(1.10)(1.12)(1.09)(1.06) ≈ $21,351.79 after 4 years, and
$15,000(1.10)(1.12)(1.09)(1.06)(1.15) ≈ $24,554.56 after 5 years.

In tabular form, you have:

Year	Appreciated Value
0	$15,000.00
1	$16,500.00
2	$18,480.00
3	$20,143.20
4	$21,351.79
5	$24,554.56

Table 1.3—Appreciation of Tony & Carol's Investment

∎

Exercises 1.4

1.4.1 Linda's laptop was worth $2,499 when it was new. Now, three years later, it is worth $945. At what rate has Linda's laptop depreciated, assuming a constant annual rate of depreciation?

1.4.2 Samar bought a new moped for $3,040. If it loses 35% of its value each year, when will it be worth $750?

1.4.3 Anibal bought a recumbent bike for $1,499. It lost 32% of its value by the end of the first year, 16% of that value in the second year, 8% of that value in the third year, and 4% of that value in the fourth year. How much is the bike worth after four years? *Hint:* See Examples 1.11 and 1.12.

1.4.4 A certain powerful super-computer costs $134,500 when it is new. At the end of each year, the owner deducts a certain fixed percentage of its value for depreciation. If the scrap value is $10,400 in 5 years, what is its value at the end of 3 years?

1.4.5 A large population has been decreasing in size by the same percentage each year. Ten years ago the size of the population was 21,139,656. Four years ago, its size was 12,171,974.

a) What is the current size of the population?

b) What is the annual rate of decrease of the size of the population?

1.4.6 Given the depreciation Formula 1.5, $D = B(1 - \delta)^n$, find formulas for each of the other three variables, B, δ, and n in turn.

1.4.7 Terry's deluxe table saw cost \$985 when new and should have a working life of 10 years. Its scrap value will be \$150 at the end of its working life.
a) What is the annual rate of depreciation?
b) What will the saw be worth at the end of 5 years?

1.4.8 Lasse's Espresso machine cost him \$345 new. If it loses 40% of its value by the end of every year, what will its value be at the end of 4 years?

1.4.9 A car costing \$31,280 loses 15% of its value each year. What will the value of the car be when its extended warranty runs out at the end of 6 years?

1.4.10 A certain laser printer costing \$2,345 loses 20% of its value each year. Make a table containing its value at the end of each of the first 5 years.

1.5 Summary of Formulas

Log Rule

$$a^m = b \Rightarrow m = \frac{\log(b)}{\log(a)}$$

The following four formulas are equivalent

$$\begin{aligned} &i) \quad A = P(1+i)^n \\ &ii) \quad P = A(1+i)^{-n} \\ &iii) \quad i = \left(A/P\right)^{1/n} - 1 \\ &iv) \quad n = \frac{\log(A/P)}{\log(1+i)} \end{aligned}$$

Depreciation

$$D = B(1-\delta)^n$$

Miscellaneous Exercises

1M1. Amjab's motor scooter cost 1,590€ (€ = euro) new. Two years later, it was valued at 950€ Assuming a constant annual rate of depreciation, how old will the scooter be when you should be able to buy it for 500€? What was the constant annual rate of depreciation, δ?

1M2. A payment of $30,000 5 years from today will be worth $22,500 at a certain time before the 5 years. If money is valued at 8% per annum, what is that earlier time relative to today?

1M3. Connie heard through the son of an acquaintance's girlfriend's brother-in-law that there was a new investment that would double her money in six months. What annual rate of interest does that imply?

1M4. In the previous exercise, what is the payment of $30,000 in 5 years worth today?

1M5. Ewan was to repay a loan of $4,000 from Frema with a payment of $4,220 one year later. Unable to repay the loan on time, Ewan asks Frema for a one year extension. Frema agrees, with the stipulation that interest the second year be 50% higher than the rate she was earning during the first year. How much must Ewan pay at the end of two years?

Chapter 2—Mixing it Up: Nominal Rates of Interest, Multiple Deposits and Withdrawals, and More

2.1 Nominal Rates

We will now look at several more examples that illustrate the concepts contained in Equations 1.1—1.4, but we will also expand on the ideas so that we can solve somewhat more complicated problems. The first thing we need to do is to point out that the year is not the only possible interest period—an impression you might have from the examples of Chapter 1. In addition to annual compounding, money is often compounded:

Semiannually—twice per year
Quarterly—four times per year
Monthly—twelve times per year
Daily—365 times per year.

Money is sometimes even compounded *continuously*, but we will not consider *continuous compounding* in this book.

From now on, the notation i will refer to the *interest rate per period*, whatever period that may be. We will need to use the interest rate per period, i, in all of our calculations involving interest. Also, n will almost always refer to the *number of periods* in a given problem.

We'll start with an example that shows the essential elements of how compounding more than once per year works.

Example 2.1: Sam borrowed some money from Emily at the nominal annual rate of 6% compounded monthly (I will explain this just below.). He promises to repay the loan with a payment of $2,817.90 after two years. How much did Sam borrow?

Solution: What does *'the nominal annual rate of 6% compounded monthly'* mean? It means that the actual interest rate is not 6% at all, but $\frac{6\%}{12}$ = 1/2% per month. That is, *the period is the month*, and the *interest rate is* $i = 1/2\% = .005$ *per month*. In this problem, there are $n = 2 \times 12 = 24$ periods (months). Look at the following time diagram for this problem:

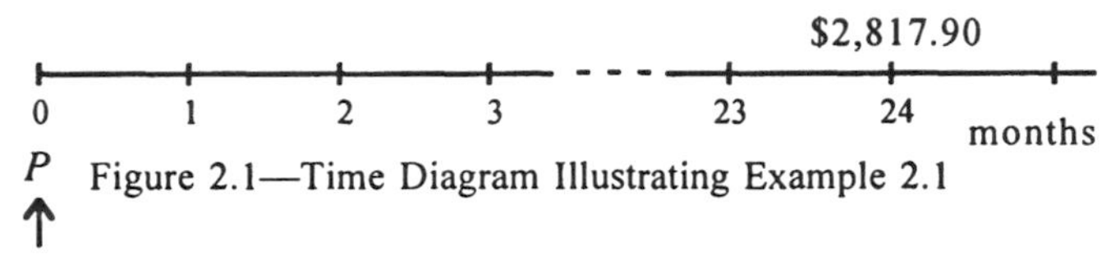

Figure 2.1—Time Diagram Illustrating Example 2.1

Notice that we are trying to find the *PV* of $2,817.90 payable in 24 months at the monthly rate of $i = .005$. This calls for Equation 1.2:

$$P = \$2{,}817.90\ (1.005)^{-24} = \$2{,}500.00.$$

So, Sam borrowed $2,500 from Emily. Note also that the focal date is at the beginning of the diagram, at time 0. ∎

Notation: We will use the special notation j_m for a *nominal annual rate of interest compounded m times per year.* In Example 2.1, we had $j_{12} = 6\% = .06$, but the rate of interest per month is $i = j_{12}/12 = 1/2\% = .005$. The thing to remember is that if the nominal annual rate of interest compounded m times per year is j_m, then the *rate of interest per period*, the real rate of interest for doing calculations, is

$$i = \frac{j_m}{m}.$$

Example 2.2: Joe wants to know to how much his $3,000 will accumulate in 2½ years if left in an account earning interest at the nominal annual rate of 8% compounded quarterly.

Solution: The transaction is illustrated on the time diagram below.

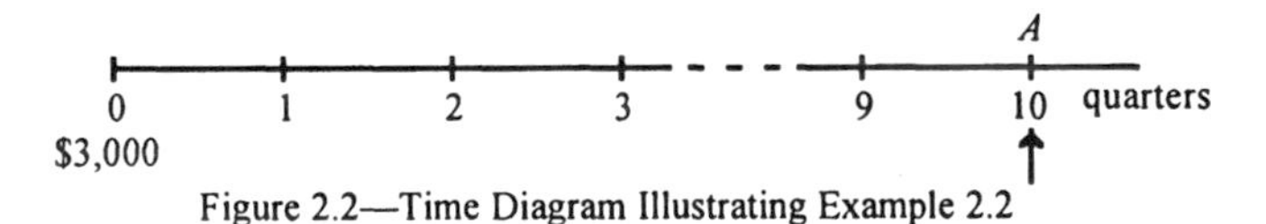

Figure 2.2—Time Diagram Illustrating Example 2.2

Here, the period is a quarter year = 3 months, and the number of quarters in 2½ years is $n = 2½\cdot 4 = 10$. The nominal annual interest rate compounded quarterly is $j_4 = 8\% = .08$, so the interest rate per quarter is $i = j_4/4 = 8\%/4 = 2\% = .02$. Applying Equation 1.1 with $P = \$3{,}000$, $i = .02$, and $n = 10$, we see that Joe will have

$$A_{10} = \$3{,}000(1 + 0.08/4)^{10} = \$3{,}000(1.02)^{10} \approx \$3{,}656.98$$

in 2½ years. Note that the focal date is at the end of the diagram, at time 10. ∎

Example 2.3: Researchers S. Jay Olshansky of the University of Illinois at Chicago School of Public Health and Steven Austad of the University of Idaho bet each other \$150 that within 150 years, the oldest living person would be at least 150 years old. Since Olshansky and Austad would probably not be around to collect, they observed that one of the two would have heirs collecting around \$500,000,000. What nominal annual interest rate, compounded monthly, were they assuming?

Solution: Olshansky and Austad bet a total of P = \$300. This will accumulate to A = \$500,000,000 in $n = 150 \times 12 = 1{,}800$ months. A time diagram illustrating this is as follows:

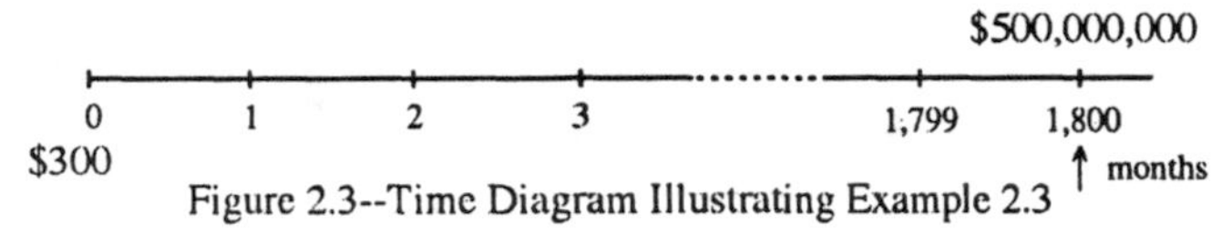

Figure 2.3--Time Diagram Illustrating Example 2.3

The equation of value is $\$300(1 + i)^{1800} = \$500{,}000{,}000$, where $i = j_{12}/12$. The solution is provided by Equation 1.3 with the period being months instead of years,

$$i = j_{12}/12 = (500{,}000{,}000/300)^{1/1800} - 1 = .00799083330\ldots,$$

so that $j_{12} = 12(.00799083330\ldots) \approx .0958899996$ or about 9.589%. ∎

Example 2.4: Barbara has an account that is earning the healthy nominal annual rate of 21% compounded semiannually. How many years will it take for the amount in her account to quadruple (i.e., become four times its original amount)?

Solution: We want n, the number of half-years. The amount, P, in Barbara's account is not specified, but it would cancel out of our equation in any case. We want to know how long this will take at $i = j_2/2 = .21/2 = .105$ per half-year. The time diagram below shows the details pictorially.

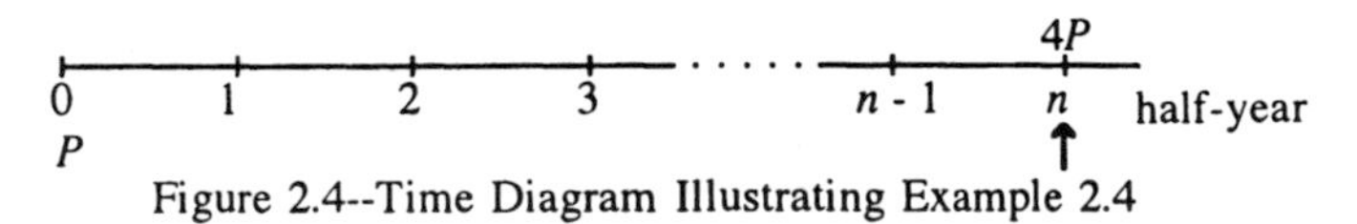

Figure 2.4--Time Diagram Illustrating Example 2.4

The equation of value is $P(1.105)^n = 4P$. From Equation 1.4, we have

$$n = \frac{\log(4/1)}{\log(1.105)} \approx 13.884$$

half-years or $(1/2)\cdot 13.884 = 6.942$ years. So, in about 6 years and 11 months, Barbara's money will have quadrupled in value. ∎

Exercises 2.1

2.1.1 Zep is wondering how much his \$1,000 will earn in five years at the nominal rate of 8% depending on how frequently the interest is compounded. Complete the following table for him.

Number of Compoundings per year (m)	Accumulated value of \$1,000 after 5 years, with compounding m times per year
1	
2	\$1,480.24
4	
12	\$1,489.85
365	
8,760 (hourly)	

2.1.2 Ruth Ellen received a check for \$20,000 on her 21st birthday from a deposit her parents made on the day she was born. How much did Ruth Ellen's parents deposit if their money earned

a) $j_1 = 6\%$?

b) $j_2 = 8\%$?

c) $j_{12} = 12\%$?

2.1.3 Einar made an investment of \$45,000 that has since tripled in value. The investment earned interest at $j_4 = 11.4\%$. How long did it take Einar's investment to triple?

2.1.4 Mimi's great-great-grandmother opened a bank account in the amount of \$200 earning interest at the nominal annual rate of 2.75% compounded quarterly. That was 145 years ago. If left to earn interest, what would the account now be worth today?

2.1.5 Helen passed by a bank that advertised that their nominal annual rate of interest was 5% compounded daily. How much would Helen have in one year if she deposited \$10,000 in that bank? What if their nominal annual rate of interest was 5% compounded hourly?

2.1.6 An investment that Valerie made 5 years ago has now doubled in value at a certain nominal annual rate compounded semiannually. What nominal annual rate compounded semiannually is Valerie getting and when will her investment be worth triple its original value?

2.1.7 Astrid invested $10,000 for 2 years at a nominal annual rate of 10%. Find her accumulated value if the interest is compounded
a) annually.
b) monthly.
c) daily (365 days per year).
d) hourly.

2.1.8 If Jesper invests $5,000 at $j_{12} = 9.6\%$, how much interest will he have earned by the end of 3 years?

2.1.9 An investment Nat made 10 years ago at $j_6 = 12\%$ is presently worth $121,719.74. How much did Nat invest?

2.1.10 How many years will it take Björn's investment of $15,000 to double at interest rate $j_2 = .085$?

2.1.11 What nominal annual rate of interest compounded monthly did Waldron earn if his investment of $100,000 earned $31,119.78 of interest over three years?

2.2 Multiple Deposits/Withdrawals

All of our examples so far have dealt with a single deposit leading to a single withdrawal. But, what if there is more than one deposit and/or withdrawal?

Example 2.5: David borrowed $3,000 from Jonathan. David promised to repay the loan with two equal payments, one after six months, and the other after one year. If the loan was made with interest at $j_{12} = 3\% = .03$, what is the amount of each of David's payments to Jonathan?

Solution: Since $j_{12} = .03$, $i = j_{12}/12 = .0025$ is the interest rate per month. The next two time diagrams should help you understand the problem better.

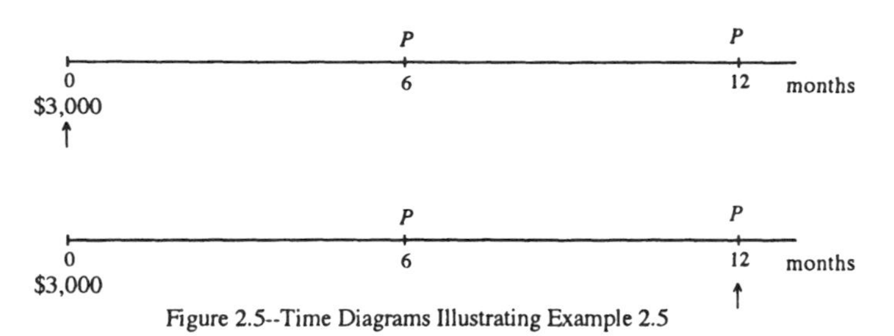

Figure 2.5--Time Diagrams Illustrating Example 2.5

In the first time diagram, I put the focal date at time 0. At time 0 months, the value of the loan is \$3,000. The sum of the present values of the two payments must also be \$3,000. The *PV* of the payment at six months is $P(1.0025)^{-6}$ (remember, the period is the month, and we need the value of *P* six months earlier), and the *PV* of the payment at one year is $P(1.0025)^{-12}$. Both of these are applications of Equation 1.2. The equation of value for the problem is therefore

$$3,000 = P(1.0025)^{-6} + P(1.0025)^{-12} = P\left((1.0025)^{-6} + (1.0025)^{-12}\right)$$
$$= P(.9851303799...) + P(.9704818654...) = P(1.955612..),$$

so $P \doteq \$3,000/1.955612.. \approx \$1,534.05$ is the value of David's two equal payments.

For completeness, let's solve the problem with the focal date at time 12 to show that we get the same answer either way. Now, we calculate the value of the loan after 12 months, which is $\$3,000(1.0025)^{12}$, and the future values of the two payments at 12 months. The first accumulates only 6 months, so its value at the focal date is $P(1.0025)^{6}$. The second does not accumulate at all since it is already at 12 months. The equation of value associated with the focal date at 12 months is

$$3,000(1.0025)^{12} = P(1.0025)^{6} + P(1.0025)^{0}$$
$$= P((1.0025)^{6} + (1.0025)^{0}) = P(2.015094...)$$

and $3,000(1.0025)^{12} = 3,091.25$, so

$$3,091.25 = P(2.015094...),$$

and, finally, $P = \$3,091.25/2.015094... \approx \$1,534.05$. Either way, David makes two payments of \$1,534.05 to Jonathan. ▮

The important thing to take away from this exercise is how we bring multiple payments to a given focal date. You evaluate the payments separately at the unique focal date and then add up the values of the deposits at the focal date, add up the values of the withdrawals at the focal date and then set them equal.

Example 2.5 illustrates something else very important about compound interest: for a constant rate of compound interest, equations of value are equivalent at any focal date. What this means in terms of Example 2.5 is that we could have put the focal date *anywhere* and still have found that David's payments to Jonathan were $1,534.05 each. See Exercise 2M7.

Example 2.6: Henrik made a deposit of $800, followed by second deposit of $1200 one year later. If money earns $j_4 = 6.4\%$, how much is in Henrik's account three years after the first deposit?

Solution: Consider the following time diagram:

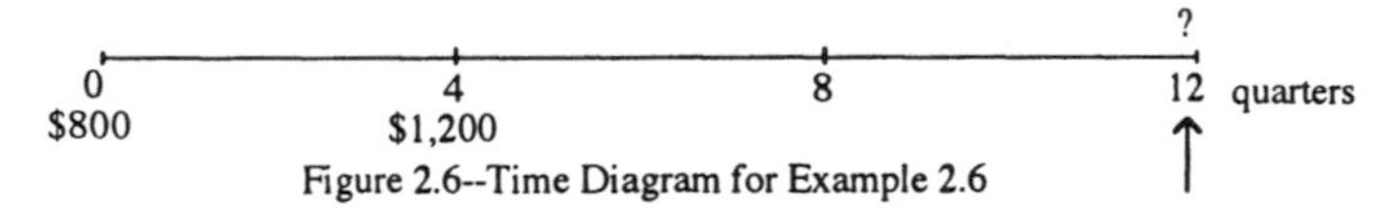

Figure 2.6--Time Diagram for Example 2.6

The interest rate per quarter is $i = j_4/4 = .064/4 = .016$. At the end of three years (12 quarters), the $800 deposit grows to $\$800(1.016)^{12} \approx \$967.86$ while the $1,200 deposit grows to $\$1,200(1.016)^8 \approx \$1,362.48$, so Henrik has $967.86 + $1,362.48. . . ≈ $2,330.34 at the end of three years. ∎

Example 2.7: Isaac talked his father-in-law into selling him his classic 1948 pick-up truck. Isaac will make three equal payments of $1,000: the first now, the second in six months, and the last in one year. If money earns $j_2 = .088$, what single payment immediately would be equivalent to the three-payment plan? Put another way, for what price was Isaac's father-in-law selling the truck?

Solution: Each payment is for $1,000, and the interest rate per half-year is $i = j_2/2 = .088/2 = .044$. With the help of the following time diagram,

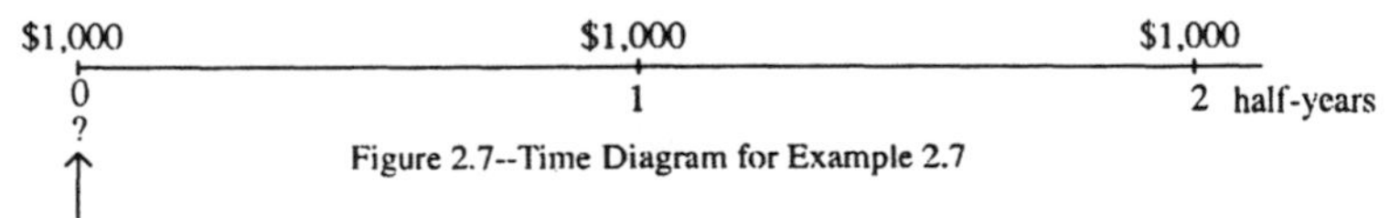

Figure 2.7--Time Diagram for Example 2.7

we see that we can write the equation of value,

$$\$1000 + \$1000(1.044)^{-1} + \$1000(1.044)^{-2} \approx \$2{,}875.34.$$

The first payment is not discounted at all, the second is discounted one period (i.e., one half-year), and the third is discounted two periods. ∎

Exercises 2.2

2.2.1 Marjorie is to repay a debt with a payment of $10,000 in 5 years. If interest is at $j_4 = 6\%$, what equivalent amount could she pay in 2½ years to liquidate her debt?

2.2.2 On February 1, Sara borrowed $50 from Caroline. She borrowed another $75 on May 1. Sara will pay back $60 on August 1 and one more payment on November 1. If Sara and Caroline agree on a nominal annual interest rate of 5% compounded quarterly, what will be the amount of Sara's final payment? (*Hint*: Note that all payments are one quarter apart and that interest is compounded quarterly. I suggest that you put the focal date of your time diagram at November 1, but it would work anywhere.)

2.2.3 Ten years ago, Denny made an investment of $10,000 in an account earning interest at 6.15% annually. He will receive five annual payments from his investment starting now, emptying his account. What will the amount of each payment be?

2.2.4 To pay for some of his senior year tuition and expenses, Stanley borrowed money from his parents at the nominal annual rate of 6% compounded monthly. His plan was to make three annual payments of $3,000, with the first payment 15 months after the loan.

a) How much did Stanley borrow from his parents?

b) Before the first payment is due, Stanley decides that he would rather make a single payment of $9,000 instead of the three payments of $3,000. How long after the loan should Stanley make his lump-sum payment so that it will be equivalent to the original plan?

2.2.5 Valeriano borrowed $10,000 one year ago and another $8,000 six months ago. He will repay the loan with interest at the nominal annual rate of 9.1% compounded monthly. He will make four equal monthly payments beginning six months from now. What will be the amount Valeriano's monthly payments?

2.2.6 Babe is saving up for a new trumpet that has a price tag of \$745. At the end of every month, she makes equal deposits into an account earning at the nominal annual rate of 3.75% compounded monthly. At the end of six months, she has accumulated \$745. What is the amount of her equal monthly deposits?

2.2.7 Chon borrowed money from Jorge and promised to repay him with two payments. She will make the first payment in the amount of \$2,000 in 10 months and the second payment in the amount of \$1,800 in 18 months. If interest is at $j_{12} = 5.2\%$, how much did Chon borrow from Jorge?

2.3 A More General Look at Equations of Value and Focal Dates

Suppose that a set of r deposits, D_1, D_2, ..., and D_r, made at times u_1, u_2, ..., and u_r, respectively, is to be equated with a set of s withdrawals, W_1, W_2, ..., and W_s, made at times v_1, v_2, ..., and v_s, respectively, where the interest rate is assumed to be fixed at i per period. To express the fact these two sets of payments are equivalent to one another in value, we construct an equation of value, and for that, we need a focal date, d. Although d may be anywhere on the time line, in many problems, one choice is often more convenient or logical than others from the point of view of solving the equation for some quantity. Study the following time diagram which contains an arbitrary focal date, one typical deposit, D_j, and one typical withdrawal, W_k.

The values of D_j and W_k moved to time d are $D_j(1+i)^{d-u_j}$ and $W_k(1+i)^{d-v_k}$, respectively. Notice that where I placed d has no effect on what these two expressions look like.

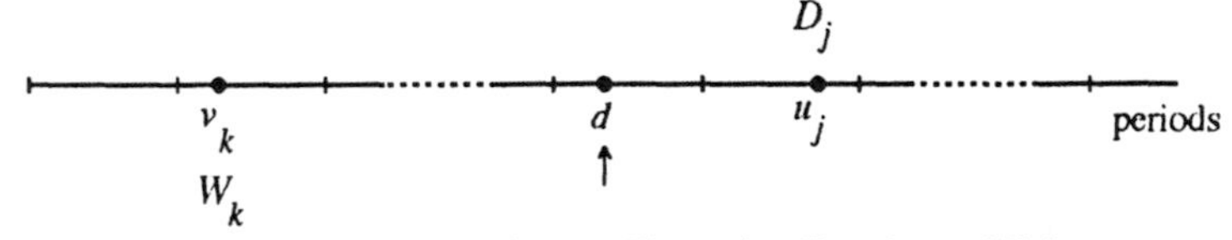

Figure 2.8--Time Diagram Illustrating Equations of Value

In the time diagram, $d - u_j$ is negative since we are moving the value of D_j *back* to time d while $d - v_k$ is positive since we are moving the value of W_k *forward* to time d.

Now, the equation of value for all the deposits and withdrawals is

$$D_1(1+i)^{d-u_1} + D_2(1+i)^{d-u_2} + \ldots + D_r(1+i)^{d-u_r} =$$
$$W_1(1+i)^{d-v_1} + W_2(1+i)^{d-v_2} + \ldots + W_s(1+i)^{d-v_s}. \qquad (2.1)$$

Remark: The fact that the placement of d has no effect on the equality in Formula 2.1 becomes clear when you realize that if you divide both sides of the equation by $(1 + i)^d$ then d disappears entirely from the equation (See Exercise 2.3.2.).

You can use Formula 2.1 as a reference for the equations of value in this book, but after you get enough practice, you should not have to refer back to it.

Example 2.8: Amy borrowed money from her mother-in-law, Hilda, and agreed to repay the loan with an interest rate of $j_{12} = 5\%$. This produced payments of \$672.23 at the end of each of the next 3 months. Amy soon realized that due to some unforeseen expenses she would not be able to make the first payment. So Hilda and Amy agreed that Amy would make two equal payments instead—one at the end of 2 months and one at the end of 3 months.

a) What is the amount of Amy's two payments?

b) How much did Amy borrow from Hilda?

Solution: The period is the month, so the interest rate per month is $i = j_{12}/12 = .05/12 = .0041666... = .0041\bar{6}$ (where the bar over the 6 indicates a repeating decimal). The following time diagram describes the setup of the problem.

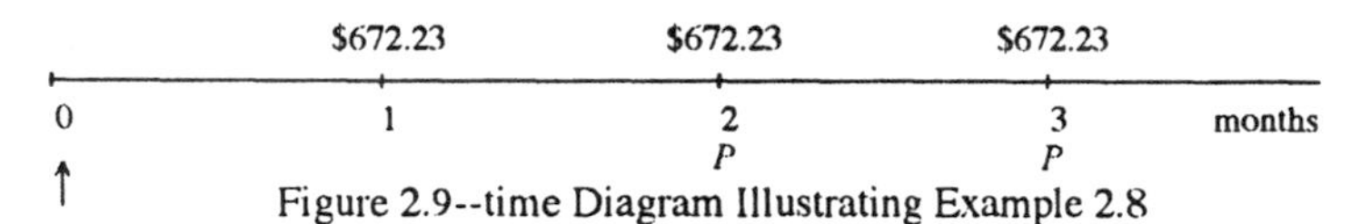

Figure 2.9--time Diagram Illustrating Example 2.8

To apply Formula 2.1, we have $r = 3$ and $D_1 = D_2 = D_3 = \$672.23$, at times $u_1 = 1$, $u_2 = 2$, $u_3 = 3$, $s = 2$, $W_1 = W_2 = P$ (to be determined), $v_1 = 2$, $v_2 = 3$, and $d = 0$.

a) Amy will make two payments of P (yet to be determined), the first at 2 months and the second at 3 months, instead of the originally planned three payments of \$672.23 at the end of each of the first three months. I have put the focal date at 0 months, but, remember, we could put it anywhere and get exactly the same solution. Formula 2.1 gives the equation of value

$$\$672.23(1.0041\overline{6})^{-1}+\$672.23(1.0041\overline{6})^{-2}+\$672.23(1.0041\overline{6})^{-3}= P(1.0041\overline{6})^{-2}+P(1.0041\overline{6})^{-3}.$$

After a bit of calculation, we get \$2,000 = P(1.979321908) so P = \$1,010.45. Amy makes two payments of \$1,010.45 at the end of two and three months instead of three payments of \$672.23 at the end of each of the first three months.

b) The left-hand side of the equation of value above is the amount of the loan—namely, \$2,000. ∎

Example 2.9: Sy has decided to deposit \$300 at the end of each of the next four months into an account paying interest at the nominal annual rate of 5.7% compounded monthly. Beginning one month after the last deposit, Sy will make three equal monthly withdrawals emptying his account. What will be the amount of the withdrawals?

Solution: Let's look at a time diagram for this:

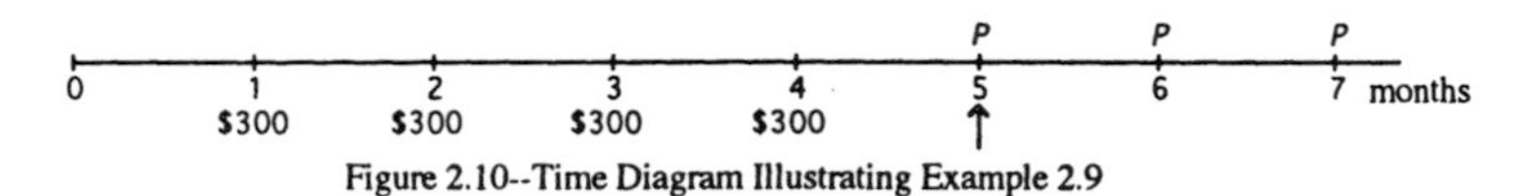

Figure 2.10--Time Diagram Illustrating Example 2.9

Although I have placed the arrow at 5 months, the solution would be the same wherever I put it. The interest rate per month is $i = j_{12}/12$ = .057/12 = .00475. At the arrow, the four \$300 payments are worth

$$\$300(1.00475)^4+\$300(1.00475)^3+\$300(1.00475)^2+ \$300(1.00475)^1=\$1{,}214.317848....$$

The total value of the three withdrawals at 5 months is

$$P+P(1.00475)^{-1}+P(1.00475)^{-2}=P(2.98583971718...),$$

and these two expressions are to be equated. When we do this, we get P = \$406.69. ∎

Exercises 2.3

2.3.1 Make up a reasonable story to go along with the following equation:

$$\$23{,}000 = \$7{,}177.45(1.095)^{-1} + \$7{,}177.45(1.095)^{-2} + \$7{,}177.45(1.095)^{-3} + \$7{,}177.45(1.095)^{-4}.$$

2.3.2 Use the remark following the Formula 2.1 to rewrite the formula in such a way that the focal date, d, does not appear. Explain in your own words what this means.

2.3.3 Quique made a loan from his bank at the rate of $j_{12} = 9\%$ and will repay the loan with a payment of \$250 in 2 months, \$300 in 5 months, and \$450 in 10 months. Make a time diagram, and write the equation of value for this loan and its repayment with the focal date at:

a) time 0.

b) time 6 months.

c) time 12 months.

d) Find the amount of the loan using each version of the equation of value

2.3.4 Krissie is supposed to pay Lanie \$600 in 6 months and \$800 in 8 months. If interest is at $j_{12} = 6\%$, what equivalent single payment could Krissie make in 7 months to liquidate her debt?

2.3.5 Marcus made a deposit of \$1,000 on March 1, 2009, a withdrawal of \$800 on September 1, 2009, and a deposit of \$950 on February 1, 2010. If $j_{12} = 5.1\%$ and there were no other deposits or withdrawals, how much remained in Marcus's account on August 1, 2011 for him to withdraw?

2.3.6 Siomak borrowed some money from his parents to pay for computer training. He promised to repay the loan with payments of \$1,000 at the end of each of the next four quarters plus a final payment of \$5,000 at the end of two years. If they agreed on an interest rate of $j_4 = .05$, what was the amount of Siomak's loan from his parents?

2.4 Effective Rate, Equivalent Rates, Comparing Nominal Rates

An important benchmark in dealing with interest calculations is the *effective* rate of interest, which is defined to be *the amount of money each dollar of deposit or investment earns in one year.* Note that we already have a symbol for the effective rate: j_1.

Not all deposits/investments get compounded the same number of times per year, so we need some way of comparing different compounding schemes. One logical way to do this is to calculate the effective rate equivalent to each of the two nominal rates. Then we know how much each dollar earns in a year for each of the rates.

If a deposit is earning interest at the nominal annual rate of j_m compounded m times per year, then the interest rate per period is, as we have seen, j_m/m. So, after m compoundings (which makes one year), each dollar has become

$$\left(1+\frac{j_m}{m}\right)^m = 1 + j_1.$$

Thus, the effective rate, the amount earned by each dollar, is

$$j_1 = \left(1+\frac{j_m}{m}\right)^m - 1. \qquad (2.2)$$

In many problems you encounter in this book, the effective rate will be given. In others, you may be given a nominal rate and you need to convert it to the effective rate. What you actually need to do will be dictated by the particular problem.

Example 2.10: Jeff's bank offers a nominal annual interest rate of 6.1% compounded monthly. Adam's bank offers an effective rate equivalent to Jeff's bank. Aila's bank offers a nominal annual interest rate that is equivalent to both Jeff and Adam's banks, but her bank compounds quarterly. What is the effective rate common to all three banks? What nominal annual rate compounded quarterly does Aila's bank offer?

Solution: For Jeff's bank, we are given that j_{12} = 6.1% = .061. We are not given Adam or Aila's bank's rates, but we are told that they are equivalent to Jeff's. What 'equivalent' means here is that all three banks will produce equal yields on equal investments over one year. After one year, each of Jeff's dollars becomes

$$\left(1+\frac{.061}{12}\right)^{12} = 1.06273...$$

dollars. The excess over $1 is the amount earned by his dollar and that is, from Equation 2.2, $\left(1+\frac{.061}{12}\right)^{12} - 1$ = .06273... dollars. So, the effective rate equivalent to j_{12} = 6.1% is j_1 = 6.273...%. This is the effective rate that Adam's bank pays.

Now, we do not yet know the nominal rate, j_4, of Aila's bank, but we do know that each of her dollars becomes $(1 + j_4/4)^4$ dollars in one year. To be equivalent to Jeff and Adam's yields, this must equal 1.06273. That is

$$(1 + j_4/4)^4 = 1.06273\ldots,$$

and we must solve this for j_4. Using Formula 1.3 with $i = j_4/4$, we find

$$j_4/4 = (1.06273\ldots)^{1/4} - 1 \approx .01532\ldots,$$

so that, $j_4 = 4(.01532\ldots) = .06131\ldots \approx 6.131\%$.

So all three banks have the same yield on single investments over one year, but due to the different compounding schemes, they advertise different but equivalent nominal rates. Summing up here,

Jeff's nominal rate compounded monthly, $j_{12} = 6.1\%$.
Aila's nominal rate compounded quarterly, $j_4 = 6.131\%$,.
The effective rate for Jeff, Aila, and Adam is $j_1 \approx 6.2373\%$, and are all equivalent, producing the same yield at the end of the year.

Formula 2.2 relates the effective rate of interest to a nominal annual rate. There is a useful formula relating two different nominal rates. To determine the nominal annual rate, j_n, compounded n times per year, equivalent to another nominal annual rate, j_m, compounded m times per year, we convert each to the equivalent effective rate and set the two versions equal Thus, we set

$$j_1 = \left(1 + \tfrac{j_n}{n}\right)^n - 1$$

equal to

$$j_1 = \left(1 + \tfrac{j_m}{m}\right)^m - 1$$

and solve for j_n. When we do this, we get

$$j_n = n\left\{\left(1 + \tfrac{j_m}{m}\right)^{m/n} - 1\right\}. \tag{2.3a}$$

Accordingly, the interest rate *per period* is

$$i = j_n/n = \left(1 + \tfrac{j_m}{m}\right)^{m/n} - 1. \tag{2.3b}$$

Using separate equations of value, we can use the concept of an effective rate to compare the relative virtues of different nominal annual rates. You will be asked to verify Equations 2.3a and 2.3b in the Exercises.

Example 2.11: Susan's bank compounds money daily at the nominal annual rate of 6.5%. Janet's bank compounds money semi-annually at the nominal annual rate of 6.6%. Whose money earns more per year?

Solution: We will find the effective rate for each and see who wins. From Formula 2.2, we find Susan's effective rate to be

$$j_{1,Susan} = \left(1+\tfrac{.065}{365}\right)^{365} - 1 \approx .06715 = 6.715\% .$$

Janet's effective rate is

$$j_{1,Janet} = \left(1+\tfrac{.066}{2}\right)^{2} - 1 \approx .06709 = 6.709\% .$$

This shows that nominal rates don't tell the whole story. Susan's rate *seems* to be lower than Janet's, but her dollars actually earn more than Janet's over the year. ∎

Example 2.12: Daniel has agreed to pay back a loan from Sarah with four consecutive monthly payments beginning in one month. The four payments are to be \$400, \$300, \$200, and \$100, respectively. The effective interest rate is j_1 = 4.074154288%. Before the first payment, Daniel and Sarah decide that instead of the four monthly payments, Daniel should make a single payment of \$1000 at a time that would make it equivalent to the four monthly payments. What is the appropriate time that would make Daniel's single payment to Sarah equivalent to the original four payments, and what was the amount of the loan?

Solution: First, let's look at a time diagram for this problem.

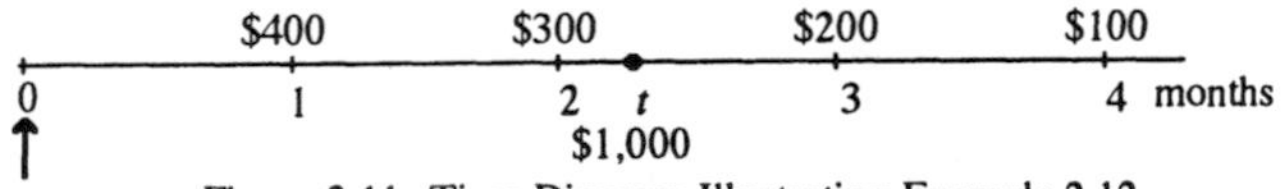

$400 $300 $200 $100
0 1 2 t 3 4 months
↑ $1,000

Figure 2.11--Time Diagram Illustrating Example 2.12

The first thing we have to do is to convert the effective rate to the actual monthly rate. We can use Equation 2.3*a* or 2.3*b* with n = 12, m = 1, and $j_m = j_1$ = .0474154288. We get $j_{12} = 12\{(1 + .04074154288/1)^{1/12} - 1\} = .04$ so $i = j_{12}/12 = .00\overline{3}$.

Notice that I have put in a time t where the single payment of \$1,000 might be made even though I really don't know where t is going to turn out to be. I have put the focal date at time 0 but putting it anywhere else would lead to the same solution. What we have to do is evaluate all of the payments at the focal date and then equate the discounted value of \$1,000 at that point with the sum of the discounted values of the four monthly payments. In other words, we need to apply the Equation of Value Formula 2.1 with $r = 4$, $D_1 = \$400$, $D_2 = \$300$, $D_3 = \$200$, $D_4 = \$100$, $u_1 = 1$, $u_2 = 2$, $u_3 = 3$, $u_4 = 4$, $s = 1$, $W_1 = \$1,000$, $v_1 = t$ (to be determined), $d = 0$, and $i = .00\overline{3}$. This gives us

$$1{,}000(1.00\overline{3})^{-t} = 400(1.00\overline{3})^{-1} + 300(1.00\overline{3})^{-2} + 200(1.00\overline{3})^{-3} + 100(1.00\overline{3})^{-4}.$$

After a bit of calculation and algebra, we get

$$(1.00\overline{3})^{t} = 1.006672207$$

This can be solved with the Log Rule with $a = 1.00\overline{3}$, $b = 1.006672207$, and $m = t$. We get

$$t = \frac{\log(1.006672207)}{\log(1.00\overline{3})} = 1.998337\text{... months.}$$

So, if Daniel makes his payment of \$1000 just before the end of two months after the loan, it will be equivalent to making payments of \$400, \$300, \$200, and \$100, respectively, at the end of the each of the first four months.

The amount that Daniel borrowed from Sarah is the value of the loan at time 0, our focal date. That is, the sum of the discounted values of the four payments. We can double-check this result by finding the discounted value of \$1,000 at the time of the loan. That is

$$\$1{,}000(1.00\overline{3})^{-1.998\ldots} = \$993.37.$$

Daniel borrowed \$993.37 from Sarah. ▮

Example 2.13: Dominík recently graduated from college and has found a good job. She thinks that she can afford a modestly priced new car. She figures that she can pay up to \$220 per month without going hungry. If she can borrow the full price of a car, plus fees and taxes, and if her bank will finance her loan at $j_{12} = 3.48\%$ for five years, how much can Dominík afford to borrow?

Solution: Let's look at a time diagram for this problem.

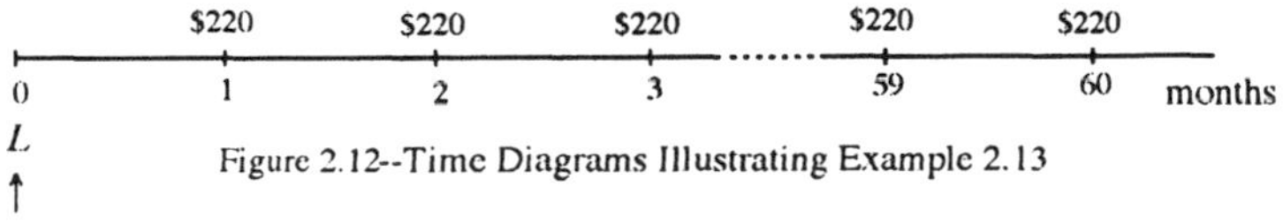

	\$220	\$220	\$220	......	\$220	\$220	
0	1	2	3		59	60	months
L ↑							

Figure 2.12--Time Diagrams Illustrating Example 2.13

Note that all of the monthly payments are at the *end* of the month. I'll say more about this below.

We are looking for the amount of Dominík's loan, L, which is the price of the car plus the usual fees and taxes. Here we have $i = j_{12}/12 =$ .0029. The equation of value sets L equal to the sum of the discounted values of all the monthly payments. That is

$$L = \$220(1.0029)^{-1} + \$220(1.0029)^{-2} + \$220(1.0029)^{-3} + \ldots$$
$$+ \$220(1.0029)^{-59} + \$220(1.0029)^{-60}.$$

The problem here is that there are 60 terms on the right hand side to sum up. We will see early in the next chapter that there is a nice formula for evaluating such sums. For now, I'll only tell you that the sum in this case is $L = \$12{,}099.35$. We will be solving problems like this in Chapter 3. ∎

Example 2.14: Hilary bought a computer with a price tag of $1,850. What convinced her to by was an advertisement that said "no payments for one year." Hilary had to pay for the computer with two payments of $1,189—the first payment one year after the purchase, and the second payment one and one half years after the purchase. What effective rate of interest was the store earning on Hilary's purchase?

Solution: Hilary borrowed $1,850 and will repay the loan with two payments of $1,189, one after one year, the second after 1½ years. Here is a time diagram for this situation:

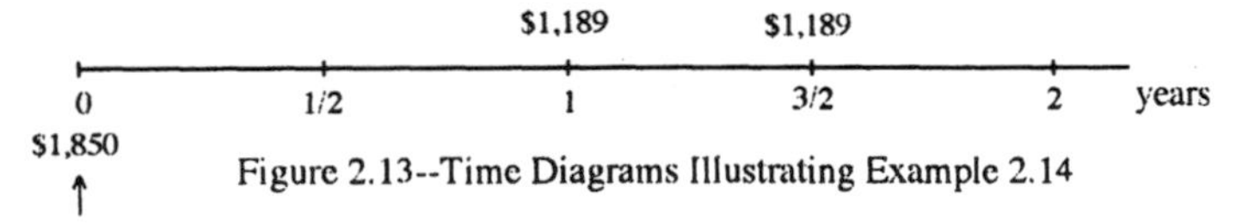

Figure 2.13--Time Diagrams Illustrating Example 2.14

We are looking for i, the effective rate of interest. The equation of value with focal date at time 0 is

$$\$1{,}189(1+i)^{-1} + \$1{,}189(1+i)^{-1.5} = \$1{,}850.$$

Although there is no problem in writing down this equation of value, this is an example of an equation that we are not equipped to solve exactly for the unknown, i. We will have two ways to attack problems like this one: problems requiring *numerical methods*.

Using Your Calculator's Solver: You should have a scientific calculator to solve problems in this book, but not all of scientific calculators have *solvers*. Many brands of calculators do have solvers, the most popular of these just now being the TI-83 and TI-84. There will be many exercises in this book where you would benefit by having such a calculator. I will show in some detail how to get the solution for i in the equation of value $\$1{,}189(1+i)^{-1}+\$1{,}189(1+i)^{-1.5}=\$1{,}850$ using the solver on the TI-83 or TI-84 calculator.

First, display the MATH menu by pressing the MATH button. Select 0: Solver to display the *equation solver*. The screen should now show:

```
EQUATION SOLVER
eqn : 0 =
```

Now, enter the equation of value, but in the form 0 = *LHS* – *RHS*, where *LHS* and *RHS* stand for left-hand side and right-hand side respectively. So, instead of entering

$$\$1{,}189(1+i)^{-1}+\$1{,}189(1+i)^{-1.5}=\$1{,}850,$$

you will enter the equation in the form

$$0=\$1{,}189(1+i)^{-1}+\$1{,}189(1+i)^{-1.5}-\$1{,}850.$$

Use the ^ button for the exponents, and suppress the commas and dollar signs. After entering the equation, press ENTER, and your screen should look like this:

```
EQUATION SOLVER
eqn : 0 =1189(1+I)^(-1) - 1189(1+I)^(-2) - 1850
I = 0
bound = {-1E99, 1E99}
```

Finally, you need a *starting value* for I. The default is I = 0, but since we will almost always be solving for an interest rate, you can feel fairly safe by I = .05. After doing this, your screen should be:

```
EQUATION SOLVER
eqn : 0 =1189(1+I)^(-1) - 1189(1+I)^(-2) - 1850
I = .05
bound = {-1E99, 1E99}
```

Now we are ready to solve for I. With the cursor as above, press ALPHA, then [SOLVE]. [SOLVE] is over the ENTER key. Your screen should now look like:

```
EQUATION SOLVER
eqn: 0 =1189(1+I)^(-1) - 1189(1+I)^(-2) - 1850
I=.22369841512...
bound={-1E99, 1E99}
left-rt=0
```

We see that Hilary's effective interest rate is i = 0.223698... ≈ 22.3698%. She is paying quite a high rate of interest on her loan.

Using Guess-and-Check: We can also approximate the solution to this problem by a method some call *Guess-and-Check* (also called *Trial-and-Error*). The idea of Guess-and-Check is to substitute values for the unknown quantity into the left-hand-side of the equation of value trying to make it as close as we can to the right-hand-side. We will show the results in a table.

As in the solver method, you can safely use $i = 0.05$ as a starting value. For any value of i, the left-hand side of the equation of value is

$$LHS(i) = \$1{,}189(1+i)^{-1} + \$1{,}189(1+i)^{-1.5}.$$

We successively substitute values if i into $LHS(i)$ until we are reasonably close to $RHS = \$1{,}850$.

Your table would be different if you made it on your own. We should all come pretty close to the correct solution for i in the end.

i	***RHS(i)***	***Comments***
.05	$2,237.47	i = .05 is much too low. Try i = .20.
.20	$1,895.34	i = .20 is still too low.
.22	$1,856.94	i = .22 is closer, but still too low
.23	$1,838.28	i =.23 is too high. We need to back up a little.
.225	$1,847.57	The solution is somewhere between .22 and .225.
.224	$1,849.44	Pretty close. We can stop here.

Table 2.1—Guess-and-Check Table for Example 2.14

Hilary was attracted by the possibility of not having to make any payments for a year, but the store was charging her at an alarmingly high effective rate of 22.4%. ∎

In subsequent chapters, other problems with solutions requiring numerical methods will arise. In a few of these, we will give detailed solutions using both of the above methods. In others, we will point the way by showing you how to get started.

Exercises 2.4

2.4.1 Rebekah goes to a bank that pays interest at the nominal annual rate of 4.75% compounded daily. What nominal rate compounded semiannually does Steve's bank offer if we know that the two rates are equivalent?

2.4.2 In 1492 Columbus sailed the ocean blue. Upon arriving in the New World, he bought a savings certificate valued at $5 in the newly opened *Banco del Nuevo Mundo*. The certificate earned interest at the effective rate of 3.25%. In 1992, some 500 years later, Lance Columbus, an heir to Christopher, discovered the certificate. What was its value in 1992?

2.4.3 Svante invested heavily in lutefisk futures. His investment amounted to $75,000. Just two years later, the value of his investment has risen to $76,000. What effective rate of interest did Svante earn on his investment?

2.4.4 When Harriet borrowed $15,000 from a neighborhood agency specializing in personal loans, she agreed to make payments of $6,900 per year at the end of each of the next three years. What effective rate of interest was she being charged? *Hint:* Use your calculator's solver if it has one, or use Guess-and-Check.

2.4.5 Monica is selling her cabin by the lake so she can travel. She would like to get $78,000 for it. She agrees to accept four annual payments of $21,000, the first payment immediately. At what effective rate of interest are these payments equivalent to price of $78,000? *Hint:* Use your calculator's solver if it has one, or use Guess-and-Check.

2.4.6 The Chelm National Bank offers interest on savings at the nominal annual rate of 4.13% compounded daily. What is the equivalent effective rate? Assume 1 year = 365 days.

2.4.7. Marshall's savings are earning at the nominal rate of 6.78% compounded quarterly. If he were to move his money to a bank compounding monthly, what nominal rate would he need to get to at least equal his present rate?

2.4.8 H. K. made an investment earning interest at $j_2 = 4\%$ for the first year, $j_4 = 6\%$ for the second year, $j_6 = 8\%$ the third year, and $j_{12} = 10\%$ the fourth and fifth years. What is the equivalent effective rate of interest over the same five-year period?

2.4.9 Verify Formulas 2.3a and 2.3b.

2.5 Summary of Formulas

Interest Rate Per Period

$$i = \frac{j_m}{m}$$

General Equation Of Value

$$D_1(1+i)^{d-u_1} + D_2(1+i)^{d-u_2} + \ldots + D_r(1+i)^{d-u_r} = W_1(1+i)^{d-v_1} + W_2(1+i)^{d-v_2} + \ldots + W_s(1+i)^{d-v_s}$$

Effective Rate From Nominal Rate Compounded m times per year

$$j_1 = \left(1 + \frac{j_m}{m}\right)^m - 1$$

Nominal Rate n Times Per Year From Nominal Rate m Times Per Year

$$j_n = n\left\{\left(1 + \frac{j_m}{m}\right)^{m/n} - 1\right\}$$

Miscellaneous Exercises

2M1. Mitch borrowed \$5,000 from Tina and she demands \$6,000 in return claiming that an effective rate of $i = 6\%$ would be appropriate. How long after the loan should Mitch pay Tina the \$6,000 to make the interest rate he charged come out to 6%?

2M2. Four years ago, when Seymour was 30, he began losing his hair. At that time he had 985,912 hairs on his head. Since then, he has been losing about 14% of his remaining hair every year. At what age will Seymour have only 10,000 hairs left on his head?

2M3. An investment that Jörgen made tripled in value in 4 years. Assuming a fixed rate of compound interest, when did Jörgen's investment double in value?

2M4. Carmen bought a large tool set for $599.99 and will pay if off with three equal monthly payments of $206.03 beginning in one month.

a) What nominal rate of interest compounded monthly is being charged? *Hint:* Use your calculator's solver if it has one, or use Guess-and-Check.

b) What effective rate of interest being charged?

2M5. Alma was going to repay a loan with two equal payments of $500, one at 2 months, and one at 6 months. Instead, she will make a single payment $1,001 at 4 months. What nominal rate of interest compounded monthly is Alma being charged? *Hint:* Use your calculator's solver if it has one, or use Guess-and-Check.

2M6. Genarro borrowed $6,000 at $j_1 = 10\%$ from his father-in-law to buy a guitar. If he pays $1,300 at the end of each of the next six years, what final amount will he need to pay at the end of seven years to liquidate his debt?

2M7. Refer to Example 2.5. Show that David's two payments to Jonathan are still $1,534.05 when you put the focal date at an arbitrary point in time *d*. Write the equation of value and note that terms involving *d* cancel out of the equation.

Chapter 3—Introduction to Annuities

3.1 Regular Annuities and Geometric Sums

An *annuity* is a finite sequence of payments made at equal intervals of time. The recipient of an annuity is called an *annuitant.* In this chapter, we will confine ourselves to annuities whose payment times coincide with interest conversion times although in practice they might not always do so. When, in addition, an annuity is made up of equal payments with a constant interest rate per period, we will call it a *regular annuity.* If, as in this and the next few chapters, annuity payments are certain to be made, the annuity may also be called an *annuity certain.* In this case, payments will be made regardless of the survival or death of the annuitant. In the case of death before the term of the annuity has expired, the remaining payments go to the estate or to the beneficiaries of the annuitant. In Part II of the book, we take up *life annuities*, where the continuance of the payments *is* dependent upon the survival of the annuitant.

Example 3.1: (Continuation of Example 2.13) Below is a repeat the time diagram from Example 2.13:

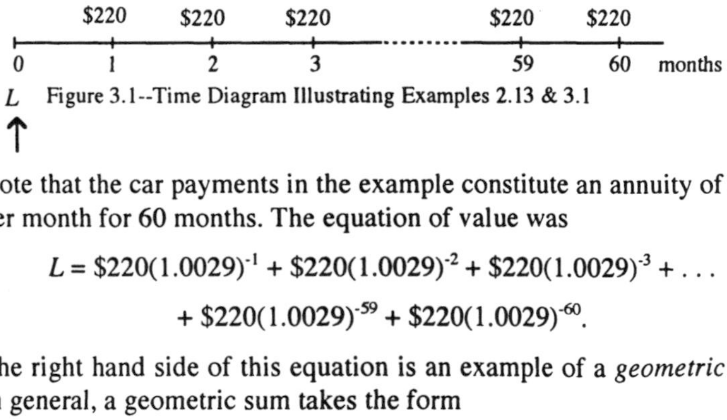

$220 $220 $220 .... $220 $220

0 1 2 3 59 60 months

L ↑

Figure 3.1--Time Diagram Illustrating Examples 2.13 & 3.1

Note that the car payments in the example constitute an annuity of $220 per month for 60 months. The equation of value was

$$L = \$220(1.0029)^{-1} + \$220(1.0029)^{-2} + \$220(1.0029)^{-3} + \ldots$$
$$+ \$220(1.0029)^{-59} + \$220(1.0029)^{-60}.$$

The right hand side of this equation is an example of a *geometric sum.* In general, a geometric sum takes the form

$$G_n = a + ar + ar^2 + ar^3 + \ldots + ar^{n-1}.$$

The individual terms, a, ar, ar^2, ar^3, ..., ar^{n-1} form what we call a *geometric progression*. There is a *first term*, a, and any later term may be calculated by multiplying the term before it by r. For example, we multiply the first term, a, by r to get the second term, ar; we multiply the second term, ar, by r to get the third term, ar^2, etc. We call r the *common ratio*. The *number of terms* in the sum is n. Evaluating G_n would involve a lot of tedious work if there weren't a simple formula for it simplifying the calculations. This formula can be written in two equivalent ways:

$$G_n = a\frac{1-r^n}{1-r}, \text{ if } r \neq 1 \tag{3.1a}$$

or

$$G_n = a\frac{r^n - 1}{r-1}, \text{ if } r \neq 1. \tag{3.1b}$$

Although the two versions, (3.1*a*) and (3.1*b*), are equivalent to one another, you might find it more comfortable to use the first if $r < 1$ and the second if $r > 1$. What this does is help you avoid running into negative numbers (not that there's anything wrong with that). The proof of Formulas 3.1*a* and 3.1*b* are left as an exercise.

When we look at a geometric sum like the equation of value we got for Dominík's car we have to identify three things before we can use Equation 3.1 to evaluate it. We need the first term, a, the common ratio, r, and the number of terms in the sum, n. Try to determine these things for the equation of value in Example 2.13 before reading on. In

$$G = \$220(1.0029)^{-1} + \$220(1.0029)^{-2} + \$220(1.0029)^{-3} + \ldots$$
$$+ \$220(1.0029)^{-59} + \$220(1.0029)^{-60},$$

the first term is $a = 220(1.0029)^{-1}$, the common ratio is $r = (1.0029)^{-1}$ because you multiply each term by this to get the next term, and the number of terms is $n = 60$. Substituting into Equation 3.1*a*, we can now write G as

$$G_n = a\frac{1-r^n}{1-r} = \$220(1.0029)^{-1}\frac{1-\left\{(1.0029)^{-1}\right\}^{60}}{1-(1.0029)^{-1}} =$$

$$\$220(1.0029)^{-1}\frac{1-(1.0029)^{-60}}{1-(1.0029)^{-1}} = \$219.3638\ldots\cdot\frac{1-.84050\ldots}{1-.99710\ldots} =$$

$$= \$219.3638\ldots\frac{0.15949\ldots}{0.00289\ldots} = \$219.3638\ldots(55.15655\ldots) \approx \$12.099.35.$$

So \$12,099.35 is what Dominík can afford to borrow to buy her new car. You will need to master calculations like these on your calculator. I suggest that you begin by consulting your calculator's manual if you have difficulty. If this does not solve your problem or if you find that you have no idea where your manual is, ask your instructor for help. ∎

Example 3.2: Paul wants to be a millionaire by the time he is 60. He is now 25. If he can put money into an investment at the beginning every quarter and earn interest at the nominal annual rate of 10% compounded quarterly, what must Paul's quarterly deposits be for him to accumulate \$1,000,000 at age 60?

Solution: Since Paul is now 25, he will have to make quarterly payments for 35 years. That is, he will make $4 \times 35 = 140$ equal payments, one at the beginning of every quarter with an interest rate per quarter of $i = j_4/4 = 2.5\% = .025$. Let's take a look at a time diagram for this.

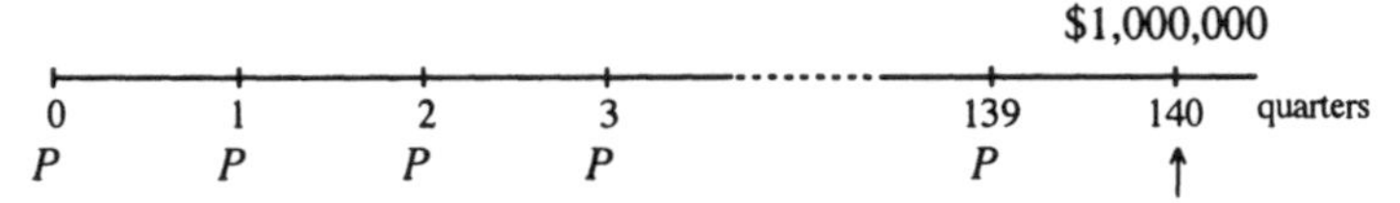

Figure 3.2--Time Diagram Illustrating Example 3.2

P is the value of each of the quarterly deposits Paul will make. That's what we'll be solving for. We will write down the equation of value, and from this, we will figure out the details of the geometric sum that results. We can write the equation of value as

$$P(1.025) + P(1.025)^2 + \ldots + P(1.025)^{139} + P(1.025)^{140} = \$1{,}000{,}000.$$

What makes this a geometric sum? It has a first term, $P(1.025)$. Each term after the first is obtained from the one preceding it by multiplying by 1.025. This makes $r = 1.025$ the common ratio. Finally, there are $n = 140$ terms.

The last payment accumulates only one quarter exposed to interest, so its value at the focal date is $P(1.025)$. That is, we *accumulate* it one quarter. The next to last payment at 138 quarters (not pictured) accumulates two quarters, so its value at the focal date is $P(1.025)^2$, and so on. Now the left-hand side of our equation of value is a geometric sum with first term, $a = P(1.025)$, common ratio, $r = 1.025$, and number of terms, $n = 140$. Using Equation 3.1*b*, we rewrite the equation of value

$$P(1.025)\frac{(1.025)^{140}-1}{1.025-1} = P(1.025)\frac{31.72058...-1}{.025} = P(1{,}259.54388...),$$

and we set this equal to \$1,000,000. That is, set

$$P(1{,}259.543886) = \$1{,}000{,}000$$

to get

$$P = \$793.938195497 \approx \$793.94.$$

For a modest \$793.94 per quarter, Paul will be a millionaire at 60. Of course, \$1,000,000 in 35 years won't be what it is today, and an interest rate of $j_4 = 10\%$ might be somewhat unrealistic as well. ∎

Now we'll do a variation on Example 3.2.

Example 3.3: Chava bought a cabin some years ago and is thinking of refinancing because the interest rate on her old loan is $j_4 = 10\%$. Her payments have been \$793.94 at the beginning of every quarter and were to go on for 35 years from the original purchase date. What was the amount of Chava's original loan, L?

Solution: Let's start with the time diagram.

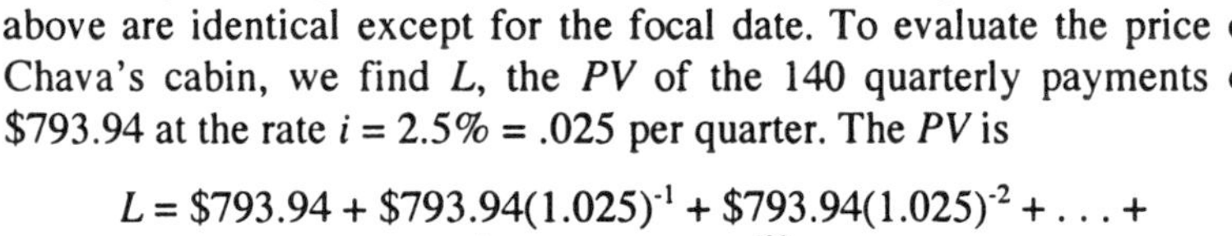

\$793.94	\$793.94	\$793.94	\$793.94	...	\$793.94		
0	1	2	3	...	139	140	quarters
L ↑							

Figure 3.3--Time Diagram Illustrating Example 3.3

If you were to fill in Paul's payments of \$793.94 on the time diagram of Example 3.2, you would see that that time diagram and the one above are identical except for the focal date. To evaluate the price of Chava's cabin, we find L, the PV of the 140 quarterly payments of \$793.94 at the rate $i = 2.5\% = .025$ per quarter. The PV is

$$L = \$793.94 + \$793.94(1.025)^{-1} + \$793.94(1.025)^{-2} + \ldots + \$793.94(1.025)^{-139},$$

a geometric sum with $a = \$793.94$, $r = (1.025)^{-1}$, and $n = 140$. Using Equation 3.1a, we calculate

$$L = \$793.94\frac{1-(1.025)^{-140}}{1-(1.025)^{-1}} = \$31{,}525.34.$$

There is a nice way we can check this answer. Since we know from Example 3.2 that the *accumulated value* of the same payments is

\$1,000,000, we should arrive at the price of Chava's house as the discounted value of \$1,000,000 for 140 quarters at 2.5% per quarter. In other words, Chava's price should be the same as

$$\$1{,}000{,}000(1.025)^{-140}.$$

Check it out. Except for a round-off error of a few cents caused by rounding \$793.938195497 to \$793.94, the answers match. Yet another way to look at the above results is to observe that if Paul were to make a *single* deposit of \$31,525.34 at 2.5% per quarter and leave it for 35 years, he would have \$1,000,000 at age 60. ▮

Exercises 3.1

3.1.1 Evaluate the geometric sum $2{,}000(.9)^3 + 2{,}000(.9)^4 + \ldots + 2{,}000(.9)^{75}$.

3.1.2 Find a, r, and n in the geometric sum $1 + 4 + 16 + 64 + \ldots + 1{,}048{,}576$. Evaluate the sum using Formula 3.1*a* or 3.1*b*.

3.1.3 Owen bought a car for \$30,000. Suppose that this car gets sold to new owners at the end of each year for 20 years for 80% of its value one year earlier. How much money has been paid for the car over the 20 years? Express the total as a geometric sum; find a, r, and n.

3.1.4 Ladislao bought a used car for \$2,400 and will pay it off with equal payments at the end of each month for two years. If the interest rate charged is $j_{12} = 8.4\%$, what is the amount of Ladislao's monthly payments?

3.1.5 Margie's grandma is very rich. She offered Margie two alternative birthday presents. The first was \$5,000,000 cash; the second was 1¢ on her birthday, 2¢ the next day, 4¢ the next day, and so on until she has received 30 payments with her money being doubled each day. Which alternative should Margie choose? Why?

3.1.6 Identify a, r, and n in the 10-term geometric progression that begins 2, -4/9, 8/81, -16/729, ...Evaluate the sum.

3.1.7 Find the sum of the first 100 terms of the geometric progression which begins $(1.05)^2, (1.05)^3, (1.05)^4, \ldots$

3.1.8 Barring repetitions, how many ancestors do you have beginning with your parents, grandparents, great-grandparents, ..., great-great-great-great-great-great-great-great-great grand-parents?

3.1.9 If Balbino deposits $1,200 at the end of each year for 10 years at $j_1 = 6\%$, how much will he have accumulated at the end of the 10 years?

3.1.10 Caleb just cut up his credit card, so he can concentrate on paying back what he owes without charging anything else. He owes $13,040.16 and wants to make equal payments that will have the total paid off at the end of two years. What will be the size of Caleb's 25 payments if the credit card company charges $j_{12} = 14.5\%$ and if he will make his first payment immediately?

3.1.11 Simon is saving up for the down payment of $2,000 on his first new car. He will make deposits at the end of each month for eight months into an account earning $j_{12} = 4\%$. One month after his last deposit, his balance is to be $2,000. How much does he have to deposit each month?

3.1.12 Derive Formulas 3.1*a* and 3.1*b*. *Hint*: Multiply G_n by r; subtract G_n from rG_n, and cancel terms. Finally, assuming that $r \neq 1$, solve for G_n. What happens to Formulas 3.1*a* and 3.1*b* if $r = 1$?

3.2 A General Formula for Regular Annuities

The above examples all deal with the simplest kind of annuity, where payments are equal, equally spaced in time, and have a constant rate of interest. There is a general formula we can use to describe any such annuity calculable at any focal date. Below is a time diagram for a regular annuity with n equal payments, P, made at times 0, 1, 2, ..., $n - 1$, to be evaluated at focal date, d.

P P P . . . P . . .
0 1 2 n - 1 n d periods

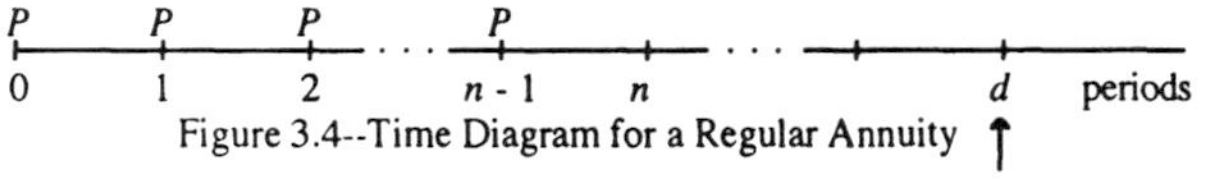

Figure 3.4--Time Diagram for a Regular Annuity ↑

The formula for the value of this annuity evaluated at focal date d is

$$S = S(P, i, n, d) = P\frac{1-(1+i)^{-n}}{i}(1+i)^{d+1}. \tag{3.2}$$

Remark: The notation $S(P, i, n, d)$ tells us to put the values of P, i, n, and d into the formula on the right-hand-side. The output is the value of the expression, *S at the focal date*. We will refer to P, i, n, and d as *inputs* to the formula. On the other hand, sometimes we will be given the value of S and three of the four *inputs* and asked to solve for the fourth input. ∎

*Derivation of Formula 3.2:**† In order to derive formula 3.2, we must find the value of each payment in turn at the arbitrary, but fixed focal date, d. The first payment is at a distance of d periods from the arrow, so its value *at* the arrow is $P(1 + i)^d$. Similarly, the second payment, at a distance of $d - 1$ periods from the arrow has the value $P(1 + i)^{d-1}$. Continuing this pattern, we can write the sum of the values of all n payments:

$$P(1+i)^d + P(1+i)^{d-1} + \ldots + P(1+i)^{d-n+1}.$$

This is a geometric sum with first term $a = P(1 + i)^d$, common ratio $r = (1 + i)^{-1}$ and n terms. From Formula 3.1*a* we have its sum:

$$P(1+i)^d \frac{1-\left\{(1+i)^{-1}\right\}^n}{1-(1+i)^{-1}} = P(1+i)^d \frac{1-(1+i)^{-n}}{1-(1+i)^{-1}}$$

$$= P(1+i)^d \frac{(1+i)}{(1+i)}\left(\frac{1-(1+i)^{-n}}{1-(1+i)^{-1}}\right) = P\frac{1-(1+i)^{-n}}{i}(1+i)^{d+1}$$

which is Formula 3.2. ∎

Strategy for Finding the Inputs for Formula 3.2: We will soon go through several examples using Formula 3.2, but first, here are a few hints on determining the inputs P, i, n and, especially, d.

a) Verify the annuity you are evaluating is regular—i.e., it is made up of equal payments with a constant interest rate per period, payment times coinciding with interest conversion times.

b) Make a time diagram showing the n payments of P. Label the time line in a natural way according to the wording of the problem. Mark the time where you want the value of the annuity—the focal date—with an upward arrow.

c) The value of the input d for Formula 3.2 is the time where the arrow is, *minus* the time where the first payment is. This number is

† More advanced items, marked with an asterix, may be omitted on first reading.

positive if the arrow is to the right of the time of the first payment; it is negative if the arrow is to the left of the time of the first payment; and it is zero if the arrow is at the time of the first payment. Placing the focal date at the time of the first payment has the effect of subtracting the time associated with the first payment from every number on the time line. The result is a zero under the first payment and the number d above the arrow. That is why I labeled the time diagram in Figure 3.4 the way I did. The four special cases below also use this strategy.

REMEMBER: However you number your time diagrams,

$d = \{\text{number at the arrow}\} - \{\text{number at the first payment}\}$

Four Special Cases of Formula 3.2:

Special Case I: The *PV* of an n-payment annuity whose payments are at the *end* of each period has focal date $d = -1$. The focal date is one period before the time of the first payment. Formula 3.2 simplifies to

$$S(P, i, n, -1) = P\frac{1-(1+i)^{-n}}{i}. \tag{3.2a}$$

Here is a time diagram for Special Case I:

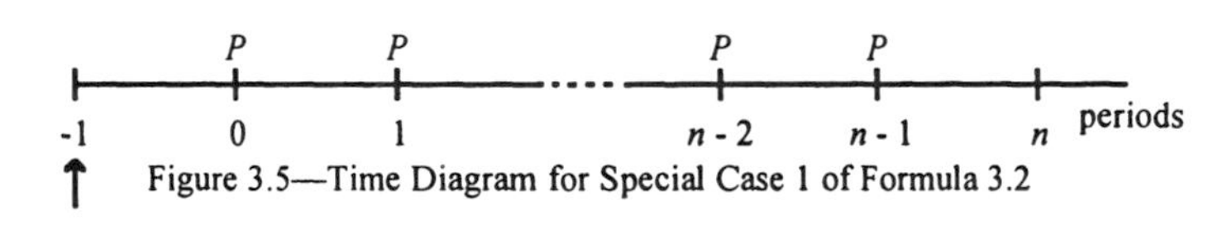

Figure 3.5—Time Diagram for Special Case 1 of Formula 3.2

Special Case II: The *PV* of an n-payment annuity whose payments are at the *beginning* of each period has focal date $d = 0$. The focal date is at the time of the first payment. Formula 3.2 simplifies to

$$S(P, i, n, 0) = P\frac{1-(1+i)^{-n}}{i}(1+i). \tag{3.2b}$$

Here is a time diagram for Special Case II:

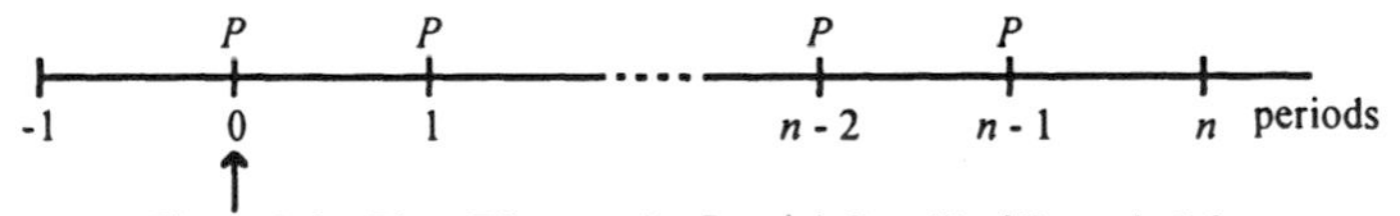

Figure 3.6—Time Diagram for Special Case II of Formula 3.2

Special Case III: The *accumulated*

value of an n-payment annuity whose payments are at the *end* of each period has focal date $d = n - 1$. The focal date is at the time of the last payment. Formula 3.2 simplifies to

$$S(P, i, n, n-1) = P\frac{(1+i)^n - 1}{i}. \tag{3.2c}$$

Here is a time diagram for Special Case III:

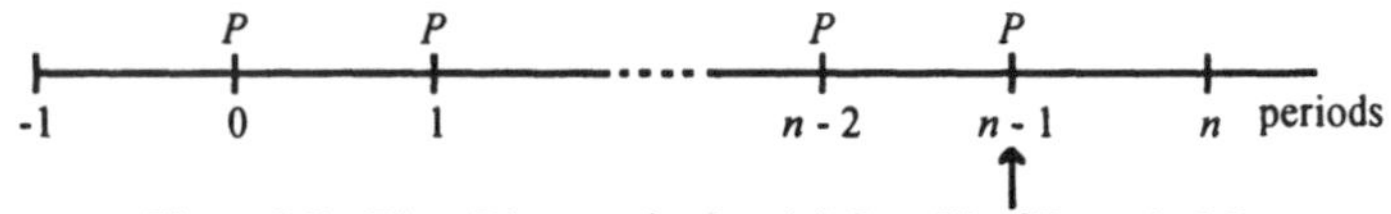

Figure 3.7—Time Diagram for Special Case III of Formula 3.2

Special Case IV: The *accumulated value* of an n-payment annuity whose payments are at the *beginning* of each period has focal date $d = n$. The focal date is one period after the time of the last payment. Formula 3.2 simplifies to

$$S(P, i, n, n) = P\frac{(1+i)^n - 1}{i}(1+i). \tag{3.2d}$$

Here is a time diagram for Special Case IV:

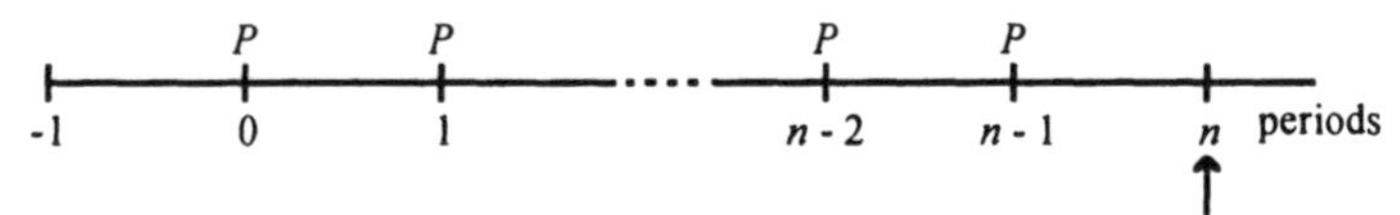

Figure 3.8—Time Diagram for Special Case IV of Formula 3.2

Convention: If in any problem, I do not specify whether payments are being made at the beginning or the end of the periods, assume that they are being made at the *end* of the periods.

Example 3.4: Fuad deposited \$100 at the end of every month for 10 years. He then stopped making deposits. If his money earned at the nominal annual rate of 5.5% compounded monthly, how much does Fuad have at the end of 20 years?

Solution: As usual, the first thing to do is to make a time diagram.

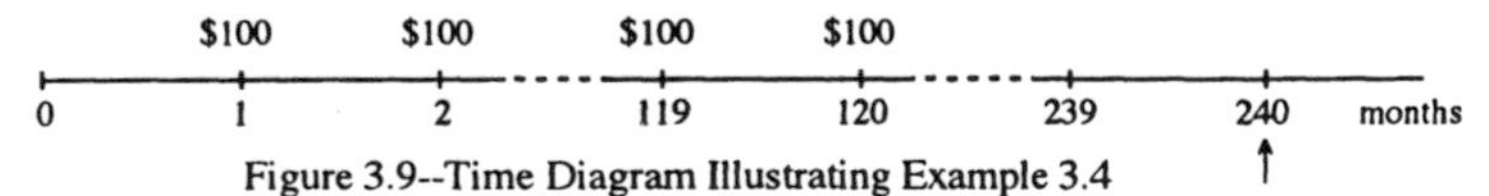

Figure 3.9--Time Diagram Illustrating Example 3.4

Fuad made 120 monthly payments and then let the balance collect interest for another 10 years. The arrow is at 240 months because this is where we want to evaluate the annuity. The difference between the number above the arrow, 240, and the time of the first payment, 1, is $240 - 1 = 239 = d$, which is the input we need for Formula 3.2. So we evaluate $S(\$100, .00458\overline{3}, 120, 239)$, because $P = \$100$, $i = .055/12 = .00458\overline{3}$, $n = 120$, and $d = 239$. I will write this out.

$$S(\$100, .00458\overline{3}, 120, 239) =$$

$$\$100\frac{1-(1.00458\overline{3})^{-120}}{.00458\overline{3}}(1.00458\overline{3})^{239+1} = \$27{,}611.98. \quad \blacksquare$$

Example 3.5: Elaine and Bill want their infant daughter Cecilia to have money for college. They want her to have $25,000 per year for four consecutive years, beginning in 18 years.

a) Supposing that their money earns 4.75% effective (i.e., $j_1 = .0475$), what single amount do Elaine and Bill need to deposit now to provide Cecilia's four payments of $25,000 beginning in 18 years?

b) If instead of making a single deposit now, Elaine and Bill make 18 equal annual deposits beginning immediately, what would be the size of those deposits?

Solution:

a) The time diagram looks like this.

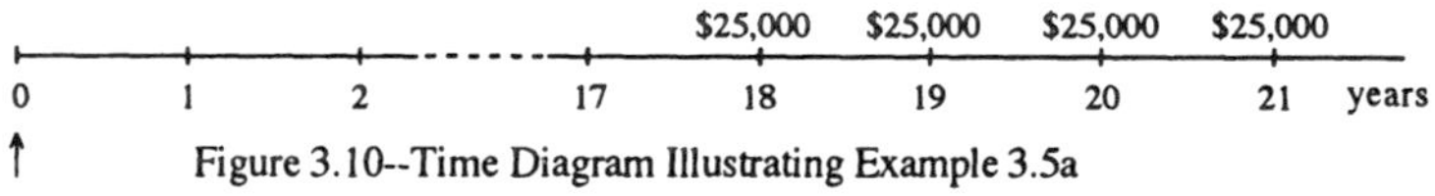

0	1	2	...	17	18	19	20	21	years
↑					$25,000	$25,000	$25,000	$25,000	

Figure 3.10--Time Diagram Illustrating Example 3.5a

The arrow is at time 0 because that is where we want the value of the annuity. To figure out the value of d, we take the number above the arrow, 0, and subtract the time of the first payment, 18, to get $0 - 18 = -18$.

Now we have the inputs for Formula 3.2. They are $P = \$25{,}000$, $i = .0475$, $n = 4$, and $d = -18$. The single deposit that Bill and Elaine have to come up with now is

$$S(\$25{,}000, .0475, 4, -18) =$$

$$\$25000\frac{1-(1.0475)^{-4}}{.0475}(1.0475)^{-18+1}=\$40,511.73.$$

b) Since Bill and Elaine might not have \$40,511.73 on hand, they could decide instead to make 18 annual deposits beginning immediately, the *PV* of which we found in part *a*), namely \$40,511.73. Note that we do not need to show the four annual payments of \$25,000 in the diagram, because we already know their value at the time of Cecilia's birth.

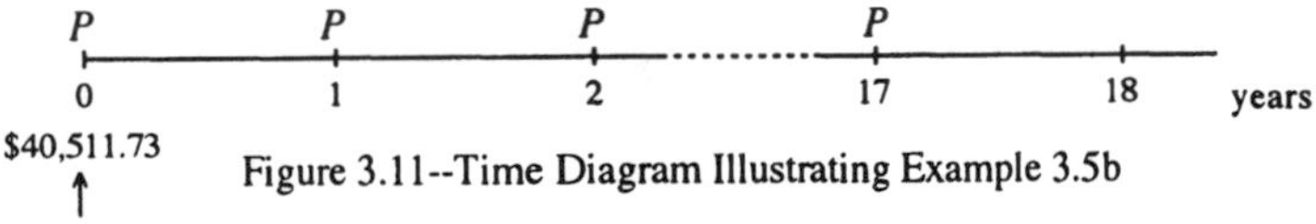

P P P P
0 1 2 .... 17 18 years
\$40,511.73
↑

Figure 3.11--Time Diagram Illustrating Example 3.5b

The first payment is at time 0, and the focal date is at time 0, so $d = 0 - 0 = 0$. We have our inputs for Formula 3.2 (or 3.2*b*): *P* to be determined, $i = .0475$, $n = 18$, and $d = 0$. We also have the value of *S*. That is, $S(P, .0475, 18, 0) = \$40,511.73$. Writing this out, we get

$$S(P,\ .0475,\ 18,\ 0)=P\frac{1-(1.0475)^{-18}}{.0475}(1.0475)=\$40,511.73.$$

After a little calculating, we find

$$P(12.4875681895) = \$40,511.73,$$

so

$$P = \$3,244.16.$$

The annual payment is a little easier to handle. ∎

Example 3.6: At age 30, Nader began contributing \$275 at the end of every month to a retirement fund. He is to do this until he is 55 years old. One month after his last contribution, he (or his estate, if he dies) will begin receiving a monthly annuity for 30 years. If his money earns $j_{12} = 7.5\%$ throughout, what monthly payments, *P*, will Nader receive?

Solution: Our time diagram will first have money going in, and later money coming out. Although, as usual, we could put the focal date anywhere, I placed it at the point where Nader is 55.

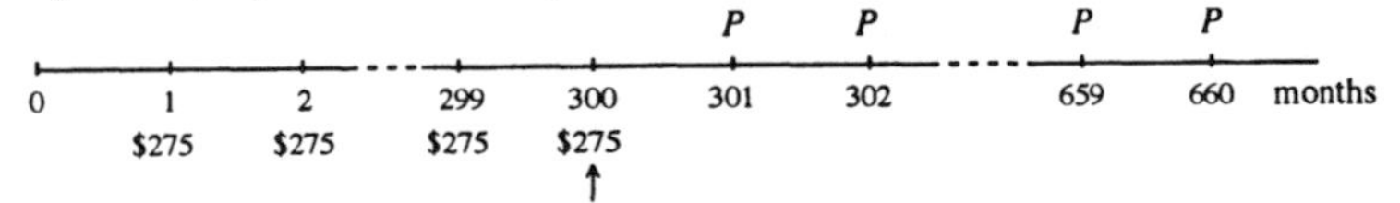

Figure 3.12--Time Diagram Illustrating Example 3.6

We have two separate annuities to evaluate here—the 300-payment annuity of \$275 per month into the fund and the 360-payment annuity of P (to be determined) per month out of the fund. These two annuities are to be set equal to one another at the chosen focal date, 300 months. This is at the end of the contributions to the fund. To find the input, d, for the first annuity, we calculate $d = 300 - 1 = 299$. The inputs for Formula 3.2 (or 3.2*c*) for the money going into the fund are $P = \$275$, $i = .075/12 = .00625$, $n = 300$, and $d = 299$, so the accumulated value of the contributions is

$$S(\$275, .00625, 300, 299) = \$275\frac{(1.00625)^{300} - 1}{.00625} = \$241{,}246.74.$$

To evaluate the 360-payment annuity that Nader will begin receiving at time 301 months, we calculate $d = 300 - 301 = -1$. The inputs for Formula 3.2 (or 3.2*a*) for the 360 payments are P (yet to be determined), $i = .00625$, $n = 360$, and $d = -1$. The value of the 360 future payments is

$$S(P, .00625, 360, -1).$$

Setting the two equal, we have

$$\$241{,}246.74 = S(P, .00625, 360, -1)$$

or

$$\$241{,}246.74 = P\frac{1 - (1.00625)^{-360}}{.00625}$$

or

$$\$241{,}246.74 = P(143.0176273)$$

or finally, $P = \$1{,}686.83$. Nader (or his estate) will receive a monthly annuity of \$1,686.83 for 30 years. ∎

Remark: A *pension* is what we call an annuity whose continuation depends on the survival of the annuitant. A pension is a so-called *life annuity*. We will deal more realistically with pensions (and other related matters) in Part II of the book. Moreover, in Part II of the book, annuities as well as other payments will, in general, not be *certain* but rather, *contingent* upon someone's survival. Therefore, as we have mentioned before, the payments and annuities we are studying in Part I are sometimes called *payments certain*, or *annuities certain*.

Example 3.7: Don deposited \$200 at the end of every month for 12 years into an account earning $j_{12} = 5.4\%$. One month after his last deposit of \$200, he began depositing \$275 per month into the same account for another 4 years. How much did Don accrue at the end of 16 years?

Solution: There is a nice way to attack this problem. Look at the two time diagrams below. The second one might suggest the way to get around the fact that Don's deposits were not constant over the 16 years.

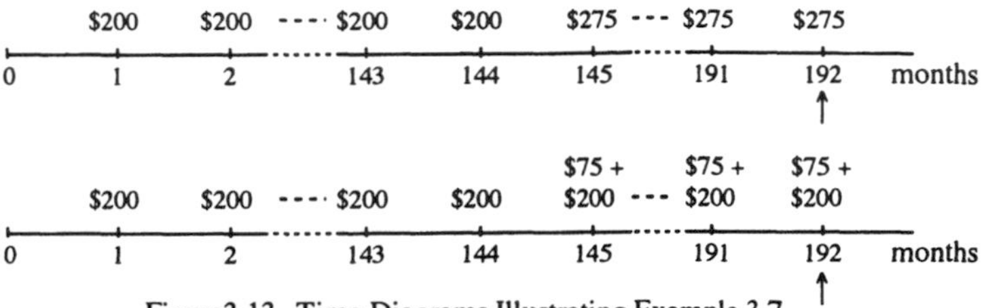

	\$200	\$200	----	\$200	\$200	\$275	---	\$275	\$275	
0	1	2		143	144	145		191	192 ↑	months

	\$200	\$200	----	\$200	\$200	\$75 + \$200	---	\$75 + \$200	\$75 + \$200	
0	1	2		143	144	145		191	192 ↑	months

Figure3.13--Time Diagrams Illustrating Example 3.7

In the second diagram, I have broken up the last 48 months of payments as though there were 16 years of monthly deposits of \$200 and 4 years of monthly deposits of \$75, the latter ones beginning at the end of month number 145. Don's accumulated amount can then be calculated with two applications of Formula 3.2 or 3.2*c*. For the 192 payments of \$200, we calculate $d = 192 - 1 = 191$. The inputs for the 192 payments of \$200 are $P = \$200$, $i = .054/12 = .0045$, $n = 192$, and $d = 191$.

To find input d for the 48-payment annuity of \$75, we calculate $d = 192 - 145 = 47$. The inputs for the second annuity are: $P = \$75$, $i = .0045$, $n = 48$, and $d = 47$. We find

$$S(\$200,.0045,192,191) + S(\$75,.0045,48,47) =$$

$$\$200\frac{(1.0045)^{192}-1}{.0045} + \$75\frac{(1.0045)^{48}-1}{.0045} =$$

$$\$60{,}801.694 + \$4{,}008.353 = \$64{,}810.05.$$

That is what Don has accumulated after 16 years. ∎

Example 3.8: Aardvark Auto Sales advertises "Up to \$4,500 off the sticker price, or 0% interest on the sticker price," on its new cars. Goldie has her eye on a new, nicely equipped convertible with a sticker price of \$31,560. On this particular model, she can choose between \$3,600 off the sticker price on a 60-month loan or the full price and $j_{12} = 0\%$ for 60 months. What would the interest rate need to be for Goldie to be indifferent about the two choices? Which option should she choose if she is offered $3,600 off the sticker price at j_{12} = 6% for 60 months?

Solution: Goldie's payments if she chooses the 0% financing are $31,560/60 = $526. To be indifferent about the options, the monthly interest rate, $i = j_{12}/12$ would need to satisfy

$$\$31,560 - \$3,600 = \$27,960 = S(\$526, i, 60, -1) = \$526\frac{1-(1+i)^{-60}}{i}.$$

Using a solver: We enter the equation $0 = \$526\frac{1-(1+i)^{-60}}{i} - \$27,960$ into the Equation Solver (See Example 2.14). After starting with a value of I = .05, and pressing [ALPHA] [SOLVE], the final screen was:

```
EQUATION SOLVER
eqn: 0 =526(1-(1+I)^(-60))/I - 27960
I = .00405989434...
bound={-1E99, 1E99}
left-rt=0
```

Goldie therefore needs a nominal annual rate of j_{12} = 12(.00405989...) ≈ 0.04872 = 4.872 % to be indifferent about the choices.

Using Guess-and-Check: In a Guess-and-Check Table I calculated values of

$$LHS(i) = \$526\frac{1-(1+i)^{-60}}{i}$$

to find i making $LHS(i)$ as close to $27,960 as I could in a few steps. Starting with i = 0.05, I found $i = j_{12}/12 \approx 0.00406$ after about seven steps. This is very close to the more precise value we got with a solver.

Goldie should decline an offer of j_{12} = 6% for 60 months. In fact, at j_{12} = 6% for 60 months, even at $3,600 below the sticker price, her monthly payments would be $540.55. ∎

Examples 3.9 and 3.10 deal with *mortgages*. You will see more on mortgages, and *outstanding principals* and *balances* in Chapter 5.

Example 3.9: Richard bought himself a little house and has been making monthly payments of $376.04 for some years. The nominal annual interest rate on his loan is 8.1% compounded monthly. Richard just made a monthly payment and realizes that he still has 149 payments to go. He would like to know how much he still owes on his house in case he wants to refinance his loan and get a lower interest rate. How much does he owe right now?

Solution: What Richard owes on his house at any point in time is called his *outstanding principal* or *outstanding balance*. Notice that we know his monthly payments, the interest rate on his loan, and the number of payments he still needs to make. Is that enough to determine his outstanding balance? We don't know how much he borrowed on the house. We will see that we have enough information because what Richard owes on his house is exactly the *present value* of all of his future payments—that is, the value of all the remaining payments right now. Here is a time diagram for this:

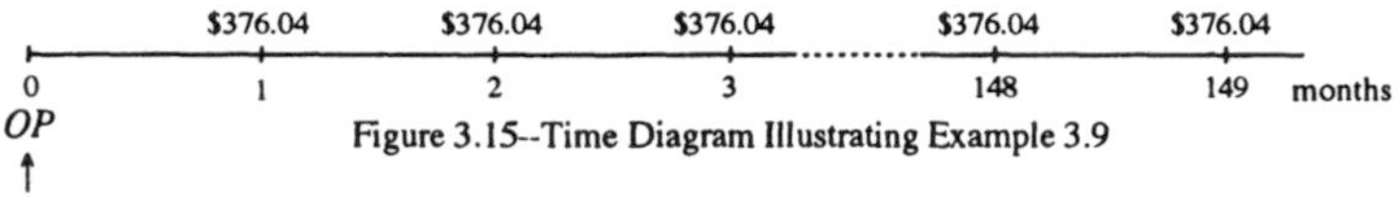

	$376.04	$376.04	$376.04		$376.04	$376.04	
0	1	2	3	......	148	149	months
OP ↑							

Figure 3.15--Time Diagram Illustrating Example 3.9

We are looking for the *OP*, the outstanding principal, at the time of Richard's latest payment. We have the inputs we need for Formula 3.2 or 3.2*a*. $P = \$376.04$, $i = .081/12 = .00675$, $n = 149$, and $d = 0 - 1 = -1$. Richard owes

$$S(\$376.04,\ .00675,\ 149,\ -1)$$

$$= \$376.04\frac{1-(1.00675)^{-149}}{.00675} = \$35,263.73. \qquad ▮$$

Note that we do not have enough information in Example 3.9 to find out the duration or the size of Richard's original loan. The information on his future payments, however, was sufficient to determine his outstanding balance. It was the *PV* of his future payments. Let's look at a different outstanding balance problem.

Example 3.10: Nadette borrowed $180,000 at j_{12} = 7.2% to buy a house. Her monthly payments are $1,221.82. How much does she still owe on her house just after her 96th payment?

Solution: We will start with a time diagram.

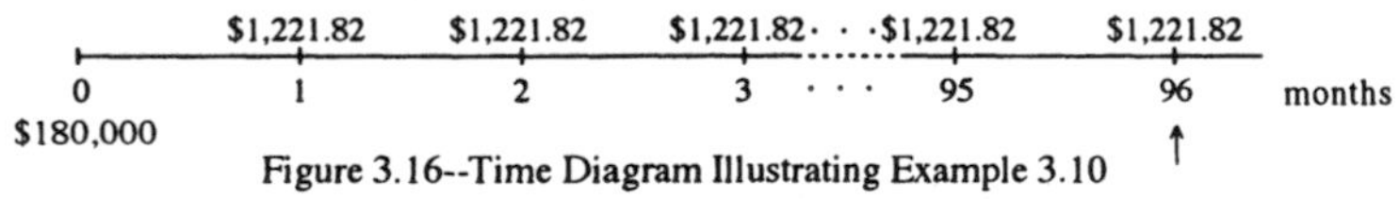

	$1,221.82	$1,221.82	$1,221.82	· · ·	$1,221.82	$1,221.82	
0	1	2	3	· · ·	95	96	months
$180,000						↑	

Figure 3.16--Time Diagram Illustrating Example 3.10

The arrow is at time 96 months because that is the time when we want to figure out what Nadette owes. If Nadette had not made any payments, she would owe $\$180{,}000(1+0.072/12)^{96} = \$180{,}000(1.006)^{96} =$ \$319,652.90. But, the bank has had Nadette's 96 payments of \$1,221.82 to credit to her account. These have accumulated to $S(\$1221.82, .006, 96, 95) = \$157{,}991.40$ after 96 months. Nadette owes the difference, \$319,652.90 - \$157,991.40 = \$161,661.50. ∎

Example 3.10 is a second example of how we can find the outstanding balance on a loan by knowing certain key facts about the original loan. In this case, we did not know, or need to know, the term of the loan.

Example 3.11: Have you ever been puzzled by the lotteries that say that someone won (let's say) \$46,000,000 and will receive the prize either as 20 annual payments of \$2,300,000 or as a lump sum of, say, \$23,748,052? That actually happened not too long ago in a Powerball Lottery. What is the explanation?

Solution: Suppose that it was you who won the \$46,000,000 lottery.

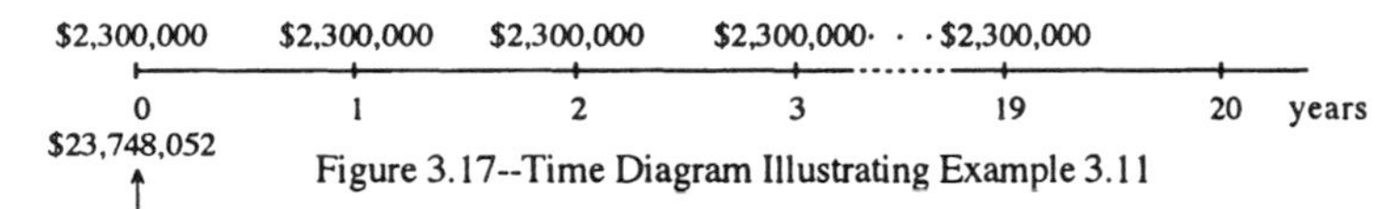

Figure 3.17--Time Diagram Illustrating Example 3.11

Claiming that you won \$46,000,000 is a bit deceptive. Your 20 payments do total \$46,000,000, but what you actually won was \$23,748,052: not that you'd want to turn that down. The way to think of it is that \$23,748,052 is, for some interest rate, i, the *PV* of 20 annual payments of \$2,300,000 made at the beginning of each year.

What the lottery folks do is add up the annual payments and announce the sum as your winnings. But take for example, the last payment of \$2,300,000 in 19 years. Is that worth the same thing as \$2,300,000 payable the day you won the lottery? The answer of course is no. That payment is worth $\$2{,}300{,}000(1+i)^{-19}$ for interest rate, i. The 20 annual payments of \$2,300,000 you won are not equivalent to \$46,000,000 now. Rather, the lottery folks assume a given interest rate, i, such that the *PV* of 20 annual payments of \$2,300,000 is \$23,748,052. Using Formula 3.2 or 3.2*b*, this looks like

$$S(\$2{,}300{,}000, i, 20, 0) = \$2{,}300{,}000\frac{1-(1+i)^{-20}}{i}(1+i)$$

$$= \$23,748,052. \qquad (3.3)$$

This is another problem requiring numerical methods to solve. The first one we saw was Example 2.14 of Chapter 2.

Using a solver: Enter the equation

$$0 = \$2,300,000\frac{1-(1+i)^{-20}}{i}(1+i) - \$23,748,052$$

into the Equation Solver. After starting with a value of I = .05, and pressing ALPHA [SOLVE], the final screen I got was:

```
EQUATION SOLVER
eqn: 0 =2300000(1-(1+I)^(-20))/I(1+I) - 23748052
I = .08412354104...
bound={-1E99, 1E99}
left-rt=0
```

The lottery people assume that their money can earn an effective rate of about 8.412%.

Using Guess-and-Check: In a Guess-and-Check Table, I calculated values of

$$LHS(i) = \$2,300,000\frac{1-(1+i)^{-20}}{i}(1+i),$$

trying to find i making $LHS(i)$ close to \$23,748,052. Starting with i = 0.05, I found $i = j_1 \approx 0.08412$ after seven steps. This is reasonably close to the more exact value we got with the solver. ∎

Example 3.12: Belle has bought a house with a 20-year mortgage on which she pays \$496 at the end of every month. According to her mortgage contract, the interest rate for the first 8 years of her mortgage will be j_{12} = 7.5% while for the final 12 years her rate will be j_{12} = 7.2%. How much did Belle borrow to buy her house?

Solution: We start with a time diagram.

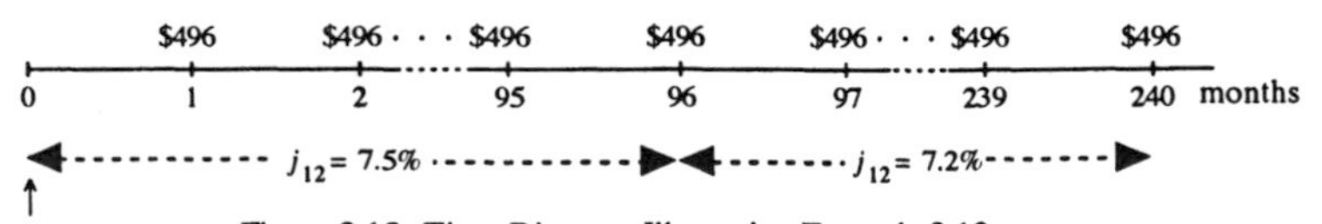

Figure 3.18--Time Diagram Illustrating Example 3.12

This problem presents a new wrinkle. The interest rate changes in the middle of the term of the loan. What we do in this case is find the value of the first 96 payments at rate $j_{12} = 7.5\%$ at time 0. Then we find the value of the second 144 payments at rate $j_{12} = 7.2\%$ at time 96 and discount this piece back to time 0 at $j_{12} = 7.5\%$. I'll go over that again in a little more detail. The value of the first 96 payments, with $d = 0 - 1 = -1$ is $S(\$496, 0.075/12, 96, -1) = \$35{,}724.91$. At time 96, the second set of payments, 144 of them, at $j_{12} = 7.2\%$, with $d = 96 - 97 = -1$, are worth $S(\$496, 0.072/12, 144, -1) = \$47{,}734.85$. But, we want the value of these 144 payments back at time 0. Since the interest rate for the first 8 years is $j_{12} = 7.5\%$ ($i = .00625$ per month), the value of the second set discounted back to time zero is $\$47{,}734.85(1.00625)^{-96} = \$26{,}246.40$. Therefore, Belle borrowed \$35,724.91 + \$26,246.40 = \$61,971.31 for her house. ∎

Example 3.13: Judith and Ted bought a small plane for which they borrowed \$110,000 at $j_{12} = 9.75\%$. They will make payments of \$1,200 at the end of every month for as long as necessary plus a smaller final payment one month after the last \$1,200 payment. How many payments of \$1,200 will they make, and what is the amount of the final smaller payment?

Solution: Here is a time diagram for this problem:

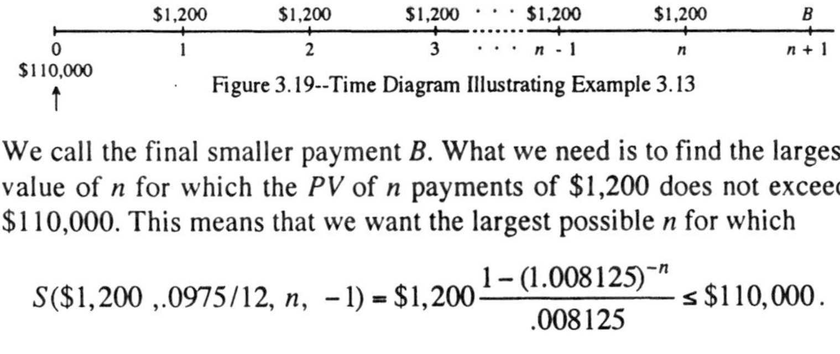

	\$1,200	\$1,200	\$1,200	· · ·	\$1,200	\$1,200	B
0	1	2	3	· · ·	$n - 1$	n	$n + 1$
\$110,000 ↑							

Figure 3.19--Time Diagram Illustrating Example 3.13

We call the final smaller payment B. What we need is to find the largest value of n for which the PV of n payments of \$1,200 does not exceed \$110,000. This means that we want the largest possible n for which

$$S(\$1{,}200\ ,.0975/12,\ n,\ -1) = \$1{,}200\frac{1-(1.008125)^{-n}}{.008125} \le \$110{,}000.$$

We will do this by finding the value of n for which

$$S(\$1{,}200,\ .0975/12\ ,n,\ -1) = \$1{,}200\frac{1-(1.008125)^{-n}}{.008125} = \$110{,}000.$$

After doing some algebra and using the Log Rule, we get $n \approx 168.765$. What tells us that $n = 168$ is too low and $n = 169$ is too high. As a check, we find that for $n = 168$, $S(\$1{,}200, 0.008125, 168, -1) =$

\$109,765.94 (too low), and for $n = 169$, $S(\$1{,}200, 0.008125, 169, -1) =$ \$110,071.61 (too high). Thus, there will be $n = 168$ equal payments of \$1,200 and a final smaller payment, B. From the time diagram, we can now write an updated equation of value with $n = 168$:

$$S(\$1{,}200,\ .008125,\ 168,\ -1) + B(1.008125)^{-169}$$

$$= \$109{,}765.94 + B(1.008125)^{-169} = \$110{,}000.$$

Solving, we get $B = (\$110{,}000 - \$109{,}765.94)\cdot(1.008125)^{169} = \$918.88$. To summarize, Judith and Ted will make 168 equal monthly payments of \$1,200 plus a final payment of \$918.88 one month later. ▮

Exercises 3.2

3.2.1 Peggy borrowed \$17,345.40 to pay for her new car. She got a 5-year loan at the nominal annual rate of 7.34% compounded monthly. If her first payment is one month after the date of the loan, what are Peggy's payments at the end of each month?

3.2.2 Mickey turned 62 one month ago, and he (or his estate) is to receive monthly retirement payments of \$2,141.09 for the next 25 years beginning immediately (one month after Mickey's 62nd birthday). To finance these payments, Mickey made deposits at the end of every month from the time he was 28 years old, the last payment, one month ago. Assuming that Mickey's money earns at the rate of $j_{12} = 8.2\%$, what were the size of the deposits into his retirement fund?

3.2.3 Karen is buying her first car. For the car she wants, she must borrow \$13,099.95. The salesman offers her a 5-year loan at $j_{12} = 7.65\%$. If her first payment is one month after the date of the loan, what will Karen's payments be at the end of each month?

3.2.4 Catherine is saving to buy an anniversary present for her parents. She is depositing \$55 at the beginning of each month into the account, and her bank account pays at the rate of $j_{12} = 4\%$. Assuming a total of 12 deposits, how much does Catherine have at the end of one year?

3.2.5 Ron and M. J. bought a house for \$125,550 eight years ago. They have been making payments of \$1,002.52 at the end of every month on a loan with a nominal annual interest rate of

8.4% compounded monthly. How much do they owe on their house eight years after buying it? *Hint:* See Example 3.10.

3.2.6 Inez is 20 and wants to accumulate $2,000,000 by the time she is 65. She found an investment which earns $j_{12} = 9.8\%$. How much does she need to deposit at the end of each month until she reaches 65 in order to reach her goal?

3.2.7 Erik is paying off a yacht with payments of $889.91 made at the end of every quarter on a 30-year loan borrowed at the nominal annual rate of 7.42% compounded quarterly. Erik just made his 35th quarterly payment. How much does Erik still owe on his yacht?

3.2.8 Ingegärd pays $250 at the end of every month on a lakefront cabin she bought. She still owes $27,049.30. The interest rate on her loan is $j_{12} = 7.5\%$. How many payments does Ingegärd still have to make?

3.2.9 Two years ago, Gene, who commutes to work by bicycle, invested $15,000 in The Abelian Group, a public interest group that promotes commuting by human power. One year ago, Gene received a payment of $9,000 and has just now cashed in his account for another $10,000. What *yield* (here, effective) *rate* did Gene receive on his investment? *Hint:* Use your calculator's solver if it has one, or use Guess-and-Check.

3.2.10 In 1977, Sandra built a house for which she borrowed $45,000 at $j_{12} = 7.5\%$. She makes payments of $350 at the end of every month, but she will make one final payment of less than $350 one month after the last regular payment of $350. How many payments of $350 will Sandra make, and what is the amount of her final payment? *Hint:* See Example 3.13.

3.2.11 Jayanti won $130,000,000 in the Big Jackpot lottery. His options were to receive $6,500,000 per year for 20 years, first payment now, or a lump sum of $49,290,168. What effective interest rate was assumed by the Big Jackpot people in calculating the lump sum? *Hint:* Use your calculator's solver if it has one, or use Guess-and-Check.

3.2.12 Anne is saving for a rainy day. She deposited $150 at the beginning of every month for the first year, $200 at the beginning of every month for the next two years, and $300 at the beginning of month for a fourth year. If the nominal annual rate of interest is $j_{12} = 6\%$, how much as Anne accumulated at the end of four years?

3.2.13 Betsy pays \$744.39 at the end of every month on the mortgage for her house. She borrowed the full price of her house at $j_{12} = 7.5\%$ and has a 20-year mortgage. What did Betsy pay for her house?

3.2.14 Greg has picked out a nice red convertible selling for \$23,130.70. If he borrows the full price of the car at an interest rate of $j_{12} = 9.99\%$ and will make payments at the end of every month for 6 years, what will the size of his monthly payment be?

3.2.15 For two years now, Neith's parents have been putting \$200 into a 'college fund' account at the beginning of every month. The account earns interest at the nominal annual rate of 5.17% compounded monthly. They will continue with these deposits for another 12 years (for a total of 14 years of deposits). How much will be in her fund at the end of the 14 years?

3.2.16 Honest Slim's Loan Company is running a 'special.' Slim will lend you \$20,000 which you will pay back with payments of \$1,950 at the end of every month for 12 months. The 'special' part is that if you pay it back on time, nobody gets hurt. Know what I mean? What nominal annual rate of interest compounded monthly is Slim charging? *Hint:* Use your calculator's solver if it has one, or use Guess-and-Check.

3.2.17 How much should Dawson deposit monthly into an account earning interest at $j_{12} = .0575$ if he wants to accrue \$15,000 at the end of 5 years?

3.2.18 At the moment, Suzanne owes Alpha Mortgage 90 equal monthly payments of \$645.74 plus a 91st payment of \$1,250 one month later. If Beta Mortgage wants to buy Suzanne's mortgage contract and earn interest at $j_{12} = 6\%$, how much should Beta pay?

3.2.19 When Lloyd was working at the Beer 'n Burger, he deposited \$150 at the end of every month into a savings account paying interest at the nominal rate of 4.45% compounded monthly. This went on for 7 years. He has not touched his savings, and it is now 14 years since he began saving. How much is in his account at this time?

3.2.20 Antonia figures she can spend up to \$240 per month for a car. If she can borrow money at the rate of $j_{12} = .089$ and has 5 years to pay, how much can she afford to borrow?

3.2.21 Roxanna has just brought home a new car. According to the conditions of her loan, she will make monthly payments of \$312.23 at the end of every month for six years to pay off her loan made at a nominal annual interest rate of 4.25% compounded monthly. How much did Roxanna borrow for her car?

3.2.22 Salvador is paying off his home mortgage with monthly payments. Including the one is just about to make, he has 89 payments still to be made. If the outstanding balance on his home loan is \$48,558 and the interest rate on his mortgage is $j_{12} = 7.12\%$, what are his monthly payments?

3.2.23 Luisa has deposited \$450 at the end of every quarter for the last three years. Her money earns at $j_4 = 6.14\%$. How much has Luisa saved at the end of the three year period?

3.2.24 Tran is saving for his daughter's college education. At the beginning of each month, he deposits \$220 into an account earning $j_{12} = 7\%$. How much will Tran have saved at the end of eighteen years?

3.2.25 Ida pays \$723.07 at the end of every month for her cabin by the lake. The cabin will be paid off in 15 years. What nominal annual interest rate compounded monthly is Ida paying if the price of the cabin was \$78,000? *Hint:* Use your calculator's solver if it has one, or use Guess-and-Check.

3.3 An Important Application of Annuities—Bonds

One way for governments and companies to raise large amounts of money is to issue *bonds* and try to sell as many as possible. Bonds are certificates bought by investors for a certain *purchase price*. In return the investor receives periodic interest payments called *coupons* until a specific *redemption date*. At the redemption date the investor is paid a specific amount, called the *redemption value*. Here is all of that again and a little more, in list form:

P: The *purchase price* of the bond to yield interest rate i per period, the period usually being the half-year.

F: The *face value* or *par value* of the bond. This is the value printed on the bond and is typically a multiple of \$100. F is multiplied by the coupon rate (see r, below) to determine the amount of each coupon paid out to the investor. F is often, but not always, the same as the redemption value.

r: The *coupon rate* when multiplied by the face value F, determines the amount of the coupon paid periodically to the investor. The coupon rate is an interest rate. The most common period for paying of coupons is the half-year, and this will be our assumption throughout.

C: The *redemption value* of a bond is the amount of money to be paid to the investor at the redemption date (= *maturity date*). Often, but not always, $C = F$. In such cases, we say that the bond is redeemable at par. We will assume that $C = F$ unless otherwise stated.

Fr: The *coupon amount*—the amount of each coupon paid to the investor.

i: The *yield rate* of a bond per interest period. This is the interest rate earned by the bondholder.

n: The number of coupon payments until the redemption date.

Here is what a time diagram for a typical bond looks like:

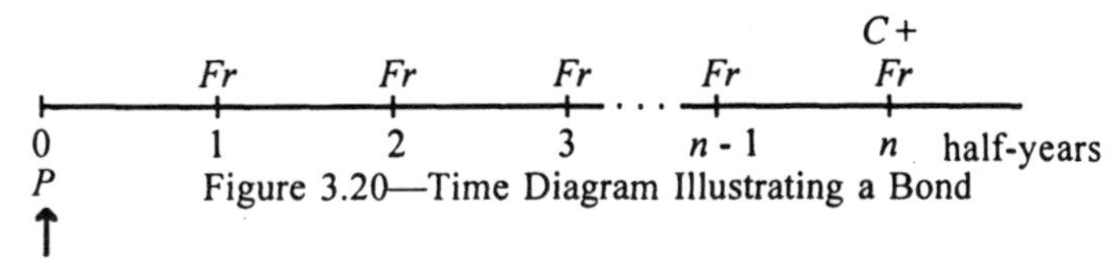

Figure 3.20—Time Diagram Illustrating a Bond

From the point of view of the investor, the money out is P, and the money in is the n coupon payments each in the amount Fr added to the redemption value, C. From what we have learned about annuities, we should be able to write down an equation of value for this bond. I have put the focal date at time 0 giving us a formula for the purchase price directly. Using Formula 3.2*a*, the equation of value we get is

$$P = S(Fr,\ i,\ n,\ -1) + C(1+i)^{-n} = Fr\frac{1-(1+i)^{-n}}{i} + C(1+i)^{-n}. \quad (3.4a)$$

With a little algebra (see Exercise 3.3.7 below), we can also write

$$P = S(Fr - Ci,\ i,\ n,\ -1) + C = (Fr - Ci)\frac{1-(1+i)^{-n}}{i} + C. \quad (3.4b)$$

Example 3.14: Diomedes buys a \$1,000 par value (face value) 20-year bond with semi-annual coupons at rate $j_2 = 8\%$ ($r = 4\%$) and redeemable at par. The bond is to have a yield of $j_2 = 10\%$ ($i = 5\%$). Find the price that Diomedes pays.

Solution: We have all of the inputs for Formulas 3.4*a* or 3.4*b*. We have $F = C = \$1,000$, $r = 0.04$, $i = 0.05$, and $n = 40$ (= 2 × 20). Thus, Diomedes pays

$$P = (\$40 - \$50)\frac{1-(1.05)^{-40}}{.05} + \$1000 = \$828.41.$$ ∎

Note that in Example 3.14, the purchase price of the bond is less than its redemption value. When this occurs, that is, when $P < C$, we say that the bond is bought at a *discount.* On the other hand, when $P > C$, when the purchase price exceeds the redemption value of the bond, we say that the bond is bought at a *premium.* The amount of the premium or discount is $|P - C|$, the magnitude of the difference between P and C. In Example 3.14, the bond was bought at a discount of $|\$828.41 - \$1000| = \$171.59$.

Example 3.15: In order to raise money to expand their franchise operation, the Eat 'n' Run fast-food chain issued 15-year \$10,000 par value bonds with semi-annual coupons at rate $j_2 = 9\%$. What price for the bond will produce a yield rate of $j_2 = 7.36\%$ for the bondholder?

Solution: The first thing we have to do is convert of $j_2 = 7.36\%$ to a semi-annual rate for use in Formula 3.4*a* or 3.4*b*. We have

$$i = j_2/2 = .0368 = 3.68\%.$$

We can use Formula 3.4*a* with inputs $F = C = \$10,000$ (remember that $F = C$ unless otherwise stated), $r = 4.5\% = .045$, $i = .0368$, and $n = 30$. The selling price for the bond is

$$P = S(\$450,\ .0368,\ 30,\ -1) + \$10,000(1.0368)^{-30} =$$

$$\$450\frac{1-(1.0368)^{-30}}{.0368}+\$10{,}000(1.0368)^{-30}=$$

$$=\$11{,}474.70.$$

This bond is bought at a premium of $1,474.70. ∎

Example 3.16: Stuart bought a $10,000 15-year bond issued by the Say Cheese Camera Company at a discount for $9,100. The coupon rate for the bond is 8% compounded semi-annually. What yield rate is Stuart enjoying?

Solution: What we know is that P = $9,100, $F = C$ = $10,000, r = 4% = .04, and n = 30. What we do not know is i, the interest rate earned per half-year. Putting all this into Formula 3.4*a*, we get

$$\$9{,}100=\$400\frac{1-(1+i)^{-30}}{i}+\$10{,}000(1+i)^{-30},$$

and this must be solved for i.

Using a solver, we enter the equation

0 = 400(1 - (1 + I)^(-30))/I + 10000(1 + I)^(-30) – 9100

into the Equation Solver. The solution for I is I = i = 0.04556181551... ≈ 4.556% per half-year. Stuart's yield rate is $j_2 \approx 9.112\%$.

Using Guess-and-Check, we find $i = j_2/2$ making

$$LHS(i)=\$400\frac{1-(1+i)^{-30}}{i}+\$10{,}000(1+i)^{-30}$$

close to $9,100. I did this and found $i \approx 0.0456$ after eight steps. This leads to $j_2 \approx 9.12\%$, an error of 0.008%. ∎

When an organization issues a series of bonds to be redeemed in installments, we call the bonds a *serial issue* of bonds. We can think of a serial issue of bonds as a sequence of separate bonds that can be analyzed individually.

Example 3.17: A serial issue bond of $100,000 with coupons at nominal rate j_2 = 8.4% will be redeemed by payments of $50,000 in 10 years, $25,000 in 15 years, and $25,000 in 20 years. Find the purchase of this issue to yield j_2 = 9.6%.

Solution: We have $r = 4.2\% = .042$, $i = j_2/2 = 9.6\%/2 = 4.8\% = .048$, and for the three separate bonds, $F_1 = C_1 = \$50{,}000$, $F_2 = C_2 = \$25{,}000$, and $F_3 = C_3 = \$25{,}000$. From Formula 3.4*b*, the individual prices are

$$P_1 = (\$2{,}100 - \$2{,}400)\frac{1-(1.048)^{-20}}{.048} + \$50{,}000 = \$46{,}197.11,$$

$$P_2 = (\$1{,}050 - \$1{,}200)\frac{1-(1.048)^{-30}}{.048} + \$25{,}000 = \$22{,}640.62,$$

and

$$P_3 = (\$1{,}050 - \$1{,}200)\frac{1-(1.048)^{-40}}{.048} + \$25{,}000 = \$22{,}354.07.$$

The purchase price for the serial issue is the sum of these prices, \$91,191.80. ∎

A bond is *callable* if the issuer has the option of redeeming it before the maturity date. For this reason, the investor should calculate the price that will guarantee at least the desired yield rate regardless of whether or not the bond is called early. If the bond is callable at par (i.e., $F = C$), it can be shown that the lowest price is at the latest possible call date if $i > r$. On the other hand, the lowest price is at the earliest possible call date if $i < r$. Even if $F \neq C$, the investor can calculate all possible choices and then offer the lowest one.

Example 3.18: Midnight Auto Sales has issued a 20-year \$1,000 bond at coupon rate $j_2 = 10\%$. The bond is callable at par after 15 years. Find the price to yield $j_2 = 8\%$.

Solution: Here, $.04 = i < r = .05$, and the bond is callable at par. We know from the comments above that the lowest price occurs if the call date is the earliest possible, namely, at 15 years. From Formula 3.4*b*, the price is

$$P = (\$50 - \$40)\frac{1-(1.04)^{-30}}{.04} + \$1{,}000 = \$1{,}172.92.$$

If you are not sure of yourself, you can also calculate the selling price at each possible call date—that is, at 30, 31, 32, ..., 40 half-years. These prices are, respectively, \$1,172.92, \$1,175.88, \$1,178.74, ..., \$1,197.93. The lowest price is for a call date 15 years after purchase. ∎

Exercises 3.3

3.3.1 Each row in the table below gives all of the necessary information to determine the price of a bond. Determine the price for each bond.

Bond	*F*	*C*	*i*	*n* (periods)	*r*	*P*
a	\$1,000	at par	5%	20	6%	
b	\$1,000	at par	6%	30	5%	
c	\$1,000	\$1,100	5%	40	4.5%	

3.3.2 Steve buys a \$5,000 par value 15-year bond with semi-annual coupons at rate j_2 = 12% for the purchase price of \$5,814.90. The bond yields j_2 = 10%. What is the redemption value of Steve's bond?

3.3.3 Rickie bought a \$10,000 20-year bond issued by the Babaloo Bongo Drum Company at a premium for \$10,980. The redemption value of the bond is \$10,138.13, and the coupon rate for the bond is 9.5% compounded semi-annually. What is Rickie's annual yield rate compounded semi-annually? *Hint:* Use your calculator's solver if it has one, or use Guess-and-Check.

3.3.4 The Trustus Manufacturing Company issues a 15-year \$5,000 bond at coupon rate j_2 = 11%. The bond is callable at par after 12 years. Find the price to yield j_2 = 14%.

3.3.5 Bob bought a 20-year \$5,000 bond with coupon rate j_2 = 6% redeemable for \$5,250 to yield j_2 = 7%. Find Bob's purchase price.

3.3.6 A serial bond issue of \$100,000 with coupons at j_2 = 8% will be redeemed with four equal payments of \$25,000—the first in 5 years, the second in 10 years, the third in 15 years, and the fourth in 20 years. If the yield is to be j_2 = 6%, find the purchase price.

3.3.7* Derive Formula 3.4*b* from Formula 3.4*a*.

3.3.8 A bond with a par value of \$50,000 has coupons at the rate of j_2 = 12%. It will be redeemed at par when it matures in 10 years. The purchaser will realize a yield rate of j_2 = 13%. Find the price of the bond.

3.3.9 Nadya bought a 20-year \$5,000 bond with coupon rate $j_2 = 6\%$ redeemable for \$5,250 to yield $j_1 = 7\%$. Find Nadya's purchase price. Compare your answer with Exercise 3.3.5.

3.5 Summary of Formulas

Geometric Sum

$$G_n = a + ar + ar^2 + ... + ar^{n-1} = a\frac{1-r^n}{1-r} = a\frac{r^n - 1}{r-1}$$

Formula For Regular Annuities where d = (number at arrow) – (number at first payment)

$$S = S(P,i,n,d) = P\frac{1-(1+i)^{-n}}{i}(1+i)^{d+1}$$

Special Cases

i) $$S(P,i,n,-1) = P\frac{1-(1+i)^{-n}}{i}$$

ii) $$S(P,i,n,0) = P\frac{1-(1+i)^{-n}}{i}(1+i)$$

iii) $$S(P,i,n,n-1) = P\frac{(1+i)^n - 1}{i}$$

iv) $$S(P,i,n,n) = P\frac{(1+i)^n - 1}{i}(1+i)$$

Price of a Bond (version *a*)

$$P = Fr\frac{1-(1+i)^{-n}}{i} + C(1+i)^{-n}$$

Price of a Bond (version *b*)

$$P = (Fr - Ci)\frac{1-(1+i)^{-n}}{i} + C$$

Miscellaneous Exercises

3M1. What amount should you invest now at $j_1 = 6.6\%$ to provide a 15-year annuity of \$25,000, first payment in 10 years?

3M2. Señor Calvo is the owner of the Hair Apparent Wig Factory. He wants to borrow \$190,000 to buy the latest wig-making machines. His bank has offered to loan him the money at the nominal annual rate of 8.12% compounded quarterly. Señor Calvo will repay the loan with equal quarterly payments of

$8,200 for as long as necessary plus a smaller final payment three months after the last payment of $8,200. Find the number of equal payments and the amount of the final smaller payment. *Hint:* See Example 3.13.

3M3. DeAnn bought a new car and borrowed $21,080.14 from the auto company's finance division to cover the car, taxes, fees, and licenses. Her monthly payments for five years are $428.44. Visiting her credit union, DeAnn saw that their rate for new car loans was $j_{12} = 7.9\%$. Should DeAnn switch her loan company?

3M4. If Padraic deposits $275 at the beginning of each month for 3 years and interest is at $j_{12} = 5.15\%$, how much will he have accumulated at the end of the 3 year period?

3M5. Medha bought a computer from a friend for $1,100 and agreed to pay if off with payments of $100 at the end of each month for 10 months plus a final payment one month after the last $100 payment. If interest is at $j_{12} = 8\%$, what will be the amount of the final payment?

Chapter 4—More General Annuities

4.1 Non-Regular Annuities

The annuities of the last chapter were *regular* annuities—payments and conversion periods coincided, and payments were usually equal. In addition, the interest rates quoted in any given example were always consistent with the payment and interest period in that example. That is, if the period was the month, the interest rate quoted was j_{12}. If the period was the quarter, the rate quoted was j_4, etc. In this chapter, we will deal with a variety of annuity problems where we are faced with some situations not encountered in Chapter 3. In each case, we shall see what is different from previous problems as well as how to handle these differences.

Example 4.1: Evaristo deposits $150 at the beginning of every month into an account earning at the nominal annual rate of 6% compounded semiannually. How much does he have at the end of 15 years?

Solution: We'll start with a time diagram.

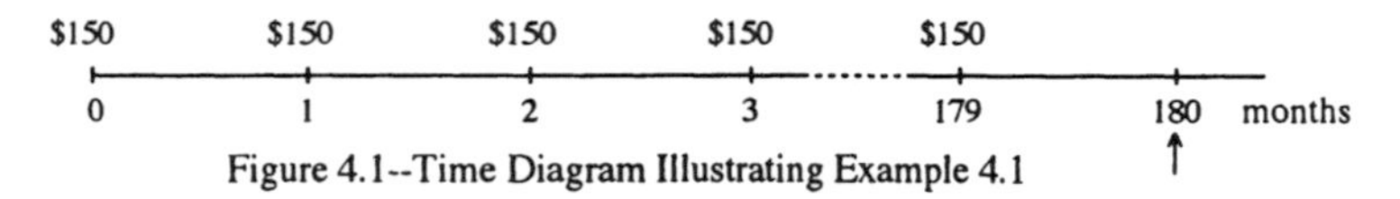

Figure 4.1--Time Diagram Illustrating Example 4.1

In this problem, the period of the interest rate quoted, half-years, is not the same as the period of the payments, months. We have to convert one to the other, and we have two ways to do this. First we will convert the semiannual interest rate to the equivalent monthly rate. We know that j_2 = 6%, but what we need in order to do the problem is the monthly interest rate $i = j_{12}/12$ equivalent to this. We go back to Formula 2.3*b* to find that

$$i = j_{12}/12 = \left(1 + \tfrac{.06}{2}\right)^{2/12} - 1 = 0.00493862203.$$

Now we have all of our inputs for Formula 3.2: P = $150, i = .00493862203, n = 180 months, and the focal date $d = 180 - 0 = 180$. Putting the inputs into Formula 3.2, we get

$$S(\$150, .004938622, 180, 180) =$$

$$\$150\frac{(1.00493862203)^{180}-1}{.00493862203}(1.00493862203)=\$43{,}564.11,$$

the money that Evaristo has accrued over 15 years.

The other way to attack the problem is to leave the semiannual rate as it is at $i = j_2/2 = .03$ and convert the six payments at the beginning of each month per half-year into an equivalent single payment at the beginning of each half year. We do this as follows: A single payment at the beginning of each six-month period equivalent to \$150 at the beginning of each month for six months is $P = \$150(1 + 1.03^{-1/6} + 1.03^{-2/6} + \ldots + 1.03^{-5/6}) = \$889.01489426\ldots$ Then we can write Evaristo's accumulated amount as

$S(\$889.01489426\ldots, .03, 30, 30) = \$43{,}564.11$, just as before. ∎

Example 4.2: Annie just won the \$800,000 'Big Jackpot' lottery paying her either an annuity of \$20,000 at the end of every six months for 20 years, or a single payment now. If the fund earns interest at the rate of $j_4 = 7.25\%$, what single payment to Annie now would be equivalent to the annuity payments?

Solution: We will start again with a time diagram.

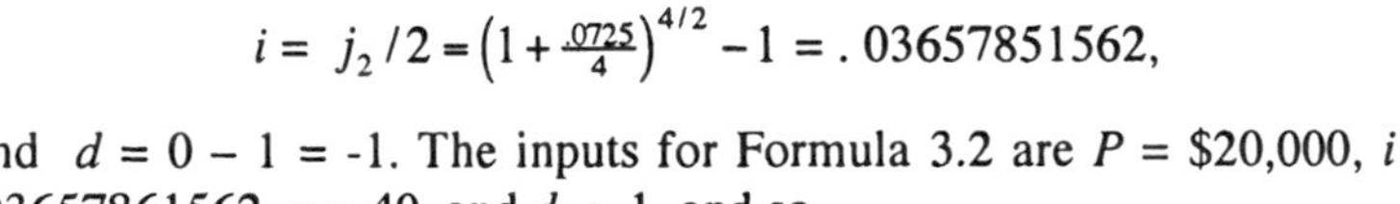

	\$20,000	\$20,000	\$20,000 · · ·	\$20,000	\$20,000	
0	1	2	3 · · ·	39	40	half-years
↑						

Figure 4.2--Time Diagram Illustrating Example 4.2

When we convert the quarterly rate to the equivalent semiannual rate, we will have a regular annuity. Using Formula 2.3*b* again, we find

$$i = j_2/2 = \left(1+\tfrac{.0725}{4}\right)^{4/2}-1 = .03657851562,$$

and $d = 0 - 1 = -1$. The inputs for Formula 3.2 are $P = \$20{,}000$, $i = .03657861562$, $n = 40$, and $d = -1$, and so

$$S(\$20{,}000,\ .03657851562,\ 40,\ -1) =$$

$$\$20{,}000\frac{1-(1.03657851562)^{-40}}{.03657851562} = \$416{,}837.14.$$

If Annie were given the option of a single payment immediately, it would be for \$416,837.14. ∎

Example 4.3: Lucho deposited \$500 at the end of every quarter for eleven years in an account earning the nominal annual rate of 6.5% compounded monthly. How much does he have at the end of 11 years?

Solution: The time diagram for this example looks like this:

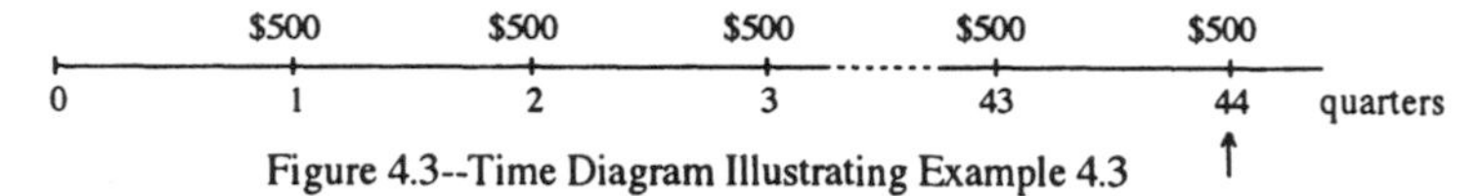

Figure 4.3--Time Diagram Illustrating Example 4.3

Here, the problem is that Lucho deposits his money quarterly, but the money is compounded monthly. We use Formula 2.3*b* again, which in this case, gives us the nominal annual interest rate per quarter

$$i = j_4/4 = \left(1 + \tfrac{.065}{12}\right)^{12/4} - 1 \approx .01633817977.$$

Now we have a regular annuity. The inputs for Formula 3.2 are P = \$500, $i = .01633817977$, $n = 44$, and $d = 44 - 1 = 43$. The accumulated value of Lucho's account is

$$S(\$500, .01633817977, 44, 43) =$$

$$= \$500\frac{(1.01633817977)^{44} - 1}{.01633817977} = \$31{,}834.83.$$ ∎

Exercises 4.1

4.1.1 Maria deposits \$150 in her bank account at the end of every month. Her bank pays interest at the nominal annual rate of 4.75% compounded daily. How much is in Maria's account after 6 years?

4.1.2 Ryan borrowed \$25,000 from his in-laws for home improvements. On the advice of the in-laws' bank, they agreed on an interest rate of 10% effective. If Ryan would like to make monthly payments to repay the loan over 5 years, what should his monthly payments be?

4.1.3 Giaconda will make equal deposits at the beginning of every month for 15 years into a fund that will provide her a 25-year annuity of \$2,500 per month. Her last deposit will be one month before she receives her first monthly check for \$2,500. What will the size of Giaconda's deposits be if her money earns 6.8% effective?

4.1.4 Evelyn saved for a vacation by depositing $100 into a savings account at the end of every week for two years. If her money was compounded monthly at the nominal annual rate of 5.4%, how much did Evelyn have in her vacation fund at the end of two years? (Assume 52 weeks per year.)

4.1.5 Alexandra bought a used car for $3,300 from her cousin. She paid $800 down and owes the rest. She will pay off what she owes with equal payments at the end of each month for two years. How much are her monthly payments if interest is to be 9% effective?

4.1.6 In order to accumulate $25,000, Amanda will deposit equal amounts at the end of each month for 5 years into an account earning interest at the nominal annual rate of 5% compounded quarterly. What monthly deposits must Amanda make?

4.1.7 Pablo bought a restaurant for a $30,000 down payment and payments of $1,100 at the end of every month for 8 years. If the interest rate on the loan was 8.6% effective, what was Pablo's purchase price?

4.1.8 At the end of each year, Madhu puts $5,000 into an account earning interest at the rate $j_{12} = 7.2\%$. How much does she have in her account after 20 years?

4.1.9 Wai-Han bought a condominium for which she paid $385,000. She made a down payment of $150,000 and borrowed the rest. The interest rate charged on her loan is $j_4 = 8\%$. If she makes payments at the end of every month for 30 years, what will be the amount of these payments?

4.1.10 Yuri wants to create a fund that will accumulate to $30,000 in three years so that he can remodel his house. He will make deposits at the end of each quarter into the fund earning at the annual rate of $j_{12} = 7.5\%$. How big are Yuri's deposits?

4.2 Annuities Whose Payments Increase/Decrease by a Fixed Percentage

In many common situations, annuities that increase or decrease by a fixed percentage each period make more sense than regular annuities. For example, if a person wants to collect annuity payments over a long period of time, it makes sense to try to have those payments anticipate the effect of inflation. The buying power of \$1,000 today would not be the same as the buying power of \$1,000 in 20 years if the inflation rate were, say, 3% annually. To match the buying power of today's \$1,000, one would need $\$1{,}000(1.03)^{20} = \$1{,}806.11$ in 20 years.

We will start with an example of an increasing annuity. Then we will derive a general formula for *increasing* and *decreasing* annuities.

Example 4.4: Rodrigo will make deposits into a retirement fund earning $j_{12} = 7.8\%$. His first deposit, payable today, will be for \$200. He wants all his deposits to be equivalent to today's \$200, assuming an inflation rate of 0.3% each month. This means that he wants next month's payment to be $\$200(1.003) = \$200.60$, his third payment in two months to be $\$200(1.003)^2 = \$201.20$, and so on. He wants these payments at the beginning of each month to continue for 35 years = 420 months. Note that Rodrigo's last deposit at the beginning of the 420th month will be $\$200(1.003)^{419} = \$701.65$. How much will be in Rodrigo's fund at the end of 35 years?

Solution: Note that the only departure from this being a regular annuity is the unequal payments. We can handle this by going back to basic principles. We will start with a time diagram for the problem.

\$200	$\$200(1.003)$	$\$200(1.003)^2$	$\$200(1.003)^3$	$\$200(1.003)^{418}$	$\$200(1.003)^{419}$		
0	1	2	3	418	419	420 ↑	months

Figure 4.4--Time Diagram Illustrating Example 4.4

We need to bring each payment forward to the focal date at 420 months at the monthly interest rate .078/12 = .0065. Then we will add them up and see what to do at that point. The value of the first payment at time 420 is $\$200(1.0065)^{420}$ because we have to accumulate \$200 for 420 months at the interest rate $i = .0065$. The value of the second payment is $\$200(1.003)(1.0065)^{419}$, the third is $\$200(1.003)^2\cdot(1.0065)^{418}$, and the last two are $\$200(1.003)^{418}(1.0065)^2$ and $\$200(1.003)^{419}\cdot(1.0065)^1$. If you look closely, you will see that what we have here is actually a geometric sum composed of $n = 420$ terms with first term $a = \$200(1.0065)^{420}$ and common ratio $r = 1.003/1.0065$. Convince yourself that this is the case. Referring back to Equation 3.1*a*, the sum in question is

$$\$200(1.0065)^{420} + \$200(1.003)(1.0065)^{419} + \$200(1.003)^2(1.0065)^{418} +$$
$$\ldots + \$200(1.003)^{418}(1.0065)^2 + \$200(1.003)^{419}(1.0065)^1 =$$

$$\$200(1.0065)^{420}\,\frac{1-(1.003/1.0065)^{420}}{1-(1.003/1.0065)}=\$671{,}723.79.$$

Rodrigo will have \$671,723.79 in 35 years. ∎

General Formula for Increasing/Decreasing Annuities: To generalize the important result above, assume that the interest rate per period, i, is constant, that there are n periods, and that the sequence of payments increases or decreases (increases if $k > 0$; decreases, if $-1 < k < 0$) at the rate k per period. We will take the payments at times $0, 1, 2, \ldots, n-1$ to be

$$P,\ P(1+k),\ P(1+k)^2, \ldots, \text{and } P(1+k)^{n-1}.$$

Here is a time diagram for this situation:

P	$P(1+k)$	$P(1+k)^2$	$P(1+k)^3$	. . .	$P(1+k)^{n-1}$		. . .		
0	1	2	3	. . .	$n-1$	n	. . .	d ↑	periods

Figure 4.5--Time Diagram Illustrating Increasing/Decreasing Annuity

I have positioned the focal date, d, to be past time n, but the formula would look exactly the same wherever I placed d. What we have to do is bring each payment to time d. The sequence of payments evaluated at time d is

$$P(1+i)^d,\ P(1+k)(1+i)^{d-1},\ P(1+k)^2(1+i)^{d-2},$$

$$P(1+k)^3(1+i)^{d-3}, \ldots, \text{and } P(1+k)^{n-1}(1+i)^{d-(n-1)}.$$

The value of all of these payments at time d is their sum,

$$P(1+i)^d + P(1+k)(1+i)^{d-1} + P(1+k)^2(1+i)^{d-2} + \ldots$$

$$+ P(1+k)^{n-1}(1+i)^{d-n+1}.$$

This is a geometric sum with n terms, the first term being

$$a = P(1+i)^d$$

and common ratio

$$r = (1+k)/(1+i).$$

With the help of Formula 3.1a, we can rewrite the sum with a new notation as

$$S^{(k)}(P, i, n, d) = P(1+i)^d \frac{1-\left(\frac{1+k}{1+i}\right)^n}{1-\left(\frac{1+k}{1+i}\right)} =$$

$$P\frac{1-\left(\frac{1+k}{1+i}\right)^n}{i-k}(1+i)^{d+1} \tag{4.1}$$

You will be asked to verify this formula in exercise 4.2.5.

Let's test this with the data from Example 4.4. We have $P = \$200$, $i = .078/12 = .0065$, $k = .003$, $n = 420$, and $d = 420 - 0 = 420$. Formula 4.1 gives us Rodrigo's accumulated value of

$$S^{(.003)}(\$200, .0065, 420, 420) =$$

$$\$200\frac{1-(1.003/1.0065)^{420}}{.0065-.003}(1.0065)^{421} = \$671{,}723.79,$$ ∎

Let's put Formula 4.1 to the test again.

Example 4.5: Dave and Maggie want to finance a retirement plan where the payments they receive will increase with inflation that they estimate will be 0.4% each month. The first payment to them in the amount of \$3,000 will be made in 25 years, and then increasing payments will continue for 30 years (to them or to their estate should they not survive the period). To make this annuity possible, Dave and Maggie will make deposits of P at the end of each month into an account earning interest at the annual rate of 6.3% compounded monthly, the last deposit in 25 years. What must the size of these monthly deposits be?

Solution: Here is a time diagram for this problem:

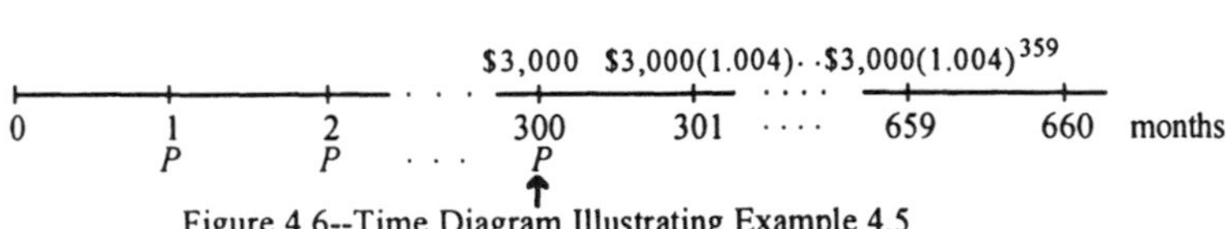

Figure 4.6--Time Diagram Illustrating Example 4.5

I have put the focal date at time 300 months. There, we can find the discounted value of the increasing annuity and set it equal to the accumulated value of Dave and Maggie's deposits. Note that the deposits form a regular annuity—not an increasing one. The inputs for these deposits are P (to be determined), $i = .063/12 = .00525$, $n = 300$, and $d = 300 - 1 = 299$. The accumulated value of the 300 deposits is therefore (using either Formula 3.2 or 3.2c)

$$S(P, 0.00525, 300, 299) = P\frac{(1.00525)^{300} - 1}{.00525} = P\cdot(725.8819222154).$$

Next, we find the PV of Dave and Maggie's increasing annuity (at time 300) and set it equal to $P(725.8819222154)$. The inputs for Formula 4.1 are $P = \$3,000$, $n = 360$, $d = 300 - 300 = 0$, and $k = .004$. This gives us

$$S^{(.004)}(\$3,000, .00525, 360, 0) =$$

$$\$3,000\frac{1 - (1.004/1.00525)^{360}}{.00525 - .004}(1.00525)^{0+1} = \$871,068.17.$$

Setting this equal to $P(725.8819222154)$ and solving for P, we get $P = \$1,200.01$. So, if Dave and Maggie deposit \$1,200.01 at the end of every month for 25 years, they will have their 30-year increasing annuity. Note that the final payment they receive at the beginning of 660th month will be $\$3,000(1.004)^{359} = \$12,575.47$. ∎

Exercises 4.2

4.2.1 Tage and Berit want to finance a retirement plan where the retirement payments increase with inflation, which they estimate at 0.25% each month. The first payment to them in the amount of 25,000 Crowns will be made in 25 years, and then monthly payments will continue for 25 years (to them or to their estate should they not survive the period). To finance this annuity, Tage and Berit will make deposits of P at the end of every month into an account earning interest at the nominal annual rate of 7.25% compounded monthly, the last deposit in 25 years. What must the size of these monthly deposits be?

4.2.2 Marcelo just inherited \$750,000 and decides to buy an annuity that will pay him (or his estate) at the end of each month for 30 years with the payments increasing by 0.6% each month. Find the first two payments and the last payment that Marcelo will receive if $j_{12} = 6\%$. Note that $i < k$!

4.2.3 Federico wants to provide himself with an increasing annuity. He wants his payments to increase by 0.35% each month. The first payment of $2,000 will be made at the end of the 241st month and then continue monthly for 40 years (to Federico or to his estate if he dies before the annuity ends). Federico or his estate will make equal deposits of P at the end of every month into an account earning interest at the nominal annual rate of 5.4% compounded monthly, the last deposit in 20 years. What must the size of these monthly deposits be?

4.2.4 To accumulate $60,000, Mool will make deposits at the end of each month for 6 years into an account earning interest at the nominal annual rate of 6% compounded monthly. His first deposit will be $600 with subsequent deposits increasing by a fixed percentage every month. By what percentage must the deposits increase each month for Mool to have his $60,000 in 6 years? *Hint:* Use your calculator's solver if it has one, or use Guess-and-Check.

4.2.5. Verify Formula 4.1.

4.2.6 Mario bought a house and was advised to buy mortgage insurance on it. Find the *PV* of all of his insurance payments if j_{12} = 8.7%, his first payment is $24 at the time of purchase, and subsequent monthly payments decrease by 0.5% each month for a total of 20 years. Assume that Mario's insurance payments are certain to be made. *Hint:* Note that $k = -.005$.

4.2.7 Bibi Titi is saving for a rainy day. She deposited $100 at the end of the first month and will increase her deposits by 0.45% each month for a total of 10 years. Her money is earning at the rate of j_{12} = 6%.

a) How large will Bibi Titi's last deposit be?

b) How much will Bibi Titi have at the end of the 10 years?

c) To accumulate the same amount you found in part b), what *equal* payments would Bibi Titi have to make at the end of each month for 10 years?

4.2.8 Zubeida is building herself a little house in the country. She borrowed $125,000 at j_{12} = 10.9% and will repay the loan with payments at the end of each month over the next 25 years. If the payments increase by 0.7% each month, find Zubeida's first and last payments.

4.2.9 Dudley wants to accumulate $2,500,000 by the time of his retirement in 30 years. To accomplish this he will make deposits at the beginning of each month for 30 years into a fund earning $j_{12} = 8\%$. If these deposits are to increase by 0.65% each month, what should Dudley's first deposit be? What will Dudley's last payment be?

4.2.10 Referring to the preceding problem, what would be the size of Dudley's payments if they were all to be equal?

4.2.11* Show that when $i = k$, $S^{(k)}(P, i, n, d) = nP(1 + i)^d$.

4.2.12 Fueni has been making deposits at the end of every month for 10 years now. Her first deposit was $125 with subsequent deposits increasing by 0.35%. If her deposits earned at the interest rate of $j_{12} = 4.2\%$, how much does Fueni have in her account today? *Hint:* See the preceding exercise.

4.3 Perpetuities

Annuities, by definition, have a finite number of payments. Eventually they stop. When a sequence of payments is to continue forever, we call it a *perpetuity*. Here is an example.

Example 4.6: Becky has done well in her career, so she wants to provide an annual scholarship of $20,000 in perpetuity beginning now. The institution managing the fund paying out the scholarship offers an effective rate of 8% on Becky's money. How much money does Becky need to put into this fund now to provide $20,000 annually forever?

Solution: At first you puzzled as to how it is possible to provide $20,000 annually forever with a finite amount of money. The answer is *interest*. We will see how to answer the question in two ways.

Call G the amount that Becky must provide. First, the fund needs the $20,000 to be paid immediately depleting it by that amount. The fund has gone from G to G - $20,000, and we are still at time 0. But it earns 8% effective, so in one year, the fund will be 8% richer and will contain (G - $20,000)(1.08), and this needs to equal G again, so that it will never run out of money. We need to solve

$$(G - \$20{,}000)(1.08) = G$$

for G. As you should check, this gives $G = \$270{,}000$. If the interest rate were lower than 8% effective, Becky would have to deposit more than $270,000 into the fund.

The second way isn't as elegant as the first, but it will allow us to handle more complicated problems later.

Aside: We need to extend the idea of the geometric sum to a *geometric series*, where the number of terms is infinite. Recall that the geometric sum

$$G_n = a + ar + ar^2 + ar^3 + \ldots + ar^{n-1}$$

can be written in the compact form

$$G_n = a\frac{1-r^n}{1-r}.$$

This is Formula 3.1*a*. What if the terms just keep going and going without end?

$$G = a + ar + ar^2 + ar^3 + ar^4 + \ldots$$

Look again at the finite version,

$$G_n = a + ar + ar^2 + ar^3 + \ldots + ar^{n-1} = a\frac{1-r^n}{1-r}.$$

If we just keep adding terms, we are essentially making n larger and larger. What happens to r^n when n gets large? If $-1 < r < 1$, which it will always is for our purposes, then r^n approaches zero. If you don't believe that (and why should you?), take a number like 0.8 and calculate 0.8^n for larger and larger values of n on your calculator. For example, $0.8^{20} = .0115\ldots$; $0.8^{100} \approx .0000000002$, $0.8^{200} = 4.1495..\times 10^{-20} \approx .000000000000000000041495$. You get the idea. If r is a number less than 1 in magnitude, then raising it to larger and larger powers makes r^n closer and closer to zero. So if r^n vanishes as n increases in the formula for G_n, then the sum of a geometric series must be

$$G = a + ar + ar^2 + ar^3 + ar^4 + \ldots = a\frac{1}{1-r} = \frac{a}{1-r}. \qquad (4.2)$$

Return to the Solution for Example 4.6: Here is what the time diagram for Becky's scholarship program looks like:

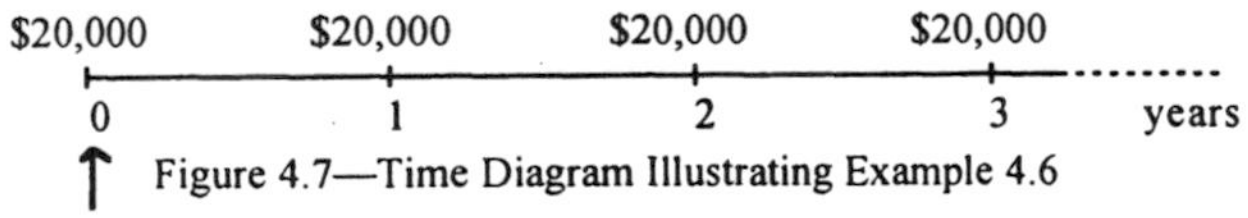

$20,000	$20,000	$20,000	$20,000	
0	1	2	3	years

↑ Figure 4.7—Time Diagram Illustrating Example 4.6

The present value of this series of payments is

$$\$20{,}000 + \$20{,}000(1.08)^{-1} + \$20{,}000(1.08)^{-2} + \$20{,}000(1.08)^{-3} + \ldots$$

This is a geometric *series* with a = \$20,000 and $r = (1.08)^{-1}$ = .925925… = $.\overline{925}$. Its sum is $a/(1 - r)$ = \$20,000/(1 - $.\overline{925}$) = \$270,000, as before. ∎

Formulas 3.2 and 4.1 can be adapted to perpetuities fairly easily. Since for $i > 0$, $(1+i)^{-n}$ approaches 0 as n grows without bound. Formula 3.2,

$$S = S(P,\ i,\ n,\ d) = P\frac{1-(1+i)^{-n}}{i}(1+i)^{d+1},$$

for perpetuities becomes

$$S(P,\ i,\ \infty,\ d) = P\frac{(1+i)^{d+1}}{i}. \qquad (4.3)$$

We will now examine some of the uses of Formula 4.3. In Example 4.6, we have P = \$20,000, i = .08, and d = 0. Substituting these into Formula 4.3, we get

$$\$20{,}000\frac{(1.08)^{0+1}}{.08} = \$270{,}000.$$

Example 4.7: Gerry wants to provide an annual scholarship of \$7,500 for worthy students at his alma mater. The first scholarship will be awarded in 5 years. How much must Gerry deposit now into a fund earning 7.2% effective to provide for this annual scholarship?

Solution: First, we'll take a look at the time diagram:

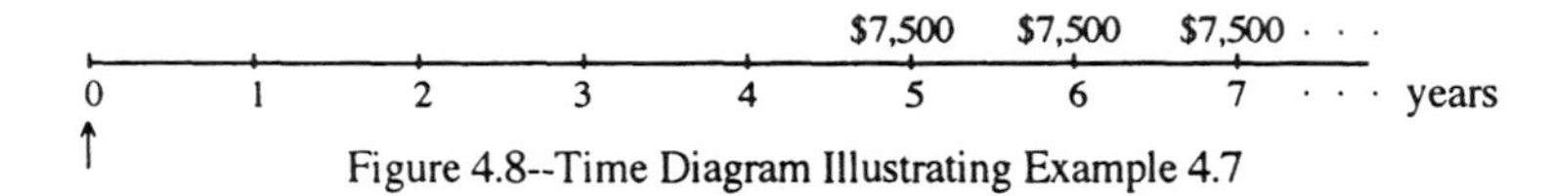

Figure 4.8--Time Diagram Illustrating Example 4.7

We have the inputs for Formula 4.3 as follows: P = \$7,500, i = .072, and $d = 0 - 5 = -5$. Substituting these, we get

$$S(\$7500,\ 0.072,\ \infty,\ -5) = \$7{,}500\frac{(1.072)^{-5+1}}{.072} = \$78{,}876.86.$$

If Gerry deposits $78,876.86 into the fund today, the fund will provide $7,500 per year in perpetuity beginning in 5 years. ∎

Exercises 4.3

4.3.1 Bev will deposit $10,000 into a fund at the beginning of every year for the next 10 years. At the end of the 10th year, an annual scholarship will be distributed and will continue forever. If the fund earns 8% effective, find the amount of the annual scholarships.

4.3.2 Professor Ümläüt, upon his retirement as CEO of Umlaut.com, deposited a sum into an account earning 6.4% annually. This was to provide an annual scholarship of $20,000 beginning immediately to the student at his alma mater who most exemplified the ideals of Umlaut.com. What was the amount of Professor Ümläüt's deposit?

4.3.3 Suzie deposited $30,000 at the beginning of every year for 10 years into a fund earning 8.6% annually. At the end of the 10th year, an endowed chair in her name was created with her fund providing a portion of the annual expenses in perpetuity. What amount of the expenses does Suzie's fund cover annually?

4.3.4 A fund of $176,891 provides an annual scholarship in the amount of $9,413 with the first award made at the time of the creation of the fund. What effective rate of interest is the fund earning?

4.3.5 A certain investment is projected to yield a dividend of $8,000 at the end of every year indefinitely. How much does a potential investor invest in order that his or her annual yield be 10% effective?

4.3.6 Boris created a $150,000 fund designed to provide an annual scholarship in perpetuity to a deserving graduate student to support field research. The first scholarship will be awarded in four years. If the fund earns interest at j_{12} = 12%, what will the size of these annual scholarships be? (Note that payments are annual, and the interest is compounded monthly.)

4.3.7 Natasha deposited \$5,000 at the beginning of each year for 20 years in a fund earning at the rate $j_1 = 9\%$. At the end of 20 years, an annual stipend to partially cover library expenses commenced. What was the amount of these stipends?

4.3.8 Vandana wishes to partially endow a chair in the Mathematics of Finance. Payments coming from her fund are to amount to \$45,000 per annum commencing in two years. If Vandana's fund earns interest at $j_2 = 10\%$, how much must she put into the fund now?

4.3.9 Verify Formula 4.2.

4.4 Perpetuities Increasing or Decreasing by a Fixed Percentage

Just as increasing annuities make sense in some cases so do increasing perpetuities. If a perpetuity provides \$10,000 annually for some purpose and if the annual inflation rate is, say, 2.5%, then in 50 years, the buying power of \$10,000 would be the equivalent of $\$10{,}000(1.025)^{-50} = \$2{,}909.42$ today, and in 150 years, \$10,000 would be the equivalent of \$246.27 today.

Example 4.8: If Gerry wants the same scholarship program as in Example 4.7 except that the payments increase by 3% every year to stave off inflation, what is the size of the fund he must create?

Solution: The time diagram from the previous example changes as follows:

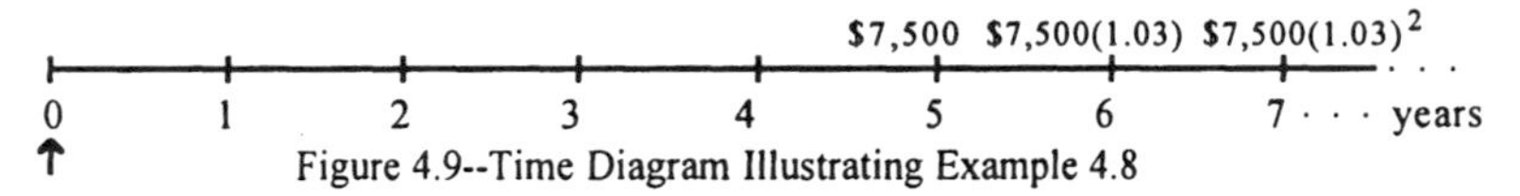

Figure 4.9--Time Diagram Illustrating Example 4.8

We can analyze this using basic principles and see if we can find a generalization of Formula 4.1 or 4.3 for increasing (or decreasing) perpetuities. The sum of the present values (at the arrow) of the first few payments is

$$\$7{,}500(1.072)^{-5} + \$7{,}500(1.03)(1.072)^{-6} + \$7{,}500(1.03)^2(1.072)^{-7} + \ldots$$

This is a geometric series whose first term is a = \$7,500(1.072)$^{-5}$ and whose common ratio is r = 1.03/1.072 because if you multiply each term by 1.03/1.072, you get the next term. The sum is therefore

$$\$7{,}500\frac{(1.072)^{-5}}{1-(1.03/1.072)} = \$135{,}217.48. \qquad ∎$$

Aside: We can express left-hand side of this equation using our usual symbols, first as

$$P\frac{(1+i)^d}{1-(1+k)/(1+i)},$$

and then after a little algebra,

$$P\frac{(1+i)^{d+1}}{i-k}.$$

Note that this is Formula 4.1 with the $\{(1 + k)/(1 + i)\}^n$ term replaced by zero. This always works out as long as $i > k$. As it turns out, the only way that an increasing perpetuity can be finite is if $i > k$, so this should not be a restrictive condition.

Result for increasing/decreasing perpetuity: Summarizing what we have now found for increasing/decreasing perpetuities, the value at comparison date d of a series of increasing or decreasing payments P, $P(1 + k)$, $P(1 + k)^2$, $P(1 + k)^3$, . . . , made at times 0, 1, 2, 3, . . . with interest rate, i per period, is

$$S^{(k)}(P, i, \infty, d) = P\frac{(1+i)^{d+1}}{i-k}, \qquad (4.4)$$

Here is another example.

Example 4.9: Beer magnate H. Paco Drools has decided to donate \$50,000,000 to endow a beer research foundation in his name. Payments for expenses, acquisitions, and salaries will begin immediately and increase by 2.5% annually. If his donation earns 9.47% effective, find the size of the first few payments.

Solution: Let's look at a time diagram.

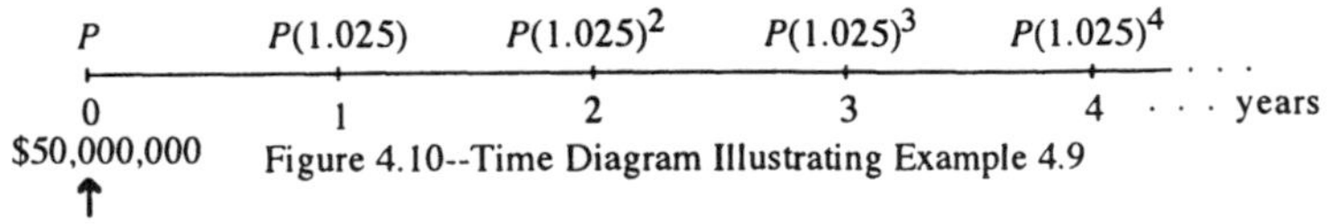

P	$P(1.025)$	$P(1.025)^2$	$P(1.025)^3$	$P(1.025)^4$	. . .
0	1	2	3	4	. . . years
\$50,000,000 ↑					

Figure 4.10--Time Diagram Illustrating Example 4.9

The inputs for Formula 4.4 are P (to be determined), i = .0947, k = .025, and $d = 0 - 0 = 0$. We also have the value of the output which is \$50,000,000.

Setting the two pieces equal to one another, we have

$$\$50{,}000{,}000 = P\frac{(1.0947)^{0+1}}{.0947-.025} \approx P(15.7058823529).$$

We get $P \approx$ \$3,183,520.60. Remember, that's only the first payment. The subsequent payments increase by 2.5% annually. The next few are \$3,263,108.61, 3,344,686.33, and \$3,428,303.49. If you are curious, the amount of the annual payment in 100 years will reach $\$3{,}183{,}520.60(1.025)^{100} = \$37{,}609{,}209.36$. ∎

Exercises 4.4

4.4.1 Continuing Exercise 4.3.1, Bev will again deposit \$10,000 into a fund at the beginning of every year for the next 10 years. This time, the annual scholarships will begin at the end of the 10th year and continue forever, but the awards will increase by 3% annually. If the fund earns 8% effective, find the amount of the first three annual scholarship awards.

4.4.2 Jan-Ove wants to provide an annual scholarship for a needy student at the table tennis academy. Payments will commence immediately with the first award in the amount of 100,000 Crowns. Subsequent awards will increase by 2.5% annually forever. If $j_{12} = 7.5\%$, how much does Jan-Ove need to provide?

4.4.3 Lois Terms wants to endow a chair in mathematics at her alma mater. The salary plus administrative expenses are estimated at \$120,000 immediately with annual increases of 3.25% in perpetuity. If the fund that Lois is creating earns 7.8% effective, what single amount must she deposit?

4.4.4 The government of town of Lepe created a fund that will provide the annual expenses for the perpetual care of the monument to its late mayor. The fund, in the amount of \$2,500,000, provides a payment of \$150,000 immediately, with annual increase at a fixed percentage. If the fund earns at 8.2% effective, by what percentage are the expense payments increasing?

4.4.5 Expenses for perpetual care for a gravesite at the local cemetery are projected to run \$300 by the end of the first year with expenses in subsequent years to increase by 3.5% annually. If money can earn interest at $j_1 = 5.3\%$, what single payment now will cover the expenses in perpetuity?

4.4.6 Zubeida will fund a new endowed chair in the Theory of Applications at the university. The expenses are predicted to be \$85,000 at the beginning of the first year and then increase at the rate of 5% annually indefinitely. What amount must Zubeida provide if her fund will earn interest at 8.5% annually?

4.4.7 A certain investment is projected to yield a dividend of \$8,000 at the end of the first year with subsequent dividends increasing by 3% annually far into the future. How much does a potential investor invest in order that his or her annual yield be 10%? Compare your result with Exercise 4.3.5.

4.4.8 Ashish invested \$400,000 to partially fund the expenses of his town's library in perpetuity. If his money earns at $j_1 = 11\%$ and annual expenses are expected to increase by 6% annually, what will the fund's first payment be immediately?

4.5 A Brief Look at Stocks

In order to raise cash to finance their business activities, corporations make *stock* offerings to investors. Stock certificates are legal documents indicating partial ownership of the corporation by the stockholder.

Many forces influence stock prices, making them quite volatile. Although a thorough treatment of stocks is outside the scope of this book, we will try to touch on a few of basic ideas.

There are two major categories of stock: *common stock* and *preferred stock.*

Common Stock

Owners of *common stock* generally enjoy certain voting rights, periodic income in the form of cash dividends, immunity from liability in the event of the bankruptcy of the corporation, and the ability to sell their stock at will.

On the other hand, dividends to holders of common stock may be small or nonexistent, as the corporation may decide to retain earnings for expansion or for other expenses. Furthermore, in the event of financial exigency or bankruptcy, common stockholders are typically the last to share in the assets of the corporation: after wage-earners, debtors, preferred stockholders, and bondholders.

We can employ the principles that we have already learned to get at least a rough idea of the value of a share of common stock. In fact, we may think of the price of common stock as the *PV* of the perpetuity consisting of all future dividends, where the interest rate used is the desired rate of return (= yield rate). Of course to determine the price, the dividends must be known, and in the case of common stock, these dividends are not constant. We give some examples where the sequence of dividends is assumed known.

Example 4.10: How much should Dolores pay per share for a common stock paying semiannual dividends the first of which is \$3.00 if her desired rate of return is $j_2 = 9\%$ and the dividend payments are to increase at the nominal annual rate of 5% compounded semiannually?

Solution: We assume the dividends are to be paid at the end of each half-year and continue indefinitely. Thus, we are dealing with an increasing perpetuity. We apply Formula 4.4 with $P = \$3.00$, $i = 0.045$, $k = 0.025$, and $d = -1$. Then the price per share of Dolores's stock should be

$$S = S^{(0.025)}(\$3.00,\ 0.045,\ \infty,\ -1) = \frac{\$3.00}{0.045 - 0.025} = \$150.00 \quad \blacksquare$$

Example 4.11: Pili is thinking of buying shares of common stock in a new corporation that manufactures prefabricated houses. To encourage investment, the corporation makes it known that it will pay semiannual dividends of \$10.00 per share for two years, followed by no dividends for two years to enable the company to finance expansion, followed by semiannual dividends of \$4.00 indefinitely. How much should Pili pay per share if she hopes to realize a rate of return of $j_2 = 12\%$?

Solution: Let's look at a time diagram for this situation.

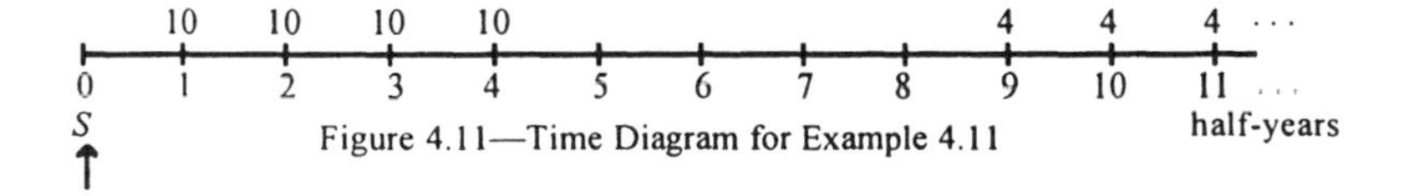

Figure 4.11—Time Diagram for Example 4.11

We can think of this as a four-payment annuity of $10.00 followed by a perpetuity of $4.00 deferred for four years (eight half-years). Pili's price is the *PV* of these payments, namely

$$S = S(\$10.00, 0.06, 4, -1) + S(\$4.00, 0.06\ \infty, -9) =$$

$$\$10.00\frac{1-(1.06)^{-4}}{0.06} + \$4.00\frac{(1.06)^{-9+1}}{0.06} = \$76.48. \qquad \blacksquare$$

Preferred Stock

Owners of *preferred stock* have certain advantages over common stockholders. Although they do not usually enjoy the voting rights afforded to common stockholders, they do typically receive fixed dividends indefinitely, unless the corporation is unable to provide them. Furthermore, preferred stockholders are always paid dividends before common stockholders, and in the event of bankruptcy, they share in the corporation's assets before common stockholders.

Example 4.12: Martti paid $75 per share for a preferred stock paying dividends of $7.00 annually. What is Martti's effective rate of return?

Solution: Assuming that dividends are paid at the end of each year, we can write down the equation of value directly, using Formula 4.3:

$$\$75.00 = S(\$7.00, i, \infty, -1) = \$7.00/i,$$

so that Martti's rate of return is $i = 0.09\overline{3} = 9.\overline{3}\%$ effective ▮

Example 4.13: Indira inherited shares of preferred stock whose per-share price was $110.00. If her nominal annual rate of return compounded semiannually is 12%, what is the amount of her fixed dividends?

Solution: We again appeal to Formula 4.3. The rate per half-year is $i = j_2/2 = 0.06$, so the equation of value is

$$\$110.00 = P/\,0.06,$$

and Indira's semiannual dividends are $P = \$6.60$ per share. ▮

Example 4.14: Write a formula for the price, S, of a share of preferred stock paying semiannual dividends of P, earning nominal annual rate of return j_2 compounded semiannually, with the first dividend being paid d half-years after purchase, where $d \geq 0$.

Solution: We already have this formula: it is Formula 4.3. The rate per half-year is $i = j_2/2$, so price is

$$S = S(P, i, \infty, d) = P\frac{(1+i)^{d+1}}{i}.$$

∎

Exercises 4.5

4.5.1 A share of a common stock earns a nominal annual rate of return of $j_2 = 9\%$ compounded semiannually. The first dividend is \$3.00, with subsequent semiannual dividends increasing by 2%. How much should a share of this stock sell for?

4.5.2 What semiannual dividends are being paid per share of preferred stock whose price was \$120.00 and whose nominal annual rate of return is $j_2 = 8\%$ compounded semiannually?

4.5.3 A share of a certain common stock is to start paying semiannual dividends after three years. The first dividend is to be \$1.50 (at 3½ years), with subsequent dividends increasing by 2.5%. If the desired annual rate of return is to be 8% compounded semiannually, what should be the per-share price?

4.5.4 What is the *effective* rate of return per share of a preferred stock whose price was \$180, and whose dividends pay \$4.50 semiannually?

4.5.5 Write the first few terms of a general formula for the price, S, of a share of common stock with semiannual dividends P_1, P_2, P_3, ... beginning in six months, if the nominal annual rate of return compounded semiannually is to be j_2.

4.5.6 A share of a certain stock pays a dividend of $P = \$12$ at the end of every two years. What price S should be paid per share if the desired effective rate of return is to be 10%? *Hint*: The period between dividend payments is two years, so find and use the interest rate, i, for that period.

4.6 Summary of Formulas

Formula for Increasing & Decreasing Annuities

$$S^{(k)}(P,\ i,\ n,\ d) = P\frac{1-\left(\dfrac{1+k}{1+i}\right)^{n}}{i-k}(1+i)^{d+1}$$

$$= S(P, i^*, n, d)(1 + k)^d,\ \ i^* = (i - k)/(1 + k)$$

Geometric Series

$$G = a + ar + ar^2 + ar^3 + \ \dots\ = \frac{a}{1-r}$$

Formula for Regular Perpetuities

$$S(P,\ i,\ \infty,\ d) = P\frac{(1+i)^{d+1}}{i}$$

Formula for Increasing & Decreasing Perpetuities

$$S^{(k)}(P,\ i,\ \infty,\ d) = P\frac{(1+i)^{d+1}}{i-k}$$

$$= S(P, i^*, \infty, d)(1 + k)^d,\ \ i^* = (i - k)/(1 + k)$$

Miscellaneous Exercises

4M1. Chauncey won the Big Bucks Lottery and settled for a lump sum payment of $2,650,899.80. If the dollar earns interest at 8% effective and if Chauncey could have chosen to have 20 annual payments beginning immediately, what would have been the amount of these payments?

4M2. You have been called upon to estimate the loss of income incurred by the widow of a 35-year-old man killed in a work-related accident. If the man was otherwise expected to live another 39 years, if his income was expected to be $35,000 one year after his death with annual increases estimated at 2.912621359%, and if money is valued at 6% per annum, what is the *PV* of the loss to his widow? Assume that he would have worked another 39 years until his death. (We will revisit this problem in Chapter 7 with a somewhat different approach.)

4M3. Deposits of $300 are made to a bank account at the end of each quarter for 5 years. Find the accumulated value of these payments given that $j_{12} = 6\%$.

4M4. Rosie just won $120,000,000 in the lottery. It can be paid in 20 equal annual installments of $6,000,000 beginning now, or she can choose the 'cash option' which is equal to the *PV* of the 20 annual payments valued at $j_1 = 6\%$. How much does she get with the cash option?

4M5. Tomoko loaned $20,000 to Satjajit as an investment. To repay the loan, Satjajit is to make semiannual payments of $1,000 to Tomoko for 5 years in addition to a lump-sum payment at the end of 5 years which is designed to yield Tomoko $j_2 = 9\%$ on her investment. What should the amount of the lump-sum payment be?

4M6.* In Formula 4.1, verify that $S^{(k)}(P, i, n, d) = S(P, i^*, n, d) \cdot (1 + k)^d$, where $i^* = (i - k)/(1 + k)$.

4M7.* In Formula 4.4, verify that $S^{(k)}(P, i, \infty, d) = S(P, i^*, \infty, d) \cdot (1 + k)^d$, where $i^* = (i - k)/(1 + k)$.

4M8. Lex paid $90.00 per share for a common stock whose first dividend, in eighteen months, was to be $4.00, with subsequent semiannual dividends increasing by 2%. Find Lex's effective rate of return. *Hint*: First use a solver or Guess-and-Check to find $i = j_2/2$, and then use Formula 2.2 to find the effective rate.

Chapter 5—Amortization and Sinking Funds

5.1 Introduction

This chapter is about repaying debts. We will discuss two ways to repay—*the amortization method* and *the sinking fund method.* When we liquidate a debt by the amortization method, we say that the debt is *amortized.* The loan is paid off by means of *installment payments* at fixed time intervals. In the sinking fund method, a original loan is repaid in a lump sum at the end of a specified term with interest on the loan being paid periodically over that term. The money for the lump sum repayment is accumulated by making deposits into a so-called *sinking fund.*

5.2 Amortization and Outstanding Principal

The techniques and formulas we will use in this chapter combine results that we developed in the preceding chapters. We already know something about repaying loans by amortization. In the previous two chapters, we looked at some problems dealing with finding the *outstanding balance* (or *principal*) on a loan. We know how to calculate present values of periodic payments and how to calculate the amount of periodic payments given the amount of the loan. We will do more of this, and we will also see how we can separate each payment into its component parts: *interest paid* and *principal repaid.*

If you have ever owned a house, you probably know from doing your taxes that in the early years of repaying your long-term debt, the *mortgage*, the larger part of each payment goes to paying interest on the debt. Only a small fraction of each early payment goes into reducing principal. On the bright side is the fact that in many cases, the interest is deductible from your taxable income.

Example 5.1: Loren and Liz bought a house 10 years ago and got an interest rate higher than what is available today. For this reason, they would like to refinance their mortgage and get a better deal. To do this, they must know how much they still owe on their house. If they borrowed $105,082.06 and got a 25-year mortgage at j_{12} = 9.74%, how much do they owe just after their 120th payment?

Solution: We will need to know the amount of Loren and Liz's monthly payments. Assuming that they make payments at the end of every month, with the aid of the time diagram below

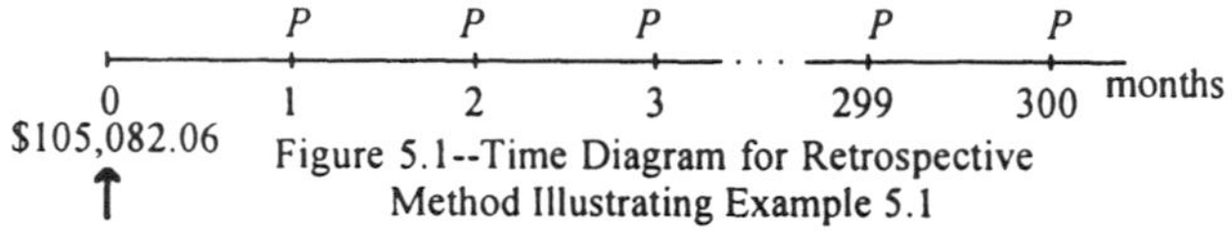

$$\begin{array}{cccccccc} & P & P & P & \cdots & P & P & \\ 0 & 1 & 2 & 3 & & 299 & 300 & \text{months} \\ \$105{,}082.06 \uparrow & & & & & & & \end{array}$$

Figure 5.1--Time Diagram for Retrospective Method Illustrating Example 5.1

we can see that we have an application of Formula 3.2 (or 3.2*a*). The inputs are P (to be determined), $i = .0974/12 = .00811\overline{6}$, $n = 25 \cdot 12 = 300$, $d = 0 - 1 = -1$, and $S = \$105{,}082.06$. We have

$$S(P, .00811\overline{6}, 300, -1) = P\frac{1-(1.00811\overline{6})^{-300}}{.00811\overline{6}} =$$

$$P\cdot(112.304355529) = \$105{,}082.06$$

so that Loren's and Liz's monthly payments are $P = \$105{,}082.06/112.304355529 = \$935.69$. The question that remains is, how much do they still owe on their house? As we saw in Examples 3.9 and 3.10, there are generally two ways to determine outstanding principal. These are called the *retrospective method* and the *prospective method.* Let's look at the retrospective method first.

With the retrospective method (looking backward in time), we know that we borrowed \$105,082.06 and that we have been paying \$935.69 at the end of each month for 10 years = 120 months and that money is worth $i = .0974/12 = .00811\overline{6}$ per month. If Loren and Liz had not made any payments for 10 years, they would owe

$$\$105{,}082.06(1.00811\overline{6})^{120} = \$277{,}219.61.$$

But they have paying the bank \$935.69 per month for 120 months, and the accumulated value of those payments 10 years after buying the house is (with $d = 120 - 1 = 119$),

$$S(\$935.69, .00811\overline{6}, 120, 119) = \$188{,}843.18\,.$$

This is the value of what they have paid back 10 years after purchase. Therefore, what Loren and Liz still owe after 10 years is the difference,

$$\$105{,}082.06(1.00811\overline{6})^{120} - S(\$935.69, .00811\overline{6}, 120, 119)$$

$$\$277{,}219.61 - \$188{,}843.18 = \$88{,}376.43.$$

With the prospective method (looking forward in time), what Loren and Liz owe after 10 years is the PV (after 120 months) of all of their future payments. Here is how that looks in a time diagram:

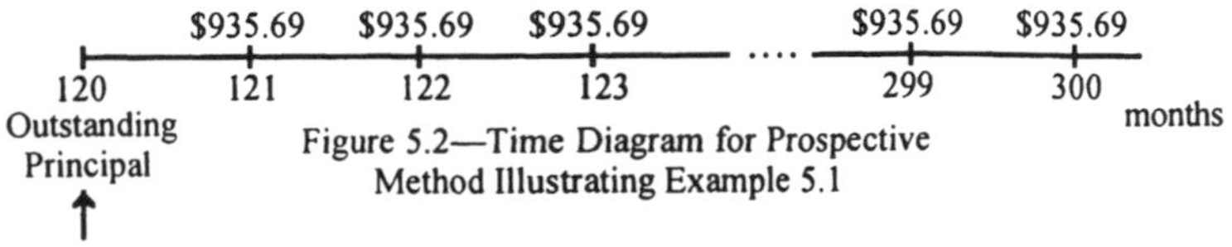

	$935.69	$935.69	$935.69	....	$935.69	$935.69	
120	121	122	123		299	300	months
Outstanding Principal ↑							

Figure 5.2—Time Diagram for Prospective Method Illustrating Example 5.1

The inputs for Formula 3.2 are $P = \$935.69$, $i = .00811\overline{6}$, $n = 12 \cdot 15 = 180$, and $d = 120 - 121 = -1$, so

$$S(\$935.69, .00811\overline{6}, 180, -1) = \$88{,}376.43,$$

the same as we got with the retrospective method. ∎

Generalization for Finding Outstanding Principal by the Retrospective Method: The data required to apply the two methods above to find the outstanding balance on a loan at a given time are not the same. Let's call OP_r the outstanding principal on our loan just after the rth payment.

For the retrospective method, we will need to know

a) the amount of the loan, L,
b) the periodic interest rate, i,
c) the number, r, of the payment just made, and
d) the amount of the periodic payment, P.

The outstanding principal then at time r is

$$OP_r = L(1+i)^r - S(P, i, r, r-1) = L(1+i)^r - P\frac{(1+i)^r - 1}{i}. \quad (5.1)$$

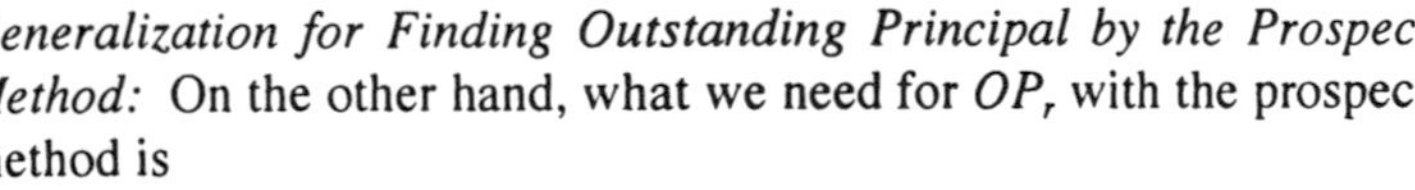

	P	P	P	....	P	P	
0	1	2	3		r - 1	r	periods
L						↑	

Figure 5.3—Time Diagram Illustrating Formula 5.1

Formula 5.1 tells us that what the borrower owes just after making the rth payment is the accumulated value of the loan for r periods, $L(1 + i)^r$, less the accumulated value of an r-period annuity of P as of time r.

Generalization for Finding Outstanding Principal by the Prospective Method: On the other hand, what we need for OP_r with the prospective method is

a) the periodic interest rate, i,
b) the number of payments still to be made, $n - r$,
c) the amount of the periodic payment, P.

Remark: In *b*) above, although we do need to know $n - r$, we do not need to know n or r. We *may* know them, but we do not need to.

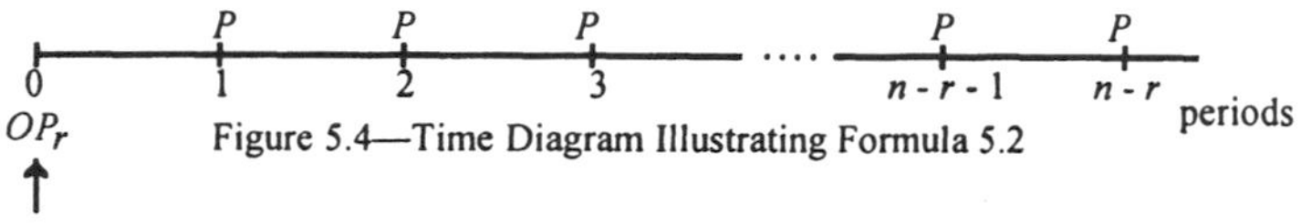

0	1	2	3	....	n - r - 1	n - r
	P	P	P		P	P

periods

OP_r ↑

Figure 5.4—Time Diagram Illustrating Formula 5.2

The outstanding principal when a payment has just been made and there are $n - r$ payments yet to be made is

$$OP_r = S(P, i, n - r, -1) = P\frac{1-(1+i)^{-(n-r)}}{i}. \tag{5.2}$$

In Exercise 5.2.15 you are asked to show that Formulas 5.1 and 5.2 are equivalent.

Remark: Note that for the retrospective Formula 5.1, we do not need to know n, and for the prospective Formula 5.2, we do not need to know L or r. This means that which of the two formulas is easier to use depends on the information you are given. In Exercise 5M9, you are asked to show that you can find n given the inputs to Formula (5.1).

Example 5.2: Bonnie and Paul borrowed $130,000 for their ranch-style home. They have just finished their sixth year of payments in the amount of $990.15 at the end of every month. The interest rate on their loan is j_{12} = 8.3974%. Find Bonnie's and Paul's outstanding principal using both methods.

Solution: We already have the information we need for the retrospective method. The inputs for Formula 5.1 are L = $130,000, i = .083974/12 = $.0069978\overline{3}$, r = 12·6 = 72, and P = $990.15. Substituting these into formula 5.1, we get

$$OP_{72} = \$130{,}000(1.0069978\overline{3})^{72} - S(990.15,\ .0069978\overline{3},\ 72,\ 71) =$$

$$= \$130{,}000(1.0069978\overline{3})^{72} - \$990.15\frac{(1.0069978\overline{3})^{72}-1}{.0069978\overline{3}} =$$

$$= \$122{,}504.10.$$

On the other hand, to use the prospective method, we need to know n in order to find $n - r$. If the term of the loan is n (months), then

$$S(\$990.15, .0069978\overline{3}, n, -1) = \$990.15\frac{1-(1.0069978\overline{3})^{-n}}{.0069978\overline{3}} = \$130{,}000.$$

To solve this for n, divide both sides by $990.15; multiply both sides by $.0069978\overline{3}$, and transpose. We then get $(1.0069978\overline{3})^{-n}$ =

.08123179990. Applying the Log Rule with $a = 1.0069978\overline{3}$, $b =$ .08123179990, and $m = -n$, we find that

$$n = -\log(.08123179990)/\log(1.0069978\overline{3}) = 360.$$

Now we know the number of payments that Bonnie and Paul still have to make: $n - r = 360 - 72 = 288$. Their outstanding principal is thus

$$OP_{72} = S(\$990.15,\ 0.0069978\overline{3},\ 288,\ -1) =$$

$$\$990.15\frac{1-(1.0069978\overline{3})^{-288}}{.0069978\overline{3}} = \$122,504.06,$$

a discrepancy of 4¢ from the retrospective method that we attribute to the rounding of Bonnie's and Paul's monthly payment from \$990.150185500 to \$990.15.

The result we find in this example certainly illustrates that the interest part of payments in the early years is very high. Only about \$7,496 of the original principal of \$130,000 has been repaid in six years of payments totaling 72·\$990.15 = \$71,290.80. ∎

Example 5.3: Cheri has just come into some money unexpectedly and would like to use it pay off the mortgage on her house. She pays \$484.25 at the end of every month for the house at an interest rate of $j_{12} = 6.89\%$. She just made a payment and has 90 payments to go at this time. How much does Cheri need to pay off her house?

Solution: Here, we have all we need for the prospective method and no more. The inputs for Formula 5.2 are $i = .0689/12 = .005741\overline{6}$, the number of payments still to be made, $n - r = 90$, and the monthly payment, $P = \$484.25$. Substituting these into Formula 5.2, we find that

$$S(484.25,\ .005741\overline{6},\ 90,\ -1) = \$484.25\frac{1-(1.005741\overline{6})^{-90}}{.005741\overline{6}} = \$33,960.36$$

is the amount Cheri needs to pay off the loan on her house. Note that we know neither n nor r in this problem. But we do know only $n - r$. ∎

Example 5.4: Ryan and Meg borrowed \$125,000 to buy their house and are considering refinancing. To do this, they need to know the outstanding balance on their loan. They pay \$998.12 at the end of each month on a loan having interest rate $j_{12} = 8.4\%$. If they have just made their 155th payment, how much do they still owe?

Solution: We do not know the term of the loan, so we do not know how many payments they have left. We would need n to use the prospective method (See Exercise 5M9.). But we do not need to find n. We have all of the ingredients for the retrospective method. Using Formula 5.1 with L = \$125,000, $i = .084/12 = .007$, $r = 155$, and P = \$998.12, we find Ryan and Meg's outstanding balance to be

$$OP_{155} = \$125{,}000(1.007)^{155} - \$998.12\frac{(1.007)^{155}-1}{.007} = \$90{,}732.63.$$ ∎

Exercises 5.2

5.2.1 What does it mean to amortize a debt?

5.2.2 Pavel is repaying a debt of \$50,000 with interest at $j_4 = 10\%$. He is to make equal payments at the end of each quarter for the next 5 years except that all but the final payment will be rounded up to the nearest dollar. The final payment will be smaller.
a) Find Pavel's equal quarterly payments.
b) Find Pavel's smaller final payment.

5.2.3 Akira borrowed \$136,000 to buy a house and is to pay off the loan at $j_{12} = 10.5\%$ over 30 years. After completing 10 years of payments, Akira found that he could refinance his loan at $j_{12} = 7.2\%$ over the next 15 years. This means that he would borrow the amount he still owed on the house after 10 years and repay this new loan at $j_{12} = 7.2\%$ over the next 15 years.
a) What were Akira's monthly payments the first 10 years?
b) What are Akira's monthly payments the next 15 years?

5.2.4 What annual payment must Silvio make in order to amortize a debt of \$16,000 with interest at $j_1 = 6\%$ in 6 years?

5.2.5 What semi-annual payment must Nadia make to amortize a loan of \$16,000 at the interest rate of $j_2 = 9\%$ in 6 years?

5.2.6 You are borrowing \$80,000 to buy a house and will amortize the loan over 30 years at interest rate $j_{12} = 9\%$. What will your monthly payments be? Make a time diagram, and write an equation of value for the problem.

5.2.7 Michiko pays \$715.12 per month on her home mortgage. She has a 15-year mortgage at $j_{12} = 7.99\%$. How much did Michiko borrow for her house?

5.2.8 Mir recently bought a house for $230,000. If his monthly payments for 20 years are $1,854.59, what nominal annual rate of interested compounded monthly is he paying? *Hint:* Use your calculator's solver if it has one, or use Guess-and-Check.

5.2.9 A few years ago, Debra bought a business and took out a loan with an interest rate of j_{12} = 9.8%. She agreed to make payments of $1,248.13 at the end of every month. After her most recent payment, she had 100 payments remaining. How much did Debra owe at that point?

5.2.10 Carl-Axel borrowed $15,559 at j_{12} = 7% to buy a car. His payments are $284.71 at the end of every month. How much does Carl-Axel still owe just after making his 32nd payment?

5.2.11 Sorie took out a loan of $30,000 at j_{12} = 12% to add a room to his house. He is making payments of $400 at the end of each month for as long as necessary plus a final smaller payment one month after his last regular payment. How much does Sorie still owe just after making his 70th payment of $400?

5.2.12 Nancy borrowed $90,000 for her house. She just made her sixtieth monthly payment of $826.15. The interest rate on their loan is j_{12} = 7.34%. Find Nancy's outstanding principal by the two methods we have studied.

5.2.13* Randy is repaying a car loan of $24,500 with equal monthly installments for 5 years. The interest rate on the loan is j_{12} = 8.8%. Find the amount of interest in the 30th payment. *Hint:* Find Randy's monthly payment, then find OP_{29}, and, finally, find the interest he owes on OP_{29}: namely, $i \cdot OP_{29}$.

5.2.14 Shira bought a silver samovar for 60,000 shekels at j_6 = 6%. If she makes equal payments at the end of every 2 months for 6 years, what is Shira's outstanding balance just after her 6th payment?

5.2.15* Show that Formulas 5.1 and 5.2 are equivalent.

5.2.16 The outstanding balance on Jamil's loan is $62,819.38. If he still has 132 monthly payments of $716.69 to make, what nominal annual interest rate compounded monthly is he paying. *Hint:* Use your calculator's solver if it has one, or use Guess-and-Check.

5.3 Amortization Schedules

If you own a house, do your own taxes and itemize your deductions, you need to know, among other things, what portion of your mortgage payments for the year consist of interest because you can deduct that amount from your taxable income. We will first go through an example showing the details.

Example 5.5: Bao took out a loan of \$30,000 to so he could open his café. He will repay the loan with five equal annual payments. The interest rate on his loan is 5.1% effective. Find Bao's annual payments, divide each payment into interest paid and principal repaid, and show the outstanding principal at the end of each year. Put the results in a table.

Solution: We'll start with a time diagram:

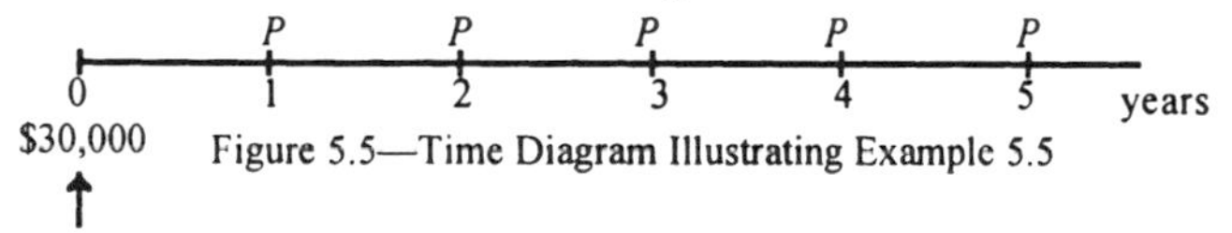

Figure 5.5—Time Diagram Illustrating Example 5.5

First we have to find Bao's annual payment, P. To do this, we need the inputs for Formula 3.2, which are P (to be determined), $i = .051$, $n = 5$, and $d = 0 - 1 = -1$. From these we get

$$S(P,\ 0.051,\ 5,\ -1) = P\frac{1-(1.051)^{-5}}{.051} = \$30{,}000$$

so $P = \$30{,}000/4.31753478312 \approx \$6{,}948.41$. We are going to construct an amortization schedule for this problem. This is what it looks like with the little we know so far:

Year	***Payment***	***Interest Paid***	***Principal Repaid***	***Outstanding Principal***
0	—	—	—	\$30,000
1	\$6,948.41			
2	\$6,948.41			
3	\$6,948.41			
4	\$6,948.41			
5	\$6,948.41			

Table 5.1—Amortization Schedule for Example 5.5

We want to see a zero in the lower right-hand cell for the amount owed after the fifth payment. We start by showing how to divide Bao's first payment of \$6,948.41 into interest paid and principal repaid. He borrowed \$30,000, so that at year's end, he owes interest on it at 5.1%.

The interest portion is \$30,000(.051) = \$1,530. That part of his first payment is the interest paid. In fact, the interest he owes at the end of each year will be 5.1% of the outstanding principal from the previous year. The remaining \$6,948.41 – \$1,530 = \$5,418.41 is the part of the first payment that repays principal. Thus, at the end of the first year, Bao's outstanding principal is \$30,000 – \$5,418.41 = \$24,581.59. These are the entries in the second row of the table.

I'll show you one more row, and then I will fill in the rest of the table for you. The interest on Bao's outstanding principle of \$24,581.59 is \$24,581.59(.051) = \$1,253.66. The principal repaid in the second year is therefore \$6,948.41 – \$1,253.66 = \$5,694.75. Finally, Bao's outstanding principal at the end of the second year is \$24,581.59 – \$5,694.75 = \$18,886.84. You should check the remainder of the entries in the amortization schedule below.

Year	*Payment*	*Interest Paid*	*Principal Repaid*	*Outstanding Principal*
0	—	—	—	\$30,000.00
1	\$6,948.41	\$1,530.00	\$5,418.41	\$24,581.59
2	\$6,948.41	\$1,253.66	\$5,694.75	\$18,886.84
3	\$6,948.41	\$963.23	\$5,985.18	\$12,901.66
4	\$6,948.41	\$657.98	\$6,290.42	\$6,611.24
5	\$6,948.41	\$337.17	\$6,611.24	\$0.00

Table 5.2—Completed Amortization Schedule for Example 5.5

Don't be surprised if the results in the lower right-hand corner of your amortization tables are a few cents off from zero. When you round the amounts in the table to two decimal places, you wind up with a bit of round-off error. This is nothing to worry about. ▮

Next, we look at a way to complete an amortization schedule the way you would do it in a 'spreadsheet' program. This does not mean that you must use a spreadsheet program, but some of the exercises are impractical without one. Exercises where we recommend use of a spreadsheet program will be accompanied by the symbol (💻).

Think of the columns of the table as being labeled *A*, *B*, *C*, *D*, and *E*, and the rows being labeled 0, 1, 2, ..., *n*, where *n* is the term of the loan. Also, *i* is the periodic interest rate, *L* is the amount of the original loan, and *P* is the periodic payment calculated from Formula 3.2.

Period (A)	*Payment (B)*	*Interest Paid (C)*	*Principal Repaid (D)*	*Outstanding Principal (E)*
0	—	—	—	L
1	P	$= i \cdot E0$	$= B1 - C1$	$= E0 - D1$
2	P	$= i \cdot E1$	$= B2 - C2$	$= E1 - D2$
3	P	$= i \cdot E2$	$= B3 - C3$	$= E2 - D3$
...	...	...	...	...
n	P	$= i \cdot E(n\text{-}1)$	$= Bn - Cn$	$= E(n\text{-}1) - Dn$

Table 5.3—Spreadsheet Style Amortization Schedule

In case you need some more explanation of how to use the above instructions, let's try to apply them to the data in Example 5.5. We had $n = 5$, $L = \$30{,}000$, and $P = \$6{,}948.41$, and these are the values we start with inside the table. Then, we fill in cell $C1 = i \cdot E0 = (.051)\$30{,}000 = \$1{,}530$. Next, $D1 = B1 - C1 = \$6{,}948.41 - \$1{,}530 = \$5{,}418.41$, and $E1 = E0 - D1 = \$30{,}000 - \$5{,}418.41 = \$24{,}581.59$. You should now see that you can complete the table almost automatically.

Now we know how to complete an amortization schedule but can we find entries in particular rows of the schedule without completing the whole table? The answer is yes. We will try to discover how to do this through an example or two.

Example 5.6: Matt is repaying a car loan of \$22,000 with payments at the end of every month for 5 years and interest at $j_{12} = 9\%$. Find the amounts of interest and principal that Matt pays in his 25th payment.

Solution: We first find Matt's monthly payment. Using Formula 3.2 with inputs P, $i = .09/12 = .0075$, $n = 60$, and $d = 0 - 1 = -1$, we find

$$S(P,\ 0.0075,\ 60,\ -1) = P\frac{1-(1.0075)^{-60}}{.0075} = \$22{,}000,$$

from which we get $P = \$456.68$. Of course, we could fill in the entire table and then pick out what we want from the row corresponding to the 25th payment, but we do not have to. Matt's outstanding principal just after his 24th payment is the PV of all of his future payments of which there are 36. His outstanding principal, OP_{24}, just after his 24th payment is therefore $S(\$456.68,\ .0075,\ 36,\ -1) = \$14{,}361.13$. The interest he owes on that amount over the 25th period is $\$14{,}361.13(.0075) = \$107.71$, and this is the interest portion of his 25th payment. The principal Matt pays from the 25th payment is $\$456.68 - \$107.71 = \$348.97$. His outstanding principal after the 25th payment is

$14,361.13 - $348.97 = $14,012.16, which is also $S(\$456.68, .0075, 35, -1)$, the *PV* of the remaining 35 payments. You can verify these calculations by obtaining the entire amortization schedule as in Exercise 5.3.6. ∎

Example 5.7: Masha is repaying her boat loan by making 10 equal semi-annual payments at $j_2 = 9\%$. If the amount of principal repaid in the 6th payment is $550, how much did Masha borrow?

Solution: Let OP_5 denote Masha's outstanding principal after her fifth payment, let L denote the amount of her loan, and let P denote the amount of her payments. By consulting Table 5.3, the amount of her 6th payment must be $P = (.045)OP_5 + \$550$. But, after her fifth payment, she still has five more to pay, so P must also satisfy

$$OP_5 = P\frac{1-(1.045)^{-5}}{.045} = P(4.38997674442).$$

Consequently, $P(4.38997674442) = (P - \$550)/.045 = 22.\overline{2}P - \$12,222.2\overline{2}$, or $17.8322454778P = \$12,222.2\overline{2}$, and so $P =$ $685.4000657.... Finally, the amount of the loan is

$$L = S(P, .045, 10, -1) = P\frac{1-(1.045)^{-10}}{.045} =$$

$$(\$685.4000657....)(7.9127...) = \$5,423.37.$$ ∎

Exercises 5.3

5.3.1 Norma borrowed $45,000 to have her house remodeled. She will make 6 equal annual payments to repay the loan. Interest on the loan is 8.5% effective. Make an amortization schedule for repaying this loan.

5.3.2 Merrick and April borrowed $265,000 to buy a house and are repaying the loan with equal payments at the end of every month for 25 years at $j_{12} = 7.1\%$. Find the amount of interest paid and principal repaid in the Merrick and April's 75th payment.

5.3.3 Sasha is repaying a debt he owes his father-in-law with 8 equal annual payments at $j_1 = 12\%$. If the amount of principal repaid in the 4th payment is $800, how much did Sasha borrow?

5.3.4 (🖳) Ken and Amy bought a house in 1977 for $30,800 amortizing the loan at $j_{12} = 7.5\%$ over 20 years.

a) How much interest did they pay in their 12th payment?

b) How much interest did they pay in their first whole year?

5.3.5 (🖳) Complete an amortization schedule for a 36-year (monthly) mortgage of $145,000 at $j_{12} = 7.95\%$. Don't print it out. Put it in a file accessible to your instructor.

5.3.6 (🖳) Verify the result in Example 5.6 by completing an amortization schedule for the entire 60 payments. Don't print it out. Put it in a file accessible to your instructor.

5.3.7 Jamila borrowed $121,719.74 at $j_{12} = 8.4\%$ to buy a duplex. The loan is to be paid off with equal monthly payments over 30 years.

a) Find Jamila's outstanding balance just after her 120th payment.

b) How much principal does Jamila repay in her 121st payment?

5.3.8 Raza will pay back a loan of $2,000 with six equal monthly payments. Make an amortization schedule assuming an interest rate of $j_{12} = 12\%$.

5.3.9 Lori is has been making monthly payments of $600.13 on her house for the past twelve years. Her original loan was for $87,500. Her current outstanding balance is $62,927.85. What nominal annual interest rate compounded monthly is she being charged? *Hint*: Use Guess-and-Check.

5.4 Sinking Fund Schedules

Another way of repaying a debt is by the *sinking-fund* method. I will introduce this method with a continuation of Example 5.5.

Example 5.8: Recall that in Example 5.5, Bao took out a loan of $30,000 to partially pay for a café. Suppose that he would rather repay the loan with a single payment at the end of 5 years by setting up a sinking fund, which will accumulate to $30,000 at 5.1% effective. He will also make five equal annual interest payments at 5.1% as they come due. Describe Bao's annual interest payments and his annual deposits into the sinking fund. Make a Sinking-Fund Schedule. Compare his total payments with those of Example 5.5.

Solution: Bao must make annual interest payments to the lender in the amount of $30,000(.051) = $1,530. By making these interest payments, the principal owed remains at $30,000 at the end of each year.

In the meantime, he must make annual deposits into a fund that will accumulate to $30,000 in 5 years, where the fund also earns at the rate of 5.1% = .051. Look at a time diagram for this situation:

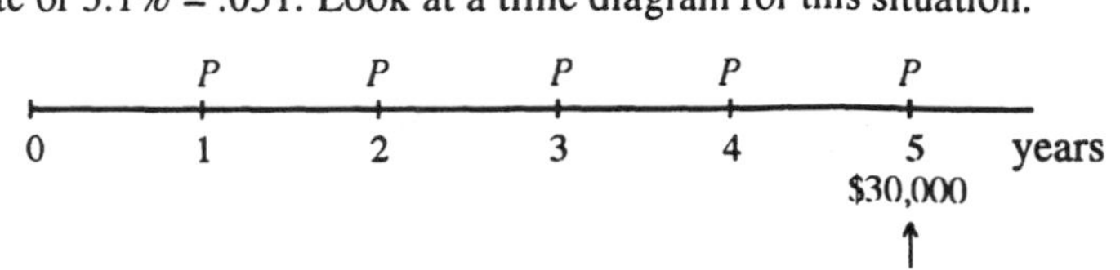

Figure 5.6--Time Diagram Illustrating Example 5.8

The 5 payments of P must accumulate to $30,000 in 5 years, money earning at the annual rate of 5.1%. To determine P (noting that $d = 5 - 1 = 4$), we need to solve

$$S(P,\ .051,\ 5,\ 4) = P\frac{(1.051)^5 - 1}{.051} = \$30,000$$

for P which gives $P\cdot(5.536680020) = \$30,000$, or $P = \$5,418.41$. Now we know that if Bao deposits $5,418.41 annually into the sinking fund, it will accumulate to $30,000 (well, $30,000.01, actually, due to round-off error) in 5 years. A *Sinking-Fund Schedule* showing the progress of Bao's fund over the 5-year period follows.

Deposit Number	*Deposit*	*Interest At 5.1%*	*Increase In Fund*	*Amount Accumulated*
0				$0.00
1	$5,418.41	$0.00	$5.418.41	$5,418.41
2	$5,418.41	$276.34	$5,694.75	$11,113.16
3	$5,418.41	$566.77	$5,985.18	$17,098.34
4	$5,418.41	$872.02	$6,290.43	$23,388.77
5	$5,418.41	$1,192.83	$6,611.24	$30,000.01

Table 5.4—Sinking Fund Schedule for Example 5.8

Let's look at how the first couple of rows of the above table were calculated. We already know that all the deposits into the fund are in the amount $5,418.41. In the last column of the first row, we have not yet accumulated any money in the fund. Therefore, the interest earned on the fund at the end of year 1 is 0. The fund therefore increases by the amount of the deposit only, $5,418.41, and that is the amount accumulated at the end of the first year. Next, 5.1% of the amount already accumulated is \$5,418.41·(0.051) = \$276.34. The fund therefore increases by \$5,418.41 + \$276.34 = \$5,694.75 to a sum of \$5,418.41 + \$5,694.75 = \$11,113.16, and so on.

Finally, we were to compare the total amount what Bao pays to the lender by both methods. In Example 5.5, the amortization method, Bao's payments total \$6,948.41·(5) = \$34,742.05. In the present example, his payments total (\$5,418.41 + \$1,530)·5 = \$34,742.05, which is the same in both cases. ∎

Remark: In reality, the interest rate paid periodically on a loan, i, is usually, but not always, higher than the interest rate earned by the deposits into the sinking fund, i'. In Example 5.6, the two were the same. That is, $i = i'$. In such cases, the total amount paid to the lender by the sinking fund method is equal to the total amount paid to the lender by the amortization method. I won't go into the details here, but when $i > i'$, the borrower pays more in total by the sinking fund method than by the amortization method. The opposite is the case when $i < i'$.

Here is how to construct a spreadsheet-style sinking fund schedule. The explanation follows the table.

Deposit Number (A)	*Deposit (B)*	*Interest Earned At i% (C)*	*Increase In Fund (D)*	*Amount Accumulated (E)*
0				\$0.00
1	P	$= i' \times E0$	$= B1 + C1$	$= E0 + D1$
2	P	$= i' \times E1$	$= B2 + C2$	$= E1 + D2$
3	P	$= i' \times E2$	$= B3 + C3$	$= E2 + D3$
...	...	...	...	...
n	P	$= i' \times E(n-1)$	$= Bn + Cn$	$= E(n-1) + D(n) = L$

Table 5.5—Spreadsheet Style Sinking Fund Schedule

The amount accumulated begins at 0 in cell $E0$. P is the amount deposited into the fund at the end of each of the n periods so that the fund will accumulate to the loan amount, L, at rate i' per period. Thus, we solve the equation $S(P, i', n, n-1) = P\{(1+i')^n - 1\}/i' = L$ for P:

$$P = L\frac{i'}{(1+i')^n - 1}. \tag{5.3}$$

The interest earned on the fund in any period is i' times the amount accumulated at the end of the previous period. The fund therefore increases by the payment amount plus the interest earned. This is added to the previous accumulated amount to give the new accumulated amount. Finally, the amount in cell E(n), namely $E(n\text{-}1) + D(n)$, will equal the amount of the loan, L.

Example 5.9: Doug borrowed $15,000 from his friend Jim to buy some electronic equipment and agreed to repay the loan in a lump sum in three years. Doug also promised to make semi-annual interest payments at the nominal annual rate of $j_2 = 8\%$. To accumulate the lump-sum repayment, Doug will make semi-annual deposits into a sinking fund earning at the nominal annual rate of $j_2 = 6.5\%$.

a) Describe Doug's interest payments and sinking fund deposits. Make a sinking fund schedule.

b) Find the total amount of Doug's payments to Jim if he were to amortize the loan over the three years at $j_2 = 8\%$, and compare the total with his sinking fund and interest payments.

Solution:

a) Note that $i = .08/2 = .04 = 4\%$, while $i' = .065/2 = .0325 = 3.25\%$. Doug will make semi-annual payments of $15,000(.04) = $600 to Jim. From Formula 5.3, his deposits into the sinking fund and his sinking fund schedule are, respectively:

$$P = \$15{,}000 \cdot 0.0325/\{(1.0325)^6 - 1\} = \$2{,}304.45 \text{ and}$$

Deposit Number	*Deposit*	*Interest @ 3.25%*	*Increase In Fund*	*Amount Accumulated*
0				$0.00
1	$2,304.45	$0.00	$2,304.45	$2,304.45
2	$2,304.45	$74.89	$2,379.34	$4,683.79
3	$2,304.45	$152.22	$2,456.67	$7,140.47
4	$2,304.45	$232.07	$2,536.51	$9,676.98
5	$2,304.45	$314.50	$2,618.95	$12,295.93
6	$2,304.45	$399.62	$2,704.07	$15,000.00

Table 5.6—Sinking Fund Schedule for Example 5.9

b) If Doug's loan were amortized at $i = 4\%$ per half-year, his semi-annual payments would be $P = \$2{,}861.43$ from solving

$$S(P,\ .04,\ 6,\ -1) = P\frac{1-(1.04)^{-6}}{.04} = \$15{,}000$$

for P. Doug's total outlay by amortization would be $6 \cdot$ $2,861.43 = $17,168.58. His total outlay by the sinking fund method was 6·($600 + $2,304.45) = $17,426.70. As pointed out above, this exceeds the amortized amount because $i = .04 > .0325 = i'$. ∎

Exercises 5.4

5.4.1 How much must Hazel deposit annually into a sinking fund earning $j_1 = 6\%$ in order to accumulate a lump sum of $40,000 in 15 years?

5.4.2 The owner of a small business borrowed $50,000 at 11% effective annually to cover start-up expenses and will make annual interest payments for 10 years. She will build up a sinking fund with annual deposits into a fund earning $j_1 = 8\%$.
a) Find her annual interest payment.
b) Find her annual sinking fund deposit.
c) Find her total annual outlay for the loan.
d) Find the annual payment she would have made if she had amortized the loan at 11% effective annually.

5.4.3 Make a sinking fund schedule for the transaction in the preceding exercise. *Hint:* Use Table 5.5.

5.4.4 Find the size of the monthly deposit into a sinking fund earning $j_{12} = 6\%$ needed to repay a debt of $100,000 due in 15 years.

5.4.5 (💻) Use a spreadsheet program and Table 5.5 to construct a sinking fund schedule for retiring a loan of $150,000 with quarterly payments for 10 years into a fund earning at $j_4 = 10\%$.

5.4.6 Koji borrowed $2,000 at $j_{12} = 10\%$ and will make interest payments for six months with six equal monthly payments. He will build up a sinking fund with monthly deposits into a fund earning $j_{12} = 8.5\%$. Make a sinking fund schedule for the transaction.

5.4.7 Nan is saving up for a nice little used car. She will accumulate $5,000 by making monthly deposits of $293.43 into a fund earning at $j_{12} = 9.99885\%$. How long will she need to make these deposits to accumulate to $5,000?

5.4.8 STNX Computer wants to borrow $2,200,000 to develop the new line of *i*STNX computers. It can repay the loan over 10 years by amortization at $j_{12} = 9\%$, or they can pay interest monthly on the loan for 10 years at j_{12} = 9.5% while making monthly deposits into a sinking fund earning j_{12} = 10.47%. Compare STNX's two options.

5.5 Yield Rates

We will introduce the idea of *yield rates* with an example although we have already done some similar problems in earlier chapters.

Example 5.10: Suppose that Lany loans her husband Lary \$10,000 now and another \$10,000 in one year as an investment. Lary will repay Lany with three annual payments of \$9,000 beginning in two years. What is the yield rate on Lany's investment?

Solution: First of all, what does *yield rate* mean exactly? In general, if Lany makes some investments (here there are the two investments of \$10,000 each) and receives some payments in return (here there are the three payments of \$9,000), then the interest rate for which the value of the investments equals the value of the payments all at the same focal date is Lany's yield rate. Another way to think of it is this: The yield rate is the interest rate that Lany is charging Lary for loaning him the two payments of \$10,000 in return for the three repayments of \$9,000. Here is a time diagram for the transaction.

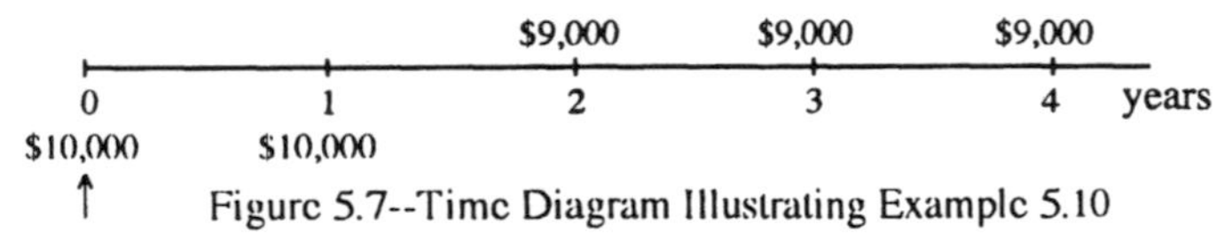

Figure 5.7--Time Diagram Illustrating Example 5.10

I have put the focal date at time 0, but (as usual) it could go anywhere. Let's translate the definition of yield rate into an equation of value. According to the definition above, the yield rate will be the value of i for which the money out equals the money in, all evaluated at the same focal date, at time 0. That is,

$$S(\$10,000,\ i,\ 2,\ 0) = \$10,000 + \$10,000(1+i)^{-1} =$$

$$\$9,000(1+i)^{-2} + \$9,000(1+i)^{-3} + \$9,000(1+i)^{-4} =$$

$$= S(\$9,000,\ i,\ 3,\ -2).$$

It is not hard to write down the equation of value that we have to solve for I, but once again, we have an equation that has to be solved by numerical methods.

Solving by Solver: After a little factoring enter into your solver

$$0 = \$10{,}000(1 + (1 + i)^{-1}) - \$9{,}000((1 + i)^{-2} + (1 + i)^{-3} + (1 + i)^{-4}).$$

Then starting with value of $i = .05$, press ALPHA and then [SOLVE]. At this point, your screen should now look like this:

```
EQUATION SOLVER
Eqn:0=10000(1+(1+I)^-1)-9000((1+I)^-2+(1+I)^-3+(1+I)^-4)
I=.12892440949...
Bound={-1E99,1E99}
Left-rt=0
```

Lany has earned an effective rate of about 12.8924% on her investment.

Solving by Guess-and-Check: In this case, since neither side of the original equation of value is a constant, we define left-hand side, $LHS(i)$, to be the difference of the two sides of the equation of value:

$$LHS(i) = \$10{,}000(1 + (1 + i)^{-1}) - \$9{,}000((1 + i)^{-2} + (1 + i)^{-3} + (1 + i)^{-4}),$$

and what we are after is the value of i that makes $LHS(i) = 0$ because then the two sides of our equation of value will be equal. I started with $i = .05$, as usual, and found $i = .1289$ after eight steps. As before, Lany's yield rate is about 12.89%. ∎

We'll do one more example on yield rates.

Example 5.11: Anders borrows \$80,000 from Bo to buy a cabin, agreeing to repay the loan with 10 equal annual payments at 7.5% effective. After 5 years, Bo sells Anders's contract (i.e. his remaining payments) to Carl for a price that is to yield Carl 9% effective.

a) What was the selling price of Anders's contract to Carl?

b) What yield rate did Bo enjoy?

Solution:

a) Here is what a time diagram looks like for the original contract.

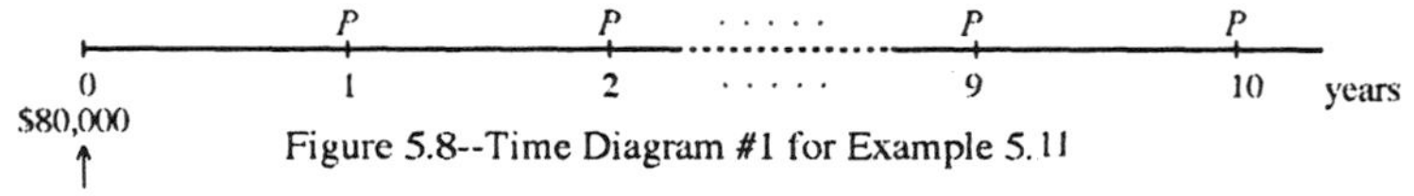

0	1	2	·····	9	10	years
\$80,000 ↑	P	P	·····	P	P	

Figure 5.8--Time Diagram #1 for Example 5.11

According to the original contract between Anders and Bo, P must satisfy

$$S(P,\ .075,\ 10,\ -1) = P\frac{1-(1.075)^{-10}}{.075} = \$80{,}000$$

from which we get $P\cdot(6.86408095599)$ = \$80,000 or $P \approx$ \$11,654.87. What Carl is buying is 5 future payments of \$11,654.87, which is to, yields him 9% effective. This means that Bo's selling price should be

$$S(\$11,654.87,\ 0.09,\ 5,\ -1) = \$45,333.38.$$

b) From Bo's point of view, he shelled out \$80,000 five years ago and handed it to Anders. In return, he got 5 annual payments of \$11,654.87 and a payment of \$45,333.38 from Carl. Here is what this looks like to Bo in a time diagram.

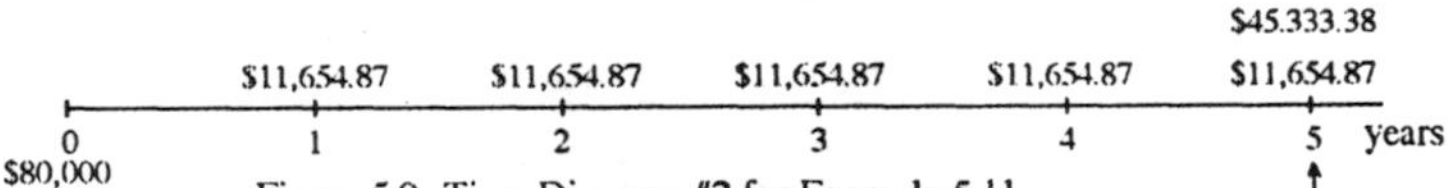

0	1	2	3	4	5 years
\$80,000	\$11,654.87	\$11,654.87	\$11,654.87	\$11,654.87	\$11,654.87 \$45,333.38 ↑

Figure 5.9--Time Diagram #2 for Example 5.11

This time I have put the focal date at 5 years, but, again, it could go anywhere. Let i denote Bo's annual yield rate. Then,

$$\$45,333.38 + S(\$11,654.87, i, 5, 4) = \$80,000(1 + i)^5,$$

or

$$\$45,333.38 + \$11,654.87\frac{(1+i)^5 - 1}{i} = \$80,000(1 + i)^5.$$

Solving by Solver: In the equation solver, enter the equation

$$0 = \$45,333.38 + \$11,654.87\frac{(1+i)^5 - 1}{i} - \$80,000(1 + i)^5,$$

start with $i = .05$, and press ALPHA and then [SOLVE].

You should now show the screen:

```
EQUATION SOLVER
Eqn: 0=45333.38+11654.87((1+I)^5-1)/I-80000(1+I)^5
I=.07039230374...
Bound={1E99,-1E99}
Left-rt=0
```

Indicating that Bo earned about 7.03923% on his transaction.

Solving by Guess-and-Check: We define the left-hand side to be

$$LHS(i) = \$45,333.38 + \$11,654.87\frac{(1+i)^5 - 1}{i} - \$80,000(1 + i)^5,$$

and we try to find a value of i for which $LHS(i) \approx 0$. I found the value $i \approx .07039$ after six steps. This is a good approximation to the actual yield rate.

By selling Anders's contract to Carl, Bo's yield rate changed from what would have been, 7.5%, to a lower 7.03923%. ∎

Exercises 5.5

5.5.1 Gordy invested $15,000 in a publishing company. In return, he received checks in the amount of $900 at the end of each six months for three years at which time he also received a final check for $16,500. What yield rate did Gordy earn on his investment? *Hint:* Use your calculator's solver if it has one, or use Guess-and-Check.

5.5.2 Rose invested $20,000 in the Kebob-a-Rebob restaurant chain. In return, she will receive semiannual payments of $1,200 for 12 years in addition to the return of her $20,000 in cash at the end of the 12 years. What is Rose's yield on her investment?

5.5.3 If Sandor loans Imre $10,000 which Imre will pay back with payments of $4,000 at the end of each of the next 3 years, what is Sandor's yield rate? *Hint:* Use your calculator's solver if it has one, or use Guess-and-Check.

5.5.4 In the preceding exercise, what annual payment would have made Sandor's yield 10%?

5.5.5 Damien loaned Tamara $5,000 four years ago. Tamara just repaid the loan with a payment of $7,320.50. What annual yield rate did Damien get on his money?

5.5.6 In the preceding exercise, what single payment in four years would have made the yield rate be 12% for Damien?

5.5.7 A $10,000 bond redeemable at par has semiannual coupons of $300 for 10 years. If the purchase price of the bond is $9,850, what is the yield rate of the bond? *Hint:* Use your calculator's solver if it has one, or use Guess-and-Check.

5.5.8 In the preceding exercise, what purchase price would have yielded $j_2 = 6.5\%$?

5.6 Summary of Formulas

Outstanding Principal—Retrospective Method

$$OP_r = L(1+i)^r - P\frac{(1+i)^r - 1}{i}$$

Outstanding Principal—Prospective Method

$$OP_r = P\frac{1-(1+i)^{-(n-r)}}{i}$$

Spreadsheet-Style Amortization Schedule:

Period (A)	Payment (B)	Interest Paid (C)	Principal Repaid (D)	Outstanding Principal (E)
0	—	—	—	L
1	P	$= i \times E0$	$= B1 - C1$	$= E0 - D1$
2	P	$= i \times E1$	$= B2 - C2$	$= E1 - D2$
3	P	$= i \times E2$	$= B3 - C3$	$= E2 - D3$
...	...	...	...	...
n	P	$= i \times E(n\text{-}1)$	$= Bn - Cn$	$= E(n\text{-}1) - Dn$

Spreadsheet-Style Sinking Fund Schedule:

Deposit Number (A)	Deposit (B)	Interest Earned On Fund At i% (C)	Increase In Fund (D)	Amount Accumulated (E)
0				$0.00
1	P	$= i' \times E0$	$= B1 + C1$	$= E0 + D1$
2	P	$= i' \times E1$	$= B2 + C2$	$= E1 + D2$
3	P	$= i' \times E2$	$= B3 + C3$	$= E2 + D3$
...	...	...	...	...
n	P	$= i' \times E(n - 1)$	$= Bn + Cn$	$= E(n\text{-}1) + D(n)$

$$\text{where } P = L\frac{i'}{(1+i')^n - 1}$$

Yield Rates:

Use Guess & Check

Miscellaneous Exercises

5M1. Leah borrowed $4,000 to pay for some root-canal work on her teeth. She agreed to make interest payments quarterly on the loan at j_4= 8% and to make a lump sum payment of $4,000 in one year. To accumulate the lump sum of $4,000, Leah will make quarterly deposits into a sinking fund earning interest at j_4 = 6%.

a) Make a sinking fund schedule including Leah's interest payments.

b) Compare Leah's total amount repaid with the sinking fund method versus what she would pay with the amortization method at j_4 = 8%.

5M2. Victor borrowed $10,000 from her bank for home improvements. He agreed to repay the loan in a lump sum in two years. He also agrees to make semi-annual interest payments at the nominal annual rate of j_2 = 6%. To accumulate the lump-sum repayment, Victor will make semi-annual deposits into a sinking fund earning at the nominal annual rate of j_2 = 5%.

a) Describe Victor's interest payments and sinking fund deposits. Make a sinking fund schedule.

b) Find the total amount of Victor's payments to the bank if he were to amortize the loan over the three years at j_2 = 6%, and compare the total with his sinking fund and interest payments.

5M3. Joshua bought a boat, for which he borrowed $28,000. He will repay the loan in a lump sum in 8 years with semi-annual interest payments at j_2 = 9%. Joshua will accumulate the $28,000 by making equal semi-annual deposits into a sinking fund paying j_2 = 6.5%.

a) Find the size of each of Joshua's deposits.

b) Find Joshua's total semi-annual outlay.

c) What would Joshua's semi-annual outlay have been if he had amortized the loan at j_2 = 9% for 8 years?

5M4. A farmer is reading what consumers have been saying about two models of tractor one of which he intends to buy. Tractor *A* will cost $99,500, will have a life of 9 years, and will cost $500 per year to maintain. Tractor *B* will cost $350 per year to maintain and will have a life of 10 years. With money worth 6% effective annually, how much should the farmer pay for Tractor *B* to make the cost of each equivalent?

5M5. Michael has just made the 120th monthly payment on his house for which he borrowed \$195,000 at $j_{12} = 7.25\%$. The term of his loan is 30 years. The outstanding principal at the time of the latest payment is called the *seller's equity*. The difference between the selling price and the outstanding principal, the part of the selling price that has already been paid, is called the *buyer's equity*. Find the seller's equity and buyer's equity for Michael's house.

5M6. Per and Ulla bought a house with a \$125,000 price tag in 1980. They were able to borrow the money at $j_{12} = 8\%$ for 25 years but felt they could not quite afford the monthly payments of \$964.77 that went along with the loan. They got the bank to agree that they pay \$925 at the end of every month except the last, at which time they would make a *balloon payment* to pay off the loan. Per and Ulla thought they could accumulate the amount required for the balloon payment gradually over the term of the loan. What would the amount of their balloon payment be 25 years later in 2005?

5M7. (Continuation) After 10 years, Per and Ulla decided to start accumulating \$40,000 in order to more than cover their balloon payment. They begin making deposits into a sinking fund earning $j_{12} = 7.2\%$ at the end of every month for the remaining 15 years of the term of the loan. These payments are to increase by 0.2% per month. What must the amount of their first payment be so that their sinking fund accumulates to \$40,000 at the end of 25 years? What is the amount of the last deposit?

5M8. Fredrik borrowed \$108,000 at $j_{12} = 7.8\%$ for 25 years to buy a house. The mortgage company charged Fredrik a fee of 3 *points* (i.e. 3% of the loan amount). What true interest rate is Fredrik paying? *Hint:* Use your calculator's solver if it has one, or use Guess-and-Check.

5M9. Show that with the inputs L, i, r, and P to Formula 5.1, we can find n, the term of the original loan. Write a formula for n.

Part II—Life Contingencies

Chapter 6—Probability & The Mortality Table

6.1 What is Probability?

If we pick a card from a well-shuffled deck in such a way that each card has an equal chance of being picked, then the *probability* of picking an ace is $4/52 = 1/13 \approx 0.076923$. Probabilities are sometimes given as percentages, so we could also say that there is about a 7.69% probability or chance of picking an ace. Similarly, the probability of picking a spade is $13/52 = 1/4 = 0.25$, and the probability of picking a black card is $26/52 = 1/2 = 0.5$. More generally, if we pick an item from a finite set S consisting of N items in such a way that each item is *equally likely* to be picked, and if A is a subset of S consisting of n items, then the probability that the selected item is a member of A is

$$P(A) = \frac{\#\text{ of items in } A}{\text{total }\#\text{ of items in } S} = \frac{n}{N}.$$

We call A an *event.*

While this is by no means the full story about probability, equally likely items are all we are going to need to know about.

Some Rules of Probability: Let S be a finite set consisting of N items.

a) For any event A, the probability of A occurring must be at least 0 and at most 1. That is, $0 \le P(A) \le 1$.
b) If $A = S$, then $P(A) = 1$. We say that S is a *certain* event.
c) If A is empty, then $P(A) = 0$. We say that A is an *impossible event.*
d) If A and B are *mutually exclusive* events (which means that they have no items in common), then the probability of at least one of the two occurring, $P(A \text{ or } B)$, is $P(A) + P(B)$.
e) The probability of A not occurring, $P(\sim A)$ satisfies $P(\sim A) = 1 - P(A)$. Equivalently, $P(\sim A) + P(A) = 1$.

Example 6.1: It turns out that there is one chance in 195,249,054 of winning the Grand Prize in the Powerball lottery with a single ticket.

a) What is the probability of winning the Grand Prize in the Powerball lottery with the purchase of a single ticket?

b) What is the probability of not winning the Grand Prize in the Powerball lottery with the purchase of a single ticket?

Solution:

a) Since there is one chance in 195,249,054 of winning, the probability of winning is $1/195,249,054 \approx 0.0000000051 2166....$ (Some might say that the probability of winning the Powerball lottery is the same whether you buy a ticket or not.)

b) From part *e*) of the General Rules above, the probability of *not* winning the Powerball lottery with the purchase of a single ticket is $1 - 1/195,249,054 = 1 - 0.0000000051 2166.... = 0.999999994$ 878.... ∎

Example 6.2: If the probability that a woman now aged 55 will survive to age 85 is 0.36448, can we find the probability

a) that she will die before reaching age 85? If yes, what is it?

b) that she will die between the ages of 75 and 85? If yes, what is it?

Solution:

a) From part *e*) of the General Remarks above, if the probability that a 55-year-old woman lives to age 85 is 0.36448, then the probability that she will not is 1 - 0.36448 = 0.63552.

b) There is no way of finding the probability that the woman will die between the ages of 75 and 85 from the information given. ∎

Example 6.3: Tex told me that the chance of rain tomorrow is 30. Does Tex know what he is talking about?

Solution: No. If Tex had said there was a 30% chance or 30% probability of rain, it would make some sense because a 30% probability is the same as a probability of .30. As we pointed out in part *a*) of the General Remarks above, probabilities must between 0 and 1.

Exercises 6.1

6.1.1 If the probability that a man now aged 28 will die between the ages of 70 and 80 is .31205, can we find the probability

a) that he will not die between 70 and 80? If yes, what is it?

b) that he will survive to 80? If yes, what is it?

6.1.2 If you toss a coin, what is the probability that it will land either heads or tails? Assume landing on its edge is impossible.

6.1.3 If the probability of a 30-year-old woman dying between the ages of 70 and 90 is 0.61313, what is the probability that she will die either before 70 or after 90?

6.1.4 Charley calculated that the probability his favorite team would win the World Series was 2.5. Could Charley be right?

6.1.5 If you bet on a horse to win the Kentucky Derby, what is the probability that it will finish 2nd and 3rd?

6.1.6 In a room of 30 randomly assembled people, the probability is about .2937 that the 30 people will all have different birthdays. What is the probability that at least two people in the group will have the same birthday?

6.2 The Mortality Table

A *mortality table* is constructed from a summary of the mortality experience of a large number of individuals. The records of many insurance companies, each having experience with many thousands of insured individuals, may be combined to form mortality tables like the ones we will be using. In this book, we use the 1980 Commissioners Standard Ordinary (CSO) Mortality Tables—one for males (Table I), and one for females (Table II). You will find the tables near the back of the book. These tables have been constructed in such a way that the last of the *survivors* will die at age 99. That is, there are no survivors to age 100. This is because, since it is a table, it must end somewhere, and by age 100 there are very few survivors in the actual population. This does not have any serious effect on the calculation of payments for insurance products, and insurance companies honor insurance contracts for people of any age.

To construct a mortality table, a number called the *radix* is chosen as the size of a hypothetical population assumed to have been born at the same time. Such a group is called a *cohort.* This radix is usually a nice round number like 100,000, 1,000,000, or, as in the case of our CSO tables, 10,000,000. For each age x, $x = 0, 1, 2, \ldots, 99$, probabilities that a random individual aged x years will die before reaching age $x + 1$ are estimated. For example, in our table for females (Table II), we find the death rate for newborns is .00289. In the context of our definition of probability for equally likely items, this means that of $N = 10{,}000{,}000$ newborn females, $n = 28{,}900$ die before reaching age one. In other words, the probability that a newborn girl will die before reaching age one is $n/N = .00289$. This, in turn, means that of the 10,000,000 originally in the cohort, there are 10,000,000 – 28,900 = 9,971,100 girls from the original 10,000,000 who survive to age one, and this is a new N for computing the probability that a one-year-old girl will die before reaching age two. For the $N = 9{,}971{,}100$ reaching age one, the probability of death before reaching two is, from our Table II, .00087 from which we could calculate that there are $n = (9{,}971{,}100)\cdot(.00087) = 8{,}675$ one-year-olds who do not survive to age two. Subtracting this number from 9,971,100, we find that $N = 9{,}962{,}425$ of the original cohort survive to age two, and so on until we reach the first age with no survivors. As I mentioned above, for the 1980 CSO tables, this age is 100.

Notation: In what follows, the symbol "(x)" is to be read "a life aged x." For ages $x = 0, 1, 2, \ldots$, the number l_x denotes the number of the original cohort of 10,000,000 who survive to age x. That is,

$$l_x = \text{the number of the original 10,000,000 surviving to age } x$$

The collection of numbers $\{l_x: x = 0, 1, 2, \ldots\}$ is a *survival function.* In Exercise 6.2.4, you are asked to make graphs of the survival functions of males and females.

At age 0, for both males and females, we have chosen the radix to be $l_0 = 10{,}000{,}000$, the number who begin life together. To give an example, if you look in Table I, you will find that the number of male survivors to age 8 is $l_8 = 9{,}893{,}156$. This also means that 10,000,000 – 9,893,156 = 106,844 of the original cohort die before reaching age 8. You can check that this number is the number who die at age 0 or 1 or 2 or 3 or 4 or 5 or 6 or 7—that is, 41,800 + 10,655 + 9,848 + 9,739 + 9,432 + 8,927 + 8,522 + 7,921 = 106,844.

The number of deaths *at* age x is denoted d_x. Therefore, if we knew the values of the survivor function, l_x, for every age x, we could construct the d_x column, because the number of deaths at age x is the number in the cohort who survive to age x but not to age $x + 1$. In symbols, this is

$$d_x = l_x - l_{x+1}$$

where x is a non-negative integer. As a reminder, in our tables, $l_{100} = 0$. In Tables I and II, you see that the column following the l_x and d_x columns is labeled q_x. This number is the probability that a person who has survived to age x dies at that age (that is, before reaching age x + 1). According to our definitions of d_x and l_x, we have

$$q_x = \frac{d_x}{l_x}.$$

Note that this agrees with our definition of probability for equally likely items, because if we select an item (here, a person) from the l_x survivors to age x, exactly d_x of these satisfy the criterion "dies at age x." Thus, P(person surviving to age x dies at age x) = $q_x = d_x/l_x$.

In actual practice, insurance companies may use different mortality tables for pricing different insurance products. For example, tables used for pricing annuities are, in general, different from those for pricing life insurance. This is because annuitants are generally a healthier population than that insured for life. In addition, mortality tables may be separated by occupation, smoker/non-smoker, etc. We will distinguish our tables only on the basis of gender.

Let's work our way through some examples using Tables I and II.

Example 6.4: Use Table II to find the probability that
a) a woman now aged 40 will die at age 40.
b) a woman now aged 40 will die in her fifties.

Solution:

a) The probability that a woman now aged 40 will die at age 40 is

$$q_{40} = \frac{\text{the number of women dying at age 40}}{\text{the number of women surviving to age 40}} =$$

$$\frac{d_{40}}{l_{40}} = \frac{23{,}100}{9{,}545{,}345} \approx .00242.$$

b) The probability that a woman now age 40 will die in her fifties is

$$\frac{\text{the number of women who reach age 50, but not age 60}}{\text{the number of women surviving to age 40}} =$$

$$\frac{l_{50} - l_{60}}{l_{40}} = \frac{9{,}219{,}130 - 8{,}603{,}801}{9{,}545{,}345} \approx .06446. \quad \blacksquare$$

Example 6.5: Find the probability that a man now aged 20 will survive to age 90.

Solution: The probability we want is

$$\frac{\text{the number of men surviving to age 90}}{\text{the number of men surviving to age 20}} =$$

$$\frac{l_{90}}{l_{20}} = \frac{645{,}788}{9{,}754{,}159} \approx .06621.$$ ∎

Probabilities like the one in Example 6.5 have a special notation. As we shall see later, they have an important place in calculating premiums for certain insurance products. In general, the probability that (x) will survive for at least another n years (that is, at least to age $x + n$) has the notation ${}_np_x$ and is defined as

$${}_np_x = \frac{\text{the number of people surviving to age } x+n}{\text{the number of people surviving to age } x} = \frac{l_{x+n}}{l_x}. \qquad (6.1)$$

Similarly, the probability that (x) will die before reaching age $x + n$ is denoted by the symbol ${}_nq_x$ and satisfies

$${}_nq_x = \frac{l_x - l_{x+n}}{l_x} = 1 - \frac{l_{x+n}}{l_x} = 1 - {}_np_x.$$

I used item *e*) of the *Rules of Probability* from Section 6.1.

Example 6.6: Explain why $l_x = d_x + d_{x+1} + d_{x+2} + \ldots + d_{99}$.

Solution: l_x is the number of people in the original population still alive at age x: They are the survivors to age x. So, their ages at death will be x or greater. In other words, d_x of them will die at age x, d_{x+1} of them will die at age $x + 1$, and so on. Thus, $l_x = d_x + d_{x+1} + d_{x+2} + \ldots + d_{99}$. ∎

Example 6.7: Find the probability that a woman now aged 35 will die within 5 years.

Solution: This is ${}_5q_{35} = 1 - {}_5p_{35}$. We calculate

$${}_5q_{35} = 1 - \frac{l_{40}}{l_{35}} = 1 - \frac{9{,}545{,}345}{9{,}637{,}125} \approx .00952.$$

There is less than 1 chance in 100 that a 35-year-old woman will die before reaching age 40. ∎

Example 6.8: According to your 1980 CSO Table, at what age is a man now aged 40 most likely to die?

Solution: The probability that a man now aged 40 will die at an age $x \geq 40$ is d_x/l_{40}. What we have to do is search the d_x column of Table I for all $x \geq 40$ and find the largest value of d_x. You can check that the largest value of d_x is $d_{78} = 329{,}936$ and that all other d_x's are less than this. A man now aged 40 is thus most likely to die at age 78. The actual probability that (40) will die at age 78 is d_{78}/l_{40} = 329,936/9,377,225 ≈ .03518. ∎

Example 6.9: Find the probability that a woman aged 25 will die *at* age 60.

Solution: We want the probability that this 25-year-old woman will survive 35 years to age 60 but not to age 61. This probability is

$$\frac{l_{60} - l_{61}}{l_{25}} = \frac{d_{60}}{l_{25}} = \frac{81{,}478}{9{,}767{,}317} \approx .00834.$$ ∎

Exercises 6.2

6.2.1 According to your 1980 CSO Table, at what age is a woman now aged 50 most likely to die? What is the probability of this event happening?

6.2.2 What is the probability that a 35-year-old woman will survive to age 70 and then die within the next two years?

6.2.3 What is the probability that a 20-year-old male will die between the ages of 75 and 85?

6.2.4 On the same set of axes using Tables I and II, make graphs showing l_x at ages 0, 10, ..., 90, 100 for both males and females. Discuss.

6.2.5* A 63-year-old man bought an insurance policy paying $100,000 to his heirs if he should die before reaching age 64. Ignoring interest and profit considerations, how much should he have paid?

6.2.6. At what age x is a man 10 times as likely to die as is a 10-year-old boy?

6.2.7. At what age x is a woman 10 times as likely to die as is a 10-year-old girl?

6.2.8. Find the probability that a man now aged 28 will die in his seventies.

6.2.9. What is the probability that a 35-year-old male will die before reaching 80?

6.2.10. What is the probability a 40-year-old woman will live at least another 40 years?

6.2.11. Find the probability that a 20-year-old male will die between 60 and 70 years of age. Would you expect the preceding probability to be higher or lower for females? Explain.

6.2.12. Find the probability that a 20-year-old female will die between 60 and 70 years of age.

6.3 Probability Distributions for Ages at Death*

Probabilities like the ones in Examples 6.5 and 6.6 have special importance in *life contingencies* and therefore have their own special notations. Specifically, the probability that (*x*) survives *t* more years, that is, to age $x + t$ and then dies at that age is

$$_{t|}q_x = \frac{\text{the number in the original population dying at age } x+t}{\text{the number in the original population surviving to age } x} =$$

$$\frac{l_{x+t} - l_{x+t+1}}{l_x} = \frac{d_{x+t}}{l_x}, \tag{6.2}$$

where t is a non-negative integer, and $x + t < 100$. Given any age x, the collection of numbers $\left\{ {}_{0|}q_x, {}_{1|}q_x, {}_{2|}q_x, \ldots, {}_{99-x|}q_x \right\}$ form the *probability distribution* for the ages at death for (*x*). More generally, if a variable, $\boldsymbol{x}$, can take on the different values $x_1, x_2, x_3, \ldots,$ and x_k with probabilities $p_1, p_2, p_3, \ldots,$ and p_k, respectively, then these probabilities are called the *probability distribution of the variable* $\boldsymbol{x}$, provided also that

i) $0 \le p_j \le 1$ for each $j = 1, 2, \ldots, k$,

and

ii) $p_1 + p_2 + p_3 + \ldots + p_k = 1$.

The term "probability distribution" is appropriate because the numbers $p_1, p_2, p_3, \ldots,$ and p_k tell us how the probability is distributed between the possible values $x_1, x_2, x_3, \ldots,$ and x_k. It might help to think of the probability distribution as follows: The variable $\boldsymbol{x}$ takes on the value x_1 $100p_1\%$ of the time; takes on the value x_2 $100p_2\%$ of the time; takes on the value x_3 $100p_3\%$ of the time; …; takes on the value x_k $100p_k\%$ of the time, and these percentages sum to 100%. For example, suppose that the variable $\boldsymbol{x}$ is the number of dots showing when you throw a fair die. Then the possible values for $\boldsymbol{x}$ are 1, 2, 3, 4, 5, and 6 with corresponding probabilities all equal to $1/6 = .1\overline{6}$. This means that in a large number of throws of the die each of the possible numbers of dots will show $100(.1\overline{6})\% = 16.\overline{6}\%$ of the time.

Example 6.10: Discuss the probability distribution of the ages at death of 40-year-old males.

Solution: Here, the variable x of the definition above is the age at death of males now aged 40. The possible values for this variable are the ages 40, 41, 42, ..., 98, 99. The corresponding probabilities are

$$_{0|}q_{40},\ _{1|}q_{40},\ _{2|}q_{40},\ _{3|}q_{40}, \ldots,\ _{59|}q_{40}.$$

We stop at $_{59|}q_{40}$ because it is the last non-zero probability. That is, $_{59|}q_{40} = d_{99}/l_{40}$, but $_{60|}q_{40} = d_{100}/l_{40} = 0$ since there are no survivors to age 100. If we graph this probability distribution, we can plot the probabilities as heights on the vertical axis against the corresponding ages on the horizontal axis. We obtain these numbers from Table I. The only arithmetic we have to do is to divide each d_x, $x = 40, 41, 42, \ldots, 98, 99$, by l_{40}. The first three are $_{0|}q_{40} = 0.0030200$, $_{1|}q_{40} = 0.0032801$, and $_{2|}q_{40} = 0.0035376$, and the last one is $_{59|}q_{40} = .0011471$. The graph of the probability distribution of $_{x|}q_{40}$, $x = 40, 41, 42, \ldots, 99$, is shown below:

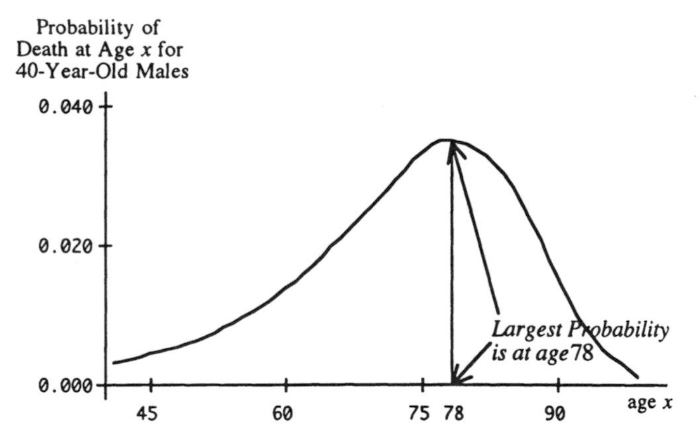

Figure 6.1--Distribution of Ages at Death for 40-Year-Old Males

Notice that the probability of death for a 40-year-old is highest at age 78 where d_{78}/l_{40} = 329,936/9,377,225 = .0351848 as found in Example 6.8.

One more thing about this distribution. With the help of Example 6.6, you can show that

$$_{0|}q_{40} + _{1|}q_{40} + _{2|}q_{40} + _{3|}q_{40} + \ldots + _{59|}q_{40} = 1.$$

(See Exercise 6.3.1 below.) ∎

Exercises 6.3

6.3.1 Show algebraically, and tell in your own words why

$$_{0|}q_{40} + _{1|}q_{40} + _{2|}q_{40} + _{3|}q_{40} + \ldots + _{59|}q_{40} = 1.$$

Hint: Make use of Example 6.6.

6.3.2 In your own words, interpret $_{40|}q_0$ and $_{0|}q_{40}$, and explain why $_{40|}q_0 < _{0|}q_{40}$. Verify that this inequality is true for both males and females.

6.3.3 What is the most likely age at death for
a) an 80-year-old woman?
b) an 85-year-old woman?
c) Explain the results of parts *a*) and *b*).

6.3.4 In Example 6.8, we saw that the most likely age for a 40-year-old man to die was 78 but that the probability of a 40-year-old man dying at that age was only .03518. Explain how the most likely age can be so unlikely.

6.3.5 For a given age x and numbers a and b with $a < b$, can you tell in general which is the larger of $_{a|}q_x$ and $_{b|}q_x$? Explain.

6.3.6 Discuss the probability distribution of the ages at death of a 95-year-old woman.

*6.4 Average Age at Death for a Life Aged x**

How would we find the average age at death for people surviving to a particular age, x? Take 40-year-old males, for example. In other words, at what age can a man surviving to age 40 expect to die on average?

Before we take up this question, let's talk about *weighted* averages. When you calculate your grade-point average, where not all of the courses carried the same number of credits, you calculate a weighted average. The weights are the numbers of credits for each course. As an example, let's say that last semester you got a 3.5 in Mathematics (1 credit), 4.0 in Music (1/2 credit), 2.5 in Physical Education (1/2 credit), 4.0 in History (1 credit), and 3.0 in English (1 credit). Your grade point average (*GPA*) for the semester was 3.4375 since

$$GPA = \frac{3.5(1.0)+4.0(0.5)+2.5(0.5)+4.0(1.0)+3.0(1.0)}{1.0+0.5+0.5+1.0+1.0} =$$

$$3.5(1/4)+4.0(1/8)+2.5(1/8)+4.0(1/4)+3.0(1/4) = 3.4375.$$

Your courses in Mathematics, History, and English each get 25% of the weight, while your courses in Music and Physical Education each get 12.5% of the weight. That's a weighted average.

Now, back to the 40-year-old males. The average age a death for (40) is a weight average. If we consider all of the l_{40} = 9,377,225 survivors to age 40, d_{40} = 28,319 of them will die at age 40, d_{41} = 30,758 of them will die at age 41, d_{42} = 33,173 of them will die at age 42, ..., d_{98} = 20,693 of them will die at age 98, and finally, d_{99} = 10,757 of them will die at age 99. The average age at death for these 40-year-olds is the sum of all their ages at death divided by the total number of people in the group. To see this, let's write out the sum of all the ages at death for the 9,377,225 survivors to age 40:

$$\overbrace{40+40+\ldots+40}^{28{,}319\text{ times}} + \overbrace{41+41+\ldots+41}^{30{,}758\text{ times}} + \overbrace{42+42+\ldots+42}^{33{,}173\text{ times}} + \ldots +$$

$$\overbrace{98+98+\ldots+98}^{20{,}693\text{ times}} + \overbrace{99+99+\ldots+99}^{10{,}757\text{ times}} =$$

$$40(28{,}319) + 41(30{,}758) + 42(33{,}173) + \ldots + 98(20{,}693) + 99(10{,}757).$$

Now, to get the average, this sum must be divided by l_{40} = 9,377,225. When we do this, the average looks like this:

$$40\left(\frac{28{,}319}{9{,}377{,}225}\right) + 41\left(\frac{30{,}758}{9{,}377{,}225}\right) + 42\left(\frac{33{,}173}{9{,}377{,}225}\right) + \ldots$$

$$+\ 98\left(\frac{20{,}693}{9{,}377{,}225}\right) + 99\left(\frac{10{,}757}{9{,}377{,}225}\right) =$$

$$40(.0030200) + 41(.0032801) + 42(.0035376) + \ldots$$
$$+ 98(.0022067) + 99(.0011471) = 73.55.$$

Notice that the numbers in parentheses are precisely the probabilities of the probability distribution of the ages at death for 40-year-olds. We can say that the average age at death for 40-year-old men is a *weighted average*, the weights being the probabilities of them dying at the different ages.

Well, actually, before we say that 73.55 is the average age at death for 40-year-old males, we need to make a small adjustment. We based the above average on the assumption that at whatever age a man dies, he does it on his birthday. This biases the calculated average downward. To make up for this, it seems reasonable to assume that the age at death, whatever it is, is on average halfway between birthdays. The assumed possible ages at death then change 40 to 40.5, 41 to 41.5, 42 to 42.5, ..., 98 to 98.5, and 99 to 99.5. The 'corrected' average age at death then becomes

$$(40.5)(.0030200) + (41.5)(.0032801) + (42.5)(.0035376) +$$
$$(98.5)(.0022067) + \ldots + (99.5)(.0011471) = 74.05,$$

exactly one-half year more than without the correction. When we say that the average age at death of 40-year-old men is 74.05, we mean that a 40-year-old man should live another 34.05 years, on average. ∎

Definition of Average Value of a Variable: If a variable, x, assumes the values $x_1, x_2, x_3, \ldots, x_k$ with corresponding probabilities $p_1, p_2, p_3, \ldots, p_k$, then the *average value* of this variable (also called the *expected value* of x), written as $E(x)$, is defined to be

$$E(x) = x_1 \cdot p_1 + x_2 \cdot p_2 + \ldots + x_k \cdot p_k. \qquad (6.3)$$

As in the previous example, we may refer to $E(x)$ as the *weighted average* of $x_1, x_2, \ldots,$ and x_k where the weights are the probabilities $p_1, p_2, \ldots,$ and p_k, respectively.

Example 6.11: Find the probability distribution and the average age at death of 95-year-old women, and show the results in a table.

Solution: Here is a table with the probability distribution of the ages at death of 95-year-old women. As pointed out above, we take the ages at death to be halfway between birthdays.

Age at Death, x	95.5	96.5	97.5	98.5	99.5
Probability of Death at Age x	${}_{0|}q_{95}$ = 0.31732	${}_{1|}q_{95}$ = 0.25651	${}_{2|}q_{95}$ = 0.20242	${}_{3|}q_{95}$ = 0.14675	${}_{4|}q_{95}$ = 0.07700

Table 6.1—Age-at-Death Probability Distribution of 95-Year-Old Women

By the definition of average value in Formula 6.3, the average age at death for 95-year-old women is

$$E(x) = (95.5)(.3173195) + (96.5)(.2565100) + (97.5)(.2024181) + (98.5)(.1467484) + (99.5)(.0770039) = 96.9096073.$$

I chose an older age to make the computations short. ∎

Generalization of Average Age at Death for a Life Aged x: Consider a person now aged x. Following the discussions above and assuming that people die, on average, halfway between birthdays, this person

will die at age $x + .5$ with probability ${}_{0|}q_x$.

will die at age $x + 1.5$ with probability ${}_{1|}q_x$.

will die at age $x + 2.5$ with probability ${}_{2|}q_x$.

⋮ ⋮ ⋮

will die at age 99.5 with probability ${}_{99-x|}q_x$.

This person's average age at death which we denote $AAD(x)$ is

$$AAD(x) = (x + .5)({}_{0|}q_x) + (x + 1.5)({}_{1|}q_x) + (x + 2.5)({}_{2|}q_x) + \ldots + (99.5)({}_{99-x|}q_x). \quad (6.4)$$

An alternate formula for calculating $AAD(x)$ is

$$AAD(x) = x + \frac{1}{2} + \frac{l_{x+1} + l_{x+2} + l_{x+3} + \ldots + l_{99}}{l_x}. \quad (6.5)$$

Remark: In some mortality tables the first age with no survivors will not be 100 as it is in our tables. The first age with no survivors is traditionally denoted by ω (the Greek letter 'omega'). This means that the last age *with* survivors is ω - 1. The effect on Formulas 6.4 and 6.5 is that you will need to change 99 to ω - 1.

If you are up for a challenge, you may verify that Formula 6.5 follows from Formula 6.4 in Exercise 6.4.6 below.

Calculations of $AAD(x)$ are quite tedious especially at the lower ages. Fortunately, this tedium has been taken over by another party (the author), and the results are tabulated for you in the fifth and tenth columns of Tables I (males) and II (females).

Expected values and probability distributions are not restricted to use with the mortality table as the next example shows.

Example 6.12: Magic Svenson pitched 150 innings of minor league softball last year. She pitched 70 scoreless innings, 40 innings with 1 earned run, 25 innings with 2 earned runs, 10 innings with 3 earned runs, 0 innings with 4, 5, 6, 7, or 8 earned runs, and 5 innings with 9 earned runs. What was Magic's ERA (earned run average) per inning?

Solution: In the definition of average value, we have the x_j and p_j values as follows:

Number of Runs (x_j)	0	1	2	3	4	5	6	7	8	9
Proportion of Innings with v_i ***runs*** (p_j)	$.4\overline{6}$	$.2\overline{6}$	$.1\overline{6}$	$.0\overline{6}$	0	0	0	0	0	$.0\overline{3}$

The proportions in the second row form a probability distribution for the numbers in the first row. Magic Svenson's ERA is therefore

$$0(.4\overline{6}) + 1(.2\overline{6}) + 2(.1\overline{6}) + 3(.0\overline{6}) + 4(0) +$$

$$5(0) + 6(0) + 7(0) + 8(0) + 9(.0\overline{3}) = 1.1$$

(average earned runs per inning). ∎

Example 6.13: Professor Fogelfroe is a 56-year-old man.

a) How long can he expect to live?

b) What is the probability that the professor will live beyond the expected age you found in part *a*)?

c) What is the probability that the professor will die between ages 70 and 80?

d) When Professor Fogelfroe was born, how long was he expected to live?

Solution:

a) From Table I, the average age at death for a 56-year-old male is 76.51, so Professor Fogelfroe should expect to live to age 76.51: another 20.51 years.

b) The probability that Professor Fogelfroe will live to at least 76 is l_{76}/l_{56} = 4,584,446/8,521,377 = .53799, and the probability that he will live to at least 77 is l_{77}/l_{56} = 4,261,105/8,521,377 = .50005. The probability that the Professor will live beyond age 76.51 should be around 51% of the way from .53799 down to .50005 or .51864. There is about a 52% chance that the Professor will live past his expected age at death. Remember, all of the calculations were made assuming that the professor is now aged 56.

c) The probability that Professor Fogelfroe will die between 70 and 80 is

$$\frac{l_{70}-l_{80}}{l_{56}}=\frac{6{,}274{,}160-3{,}274{,}541}{8{,}521{,}377}=.35201\,.$$

d) When the professor was born, he was expected to live to $AAD(0)$ = 70.83, some 5 years and 8 months less than he could expect to live once he survived to age 56. ∎

Example 6.14: Verify from both Formula 6.4 and 6.5 that the value in Table II for females, $AAD(85)$ is 90.18.

Solution: Formula 6.4 gives

$$AAD(85) = (85.5)({}_{0|}q_{85}) + (86.5)({}_{1|}q_{85}) +$$

$$(87.5)({}_{2|}q_{85}) + \ldots + (99.5)({}_{14|}q_{85})$$

while Formula 6.5 gives

$$AAD(85) \;= 85+\frac{1}{2}+\frac{l_{86}+l_{87}+\ldots+l_{98}+l_{99}}{l_{85}}.$$

Since we are doing all of this by hand, it is easiest to make a table. From Table II, we get:

x	x+.5	${}_{x-85|}q_{85}$	$(x+.5)\,{}_{x-85|}q_{85}$	l_x
85	85.5	0.116099910	9.926542264	—
86	86.5	0.114279536	9.885179835	2,885,681
87	87.5	0.110302158	9.651438840	2,512,591
88	88.5	0.104290912	9.229745699	2,152,486
89	89.5	0.096541382	8.640453651	1,812,006
90	90.5	0.087456359	7.914800500	1,496,826
91	91.5	0.077496834	7.090960341	1,211,306
92	92.5	0.067163310	6.212606219	958,301
93	93.5	0.056934237	5.323351142	739,032
94	94.5	0.047325126	4.472224366	553,158
95	95.5	0.038747958	3.700429961	398,655
96	96.5	0.031322499	3.022621124	272,154
97	97.5	0.024717326	2.409939278	169,895
98	98.5	0.017919487	1.765069467	89,200
99	99.5	0.009402968	0.935595277	30,698
	Totals	—	90.18095796	15,281,989

We can now calculate from Formula 6.5 that

$$ADD(85) = 85 + \frac{1}{2} + \frac{15{,}281{,}989}{3{,}264{,}714} = 90.18095796,$$

as advertised. This agrees with the value given in the tenth column of Table II for $ADD(85)$. ∎

Example 6.15: We saw from Table I that the expected age at death for newborn males is 70.83. What is the *median* age at death for males? That is, at what age is half the original cohort dead and the other half alive?

Solution: What we are searching for is the age x at which l_x = 5,000,000. Unfortunately, we cannot find such an integer, but we can approximate the median age. From Table I, we find that

$$l_{74} = 5{,}201{,}587$$

and

$$l_{75} = 4{,}898{,}907$$

so that the median age at death for males is somewhere between 74 and 75. Now, 5,000,000 is almost exactly two thirds of the way down from 5,201,587 to 4,898,907. Here is why: The difference between 5,201,587 and 4,898,907 is 302,680. That is

$$5{,}201{,}587 - 4{,}898{,}907 = 302{,}680.$$

The difference between 5,201,587 and 5,000,000 is 201,587:

$$5{,}201{,}587 - 5{,}000{,}000 = 201{,}587.$$

This tells us that 5,000,000 is 66.6% of the way from 5,201,587 to 4,898,907 because

$$201{,}587/302{,}680 = .6660.$$

The median is 66.6% $\approx$ 67% of the way from 74 to 75, so, we can say that the median age at death for males is 74.67. ∎

Example 6.16:

a) According to Table I, at what age has one-quarter of the original male population died and three-quarters survived? This number, Q_1, is called the *first quartile* of the distribution of ages at death.

b) At what age has three-quarters of the original male population died and one-quarter survived? This number, Q_3, is called the *third quartile* of the ages at death. In keeping with this notation, the median (discussed in Example 6.15) is Q_2, the *second quartile* of the distribution of ages at death.

Solution:

a) We would like to find x in Table I so that $l_x = 7{,}500{,}000$. Since $l_{64} = 7{,}503{,}368$ and $l_{65} = 7{,}329{,}740$, we see that Q_1 is between 64 and 65 but closer to 64. In fact, following the analysis in the preceding example, 7,500,000 is about 2% of the way from l_{64} down to l_{65}. Therefore, $Q_1 \approx 64.02$.

b) This is similar to part *a*), but now we want to find x such that $l_x = 2{,}500{,}000$. From Table I, $l_{82} = 2{,}633{,}724$, and $l_{83} = 2{,}324{,}920$, so Q_3 lies between 82 and 83 but is a little closer to 82. Here, 2,500,000 is about 43% of the way from l_{82} down to l_{83}. Therefore, $Q_3 \approx 82.43$. ∎

Exercises 6.4

6.4.1 According to your 1980 CSO Table I, newborn males can expect to live about 71 years. At what age should a 71-year-old male expect to die? Explain.

6.4.2 According to your 1980 CSO Table II, newborn females can expect to live about 76 years. At about what age is a 76-year-old female expected to die? Explain.

6.4.3 Verify from Formula 6.4 or 6.5 that the value in Table I for males *AAD*(91) really is 93.94.

6.4.4 We can see from Table II that the expected age at death for newborn females is 75.83. What is the *median* age at death for females. That is, at what age is half the original cohort dead and half alive? Find the answer to two decimals as I did in Example 6.15.

6.4.5 Following Example 6.16, find the Q_1 and Q_3 for the female population.

6.4.6 Derive Formula 6.5 from Formula 6.4.

6.4.7 According to your 1980 CSO tables how much longer can a 21-year-old woman expect to live than a 21-year-old man?

6.4.8 How old a male partner should a 30-year-old woman take in order that they could expect to die at around the same time?

6.4.9 Compute the average age at death for a 95-year-old woman.

6.5 The Force of Mortality

An interesting aspect of the study of mortality is the *force of mortality* which, as the name might suggest, is a measure of the likelihood of dying after reaching a given age. The mathematics we would need to make a rigorous study of the force of mortality is outside the scope of this book, but we can briefly cover some of the more important highlights.

We will denote the force of mortality at age x by $\mu(x)$. The letter μ is the Greek equivalent of our m. (The Greeks and some others pronounce this 'moo,' like the cow, and English speakers pronounce it 'mew,' like the cat.)

Imagine that t is a small interval of time. Then, in words, $t \cdot \mu(x)$ is the approximate probability that a person who survives to age x will die before reaching age $x + t$: that is, in the interval between the ages x and $x + t$. Put slightly differently,

$$\mu(x) = \frac{P(\text{a life aged } x \text{ dies before reaching age } x+t)}{t} = \frac{l_x - l_{x+t}}{t \cdot l_x} = \frac{1 - {}_tp_x}{t}, \tag{6.6}$$

approximately. The approximation gets more and more accurate the smaller we set t. Unfortunately, we do not have the survivor numbers except at integer ages, so we cannot evaluate $\mu(x)$ with any great accuracy. But we can get some idea of the behavior of $\mu(x)$ from the mortality table because, when $t = 1$ year, the approximation above reduces to $\mu(x) \approx q_x$. That is, for $t = 1$,

$$\mu(x) \approx \frac{l_x - l_{x+1}}{1 \cdot l_x} = \frac{d_x}{l_x} = q_x.$$

Example 6.17: Make a plot of $\mu(x) \approx q_x$ versus x, for males of ages 0 to 50, and discuss.

Solution: Below is the graph for males aged 0 to 50 based on the q_x column of Table I (the graph increases quite predictably after age 50).

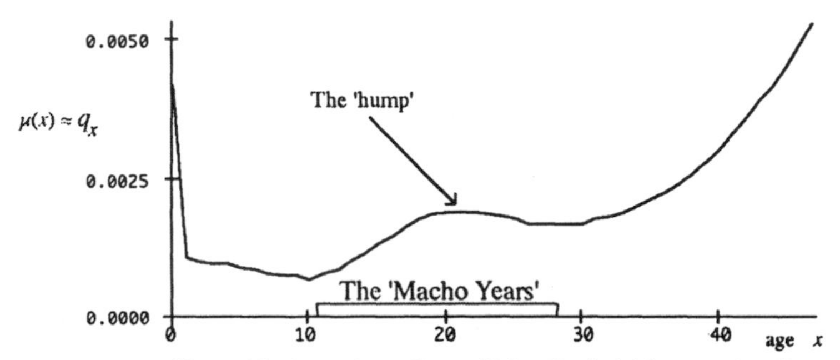

Figure 6.2--Approximate Force of Mortality for Males Ages 0 to 50

You can see the drop in mortality after the high infant mortality rate in the first year of life. More surprising is the 'hump' in the force of mortality between the ages of about 11 and 28. There is a sudden rise from about age 11 to age 21 then a drop until about age 28. I call these the *macho years* when many boys and young men consider themselves immune to danger. Notice that the force of mortality at age 21 is not reached again until about age 33. ∎

Remark: The graph of the force of mortality for females usually has a similar hump, but it appears a few years later, and is likely accounted for by an increase in mortality during the childbearing years. This hump is present in Table II but is less pronounced.

Exercises 6.5

6.5.1 Discuss the possible causes of the 'hump' between ages 11 and 28 in the force of mortality pictured in Figure 6.2.

6.5.2 At what age is the force of mortality lowest for males?

6.5.3 At what age is the force of mortality lowest for females?

6.5.4* Suppose, for the sake of argument that there is a population of 10,000,000 beings on some planet whose survival function, l_x, is given by the formula $l_x = 10{,}000{,}000\,(0.933)^x$. (By the way, this does not resemble the survival functions in either Table I or Table II.) Approximate the force of mortality at age 30 for this population. Do this using Formula 6.6 with $x = 30$, and try $t =$ .1, .01, and .001 in succession.

6.5.5 Verify that $\dfrac{1-{}_tp_x}{t}$ reduces to q_x when $t = 1$.

6.6 Summary of Formulas

Probability of Event A with Equally Likely Items

$$P(A) = \frac{\text{\# of items in } A}{\text{total \# of items outcomes}}$$

A Life Aged x (x)

Number of Survivors to Age x l_x

Number of Deaths at Age x $d_x = l_x - l_{x+1}$

Probability (x) Dies at Age x $q_x = \dfrac{d_x}{l_x}$

Probability (x) Survives At Least Another n Years

$${}_np_x = \frac{l_{x+n}}{l_x} = 1 - {}_nq_x$$

Probability (x) Survives to Age $x + t$ and Dies at That Age

$${}_{t|}q_x = \frac{d_{x+t}}{l_x}$$

Expected Value of a Variable $E(V) = v_1p_1 + v_2p_2 + \ldots + v_kp_k$

Average Age at Death for (x)

$$AAD(x) = (x+.5)_{0|}q_x + (x+1.5)_{1|}q_x + \ldots + (99.5)({}_{99-x|}q_x) = x + \frac{1}{2} + \frac{l_{x+1} + l_{x+2} + l_{x+3} + \ldots l_{99}}{l_x}$$

Force of Mortality for (x) $\mu(x) \approx \dfrac{l_x - l_{x+t}}{t \cdot l_x} = \dfrac{1-{}_tp_x}{t}$, for small t

Approximate Force of Mortality for (x)

$$\mu(x) \approx q_x \text{ when } t = 1$$

Miscellaneous Exercises

6M1. Use the definitions of q_x, l_x, d_x, and $AAD(x)$ to complete the mortality table below. *Hint:* Note the remark following Formulas 6.4 and 6.5.

x	l_x	d_x	q_x	$AAD(x)$
0	1,000			
1		255		
2	145		0.6000	
3			0.5000	
4			1.0000	

6M2. The new Watt-A-Lite 75-watt light bulb has been studied by the company's technical staff which found that the annual failure rates (i.e., death rates) were $q_0 = .2$, $q_1 = .4$, $q_2 = .6$, $q_3 = .8$, and $q_4 = 1$. Construct a mortality table for a lot of size $l_0 = 10{,}000$ bulbs. *Hint:* Note the remark following Formulas 6.4 and 6.5.

6M3.* (🖳) Females on earth have survival function given in Table II. Females on planet Q0317B have survival function

$$l_x = 10{,}000{,}000(1 - x/100),\ 0 \le x \le 100.$$

Females on planet Q0317C have survival function

$$l_x = 10{,}000{,}000\sqrt{1 - x/100},\ 0 \le x \le 100.$$

Using Formula 6.4 or 6.5 and a spreadsheet, find *AAD*(35) for both planets.

6M4.* (🖳) Using a spreadsheet, find the power n for which the survival function

$$l_x = 10{,}000{,}000(1 - x/100)^n,\ 0 \le x \le 100$$

has *AAD*(35) is 77.98, the same as for females on Earth.

Chapter 7—Life Annuities[†]

7.1 Introduction

In this chapter we study *life annuities*: annuities in which the continuation of payments is *contingent* upon the survival of a life. A life annuity may consist of payments *to* an individual, as is the case with a *pension*, or may be payments *from* an individual, as is in the case when that person pays periodic premiums for some insurance or other product. In either situation, the payments end when that person dies.

Perhaps the simplest example of a life annuity is the *whole-life* annuity

. The time diagram for the *present value* of a whole-life annuity of P paid at the end of each year to a life now aged x, looks like this:

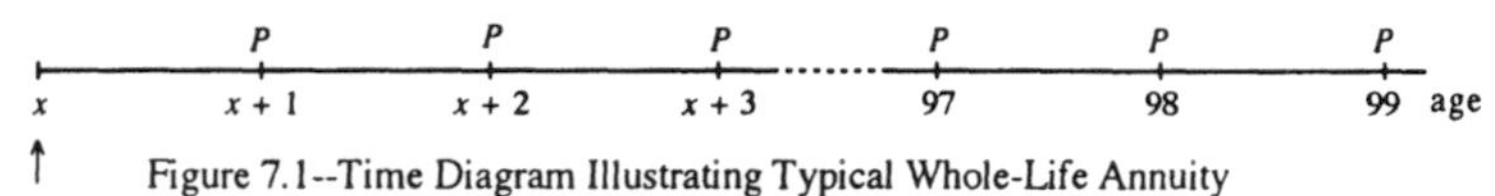

Figure 7.1--Time Diagram Illustrating Typical Whole-Life Annuity

To find the present value of this annuity, we will need to find the present values of each of the individual payments and then add them up. We will study several other types of life annuity shortly. Keep in mind that what life annuities have in common is that payments cease when the annuitant dies.

Remark: The present value of a single payment or series of payments where we ignore profit and expenses is call a *net single premium*, which we abbreviate *NSP*.

7.2 Pure Endowments

Pure endowments are the building blocks of life annuities because all life annuities are sums of pure endowments. An n-year pure endowment is the present value of a payment to a person in n years, provided that the person is alive at that time. The *NSP* for a pure endowment of each unit of currency to (x) is denoted by ${}_nE_x$, so that

---

[†] All but a few of the examples and exercises in this and succeeding chapters are tied to the Mortality Tables I & II, and to the interest rates in Tables III & IV, which run from 2% to 8% in steps of ¼%. In order to accommodate situations not covered in these tables, I have provided an interactive program at www.augsburg.edu/home/math/faculty/kaminsky/finlit.html.

the *NSP* for a payment of P in n years to (x) is $P \cdot {}_nE_x$. Here is a time diagram for an n-year pure endowment of P:

P

$x \quad x+1 \quad x+1 \quad \cdots \quad x+n-1 \quad x+n \quad$ age

NSP ↑

Figure 7.2—Time Diagram for a Pure Endowment of P

We will derive a formula for ${}_nE_x$ using the so-called *mutual fund method.* Suppose that all of the x-year olds, of which there are l_x, decide to buy an n-year pure endowment of P. They each put their net single premium, $P \cdot {}_nE_x$, into a mutual fund that will earn interest at effective rate i over the n-year period. The initial value of the fund is therefore $P \cdot {}_nE_x \cdot l_x$. After n years, the survivors to age $x+n$, of which there are l_{x+n}, divide up the accumulated value of the fund, which is $P \cdot {}_nE_x \cdot l_x \cdot (1+i)^n$. In other words,

$$P \cdot {}_nE_x \cdot l_x \cdot (1+i)^n = P \cdot l_{x+n}.$$

Using the definition of ${}_np_x = l_{x+n}/l_x$, we can solve this for ${}_nE_x$:

$${}_nE_x = (1+i)^{-n} \cdot {}_np_x \tag{7.1}$$

Example 7.1: Mary Jane is 50 years old, and recently inherited some money. She would like to put some of it to work on her behalf. A friend advised her to go to a licensed insurance agent and pay a sum now so that if and when she reached 70, she would receive a payment of, say, \$250,000. She understands that if she dies before reaching age 70, neither she nor her heirs will get a penny. If money earns 7.5% effective, what net single premium should Mary Jane pay the insurance agent for this pure endowment of \$250,000 payable at age 70?

Solution: Here is a time diagram illustrating Mary Jane's transaction.

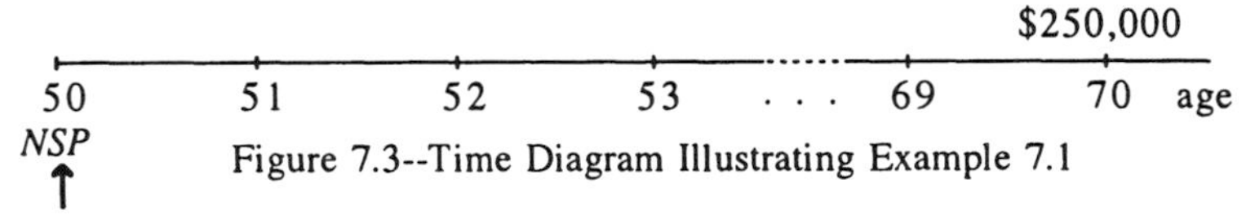

Figure 7.3--Time Diagram Illustrating Example 7.1

From Formula 7.1 and Table II we can write

$$NSP = \$250{,}000 \cdot {}_{20}E_{50} = \$250{,}000 \cdot (1.075)^{-20} \cdot {}_{20}p_{50} = (1.075)^{-20} \cdot \frac{l_{70}}{l_{50}} =$$

$$\$250{,}000\,(1.075)^{-20} \cdot \frac{7{,}448{,}818}{9{,}219{,}130} \approx \$47{,}551.93.$$

We will make the calculation above, and others soon to come, easier to carry out with the aid of so-called *commutation functions*. The first of these is the commutation function D_x. First, rewrite ${}_nE_x$ as

$${}_nE_x = (1+i)^{-n}\,{}_np_x = (1+i)^{-n}\frac{l_{x+n}}{l_x} = \frac{(1+i)^{-(x+n)}l_{x+n}}{(1+i)^{-x}l_x},$$

where we have multiplied and divided by $(1 + i)^{-x}$. We now define D_x: $D_x = (1+i)^{-x}l_x$, for $x = 0, 1, 2, \ldots$ Then we may write a second version of Formula 7.1:

$${}_nE_x = \frac{(1+i)^{-(x+n)}l_{x+n}}{(1+i)^{-x}l_x} = \frac{D_{x+n}}{D_x}. \tag{7.2}$$

These D_x's are found along with the other commutation functions defined later (N_x and M_x) in Tables III and IV. Let's use Formula 7.2 to repeat the calculation above. We will need to go to Table IV (for females) with $i = 7.5\%$. We find

$$NSP = \$250{,}000 \cdot \frac{D_{70}}{D_{50}} = \$250{,}000 \cdot \frac{47{,}151.42}{247{,}894.41} \approx \$47{,}551.92,$$

as before except for a \$0.01 round-off error. If Mary Jane's inheritance were \$50,000, she would still have a little cash left. ▮

Example 7.2: Harvey is 40 and pays a premium of \$75,000 for a pure endowment payable to him at age 65. If money earns at $i = 4\%$ annually, what pure endowment might Harvey receive at age 65?

Solution: We'll start with a time diagram.

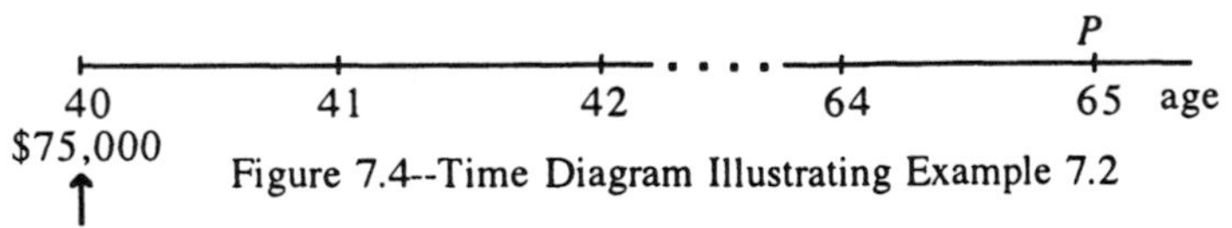

Figure 7.4--Time Diagram Illustrating Example 7.2

This time we know that NSP = \$75,000, and we need to find the contingent payment, P, at age 65. The inputs for Formula 7.2 are P = \$75,000, gender = *male*, $i = .04$, and $n = 25$. Using Table III, we have the equation

$$\$75{,}000 = P\frac{D_{65}}{D_{40}} = P\frac{572{,}692.53}{1{,}953{,}173.24} = P\cdot(.29321133337),$$

which, when solved for P gives $P = \$75{,}000/(.29321\ldots) = \$255{,}788.20$. That is what Harvey will receive at age 65 if he is alive.

Remark: Just as $(1+i)^{-n}$ and $(1+i)^n$ act as discount and accumulation factors, respectively, when dealing with payments certain, ${}_nE_x = (1+i)^{-n}\cdot{}_np_x$ and $1/{}_nE_x = (1+i)^n/{}_np_x$ act as discount and accumulation factors when dealing with contingent payments. We say that ${}_nE_x$ acts as a discount factor *with interest and survivorship*, and $1/{}_nE_x$ as an accumulation factor *with interest and* survivorship. ∎

Exercises 7.2

7.2.1 Abby wants to receive a lump sum payment when she reaches age 60, but only if she reaches that age. She is now 45. To get this payment, she pays her insurance company \$40,000.

a) Assuming that money earns 4.5% annually, how much will Abby's payment at age 60 be?

b) How much would Abby's payment be if $i = 6\%$?

7.2.2 Find the present value at interest rate $j_1 = 3.75\%$ of a payment of \$50,000 to be made in 25 years:

a) If the payment is certain to be made.

b) If the payment is contingent upon the survival of a male now aged 20.

7.2.3 Johanna inherited \$100,000 when she was 30 and immediately bought a 30-year pure endowment with the money. Assume that money earns $j_1 = 5.75\%$.

a) How much will Johanna receive when she is 60?

b) How much would Johanna or her estate receive if she paid for a payment-certain in 30 years instead of a pure endowment?

7.2.4* Redo part *a*) of Exercise 7.2.3 assuming that the survival function is given by the formula $l_x = 10{,}000{,}000(1 - x/105)$, $0 \le x \le 105$ instead of by Table II.

7.2.5* Redo part *a*) of Exercise 7.2.3 assuming that the survival function is given by the formula $l_x = 10{,}000{,}000\sqrt{1 - x/105}$, $0 \le x \le 105$ instead of by Table II.

Remark: Neither of the survival functions of the last two exercises are used in practice, but a version of the first one, $l_x = 10{,}000{,}000\,(1 - x/105)$, $0 \le x \le 105$, was introduced in the early 18th century by the mathematician Abraham de Moivre as a simple approximation to l_x.

7.2.6 Rong paid the Endowments-R-Us Company \$50,000 at age 35 in return for a pure endowment payable to her at age 60. If money is valued at 4.5% effective, what is the size of Rong's pure endowment payment?

7.2.7 Other things being equal, should a man or a woman of the same age pay more to receive the same life annuity? Explain.

7.2.8 Complete the Table to find the present value at effective rate j_1 of a payment of \$100,000 to be made in 20 years if payment is contingent upon the survival of a male now aged x.

$j_1 \backslash x$	25	45	65
3%			
5%			
7%			

7.2.9 Repeat Exercise 7.2.8 for females.

7.3 Life Annuities

In Section 7.1 we defined a whole life annuity payable to (x). We can see now that the *NSP* for this annuity may be found by treating it as a sum of pure endowments, the first payment of P at age $x + 1$, the last at age 99. That is,

$$NSP = P \cdot {}_1E_x + P \cdot {}_2E_x + \ldots + P \cdot {}_{99-x}E_x = P({}_1E_x + {}_2E_x + \ldots + {}_{99-x}E_x).$$

But, using the fact that ${}_nE_x = D_{x+n}/D_x$ for each n, we can rewrite the *NSP* as

$$NSP = P\left(\frac{D_{x+1}}{D_x} + \frac{D_{x+2}}{D_x} + \ldots + \frac{D_{99}}{D_x}\right) = P \cdot \frac{D_{x+1} + D_{x+2} + D_{x+3} \ldots + D_{99}}{D_x}.$$

Of course, adding up all of those D's is tedious, but it suggests our next commutation function, N_x. We define

$$N_x = D_x + D_{x+1} + D_{x+2} \dots + D_{99},$$

$x = 0, 1, 2, \dots, 99$, and $N_{100} = 0$. For example,

$$N_{50} = D_{50} + D_{51} + D_{52} \dots + D_{99}.$$

and

$$N_{75} = D_{75} + D_{76} + D_{77} \dots + D_{99}.$$

Note that the column following the D_x column in Tables III and IV lists this commutation function. Now we can put the net single premium for our whole-life annuity into a very compact formula.

Net Single Premium for a Whole-Life Annuity of P to (x) with first payment at age x + 1 is

$$NSP = P \frac{N_{x+1}}{D_x}. \tag{7.3}$$

Example 7.3: Rubén is 58 years old and has built up a sizable sum to provide him with a pension (that is, a whole-life annuity) for the rest of his life with payments of \$75,000 per year beginning at age 59. Assuming interest at 6.5%, how much is in Rubén's pension fund?

Solution: We are looking for the *NSP* in Formula 7.3 with inputs of P = \$75,000, gender = *male*, $i = .065$, and $x = 58$. Consulting Table III with $i = .065$, we find the other ingredients, $N_{59} = 2{,}096{,}721.29$ and $D_{58} = 215{,}659.35$. Rubén's pension fund must have totaled

$$NSP = \$75{,}000 \frac{N_{59}}{D_{58}} = \$75{,}000 \frac{2{,}096{,}721.29}{215{,}659.35} = \$729{,}178.20. \quad \blacksquare$$

Example 7.4: Jacqueline is 55 years old and will start to collect her pension next year at age 56. She has built up a pension fund of \$495,000. If money earns $i = 4.5\%$ per annum, what will her annual pension payments be?

Solution: Here we have *NSP* = \$495,000, and want to find P using Formula 7.4. We need to go to Table IV with $i = .045$ to find N_{56} and D_{55}. Then we substitute into Formula 7.4 to get

$$\$495{,}000 = P \frac{N_{56}}{D_{55}} = P \frac{11{,}115{,}091.11}{795{,}756.64} = P\,(13.9679527022).$$

Dividing both sides by 13.9679..., we have P = \$35,438.26. That is the size of Jacqueline's annual pension for the rest of her life beginning at age 56. ∎

A reminder: While there will be many instances where payments are made at the beginning of the period, the default situation is still that payments are made at the end of the period unless otherwise stated.

The *whole-life* annuity is just one of many types of life annuity. Some are common enough to be identified by name. We will look at some of these after we derive a general formula covering *regular life annuities*: that is, life annuities with equal annual payments and constant effective interest rate.

General Formula for the NSP of a Regular Life Annuity: Suppose that (x) is to receive a life annuity consisting of $b - a$ payments of P, with the first payment at age a and final payment at age $b - 1$, with $b - a \geq 1$. Suppose also that we want the value of this annuity when the annuitant is aged c. We assume a constant effective rate of interest, i, throughout. Here is a time diagram for this life annuity:

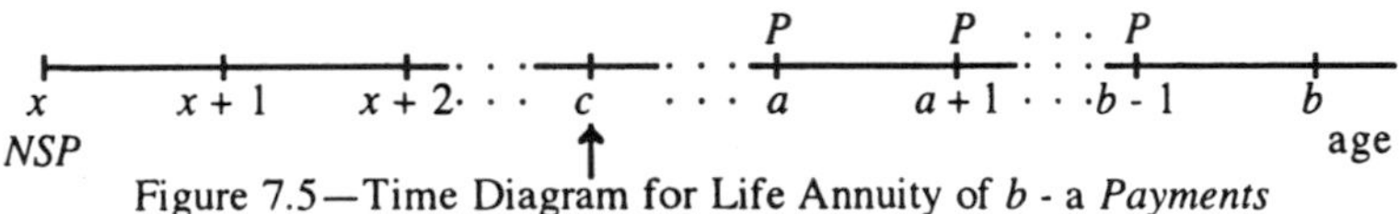

Figure 7.5—Time Diagram for Life Annuity of b - a *Payments*

We will use the symbol $\widehat{S}$ (read "S-hat") for the net single premium of regular life annuities. The general formula will have seven inputs. These are:

x—the age of the annuitant at the time of purchase;
P—the size of the annual annuity payment;
g—the gender of the annuitant: *male* or *female*;
i—the constant effective rate of interest;
a—the age of the annuitant when the first payment is made;
b—the age of the annuitant one year after the last payment;
c—the annuitant's age when the annuity is to be valued.

The short name and long for the net single premium are

$$\widehat{S} \text{ and } \widehat{S}(x, P, g, i, a, b, c).$$

In most cases we will be evaluating the annuity at the time of purchase, so that almost always, we will have $c = x$.

This life annuity consists of $b - a$ payments, and the present value of each at age x is a pure endowment. Therefore, the value of the life annuity when the annuitant is aged x is

$$\hat{S} = P \cdot \left(\frac{D_a}{D_x} + \frac{D_{a+1}}{D_x} + \ldots + \frac{D_{b-1}}{D_x} \right) =$$

$$P \cdot \left(\frac{D_a + D_{a+1} + \ldots + D_{b-1}}{D_x} \right) = P \cdot \frac{N_a - N_b}{D_x}.$$

This is the formula we want when $c = x$, but when $c \neq x$, we need to multiply the last expression by the accumulation factor $1/_{c-x}E_x = D_x/D_c$ to account for both interest and survivorship . The final version of the formula is then

$$\hat{S} = P \cdot \frac{N_a - N_b}{D_c} . \tag{7.4}$$

Note that the formula simplifies to

$$\hat{S} = P \cdot \frac{N_a}{D_c}$$

when $b = 100$, since $N_{100} = 0$.

Example 7.5: (*Deferred life annuity*) Monika is 40 years old and wants to have a yearly income of \$60,000 for life, beginning when she is 65. Assuming interest at 3% effective, what net single premium, payable today, will provide for this pension?

Solution: A deferred life annuity is one that begins making payments after a specified number of years, depending, of course, on survival. Here we have a 25-year deferred life annuity. We want to find $\hat{S}$ in Formula 7.4 with inputs $c = x = 40$, $P = \$60,000$, $g = female$, $i = .03$, $a = 65$, and $b = 100$, so we will need to get N_{65} and D_{40} ($N_b = N_{100} = 0$) from Table IV. We find that

$$\hat{S} = \$60,000\frac{N_{65}}{D_{40}} = \$60,000\frac{16,047,751.58}{2,926,190.81} = \$329,050.69.$$

Before leaving this problem, let's look at a time diagram showing what we have found.

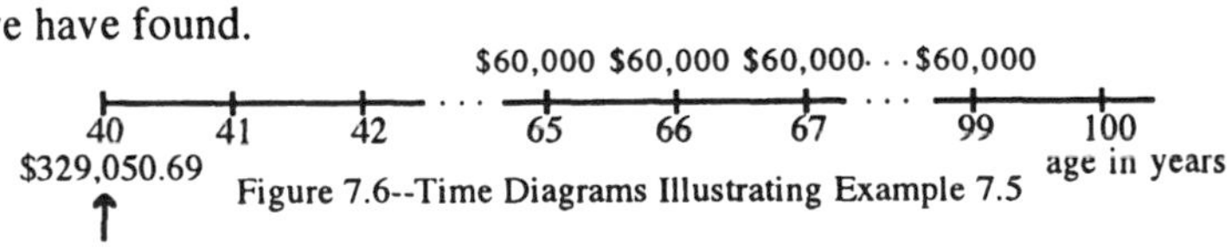

$60,000 $60,000 $60,000 . . . $60,000

40 41 42 . . . 65 66 67 . . . 99 100

$329,050.69 ↑ age in years

Figure 7.6--Time Diagrams Illustrating Example 7.5

We see from the time diagram that Monika would need \$329,050.69 right now in return for \$60,000 per year for life beginning at age 65. Note that all payments except the *NSP* are contingent on Monika's survival, so that if she were to pay the \$329,050.69 and die before reaching age 65, no money would be paid out. This is the reason that the premium is much less than the corresponding annuity-certain that would pay \$60,000 per year regardless of Monika's survival or death. You can check for yourself that the premium for that would be $S(\$60{,}000, 0.03, 35, -25) = \$634{,}217.31$.

Example 7.6: (*Temporary life annuity*) Since it would probably be difficult for someone in Monika's position to come up with \$329,050.69 immediately, it would be much more comfortable for someone in her situation to pay annual premiums into a fund until the pension was to go into effect. These premiums constitute a *25-year temporary annuity*. Let's find the *net annual premium* (*NAP*), *P*, which Monika should pay from now until age 64, that will provide her with a \$60,000 per year pension. Assume that interest remains at 3% effective.

Solution: We will start with a time diagram.

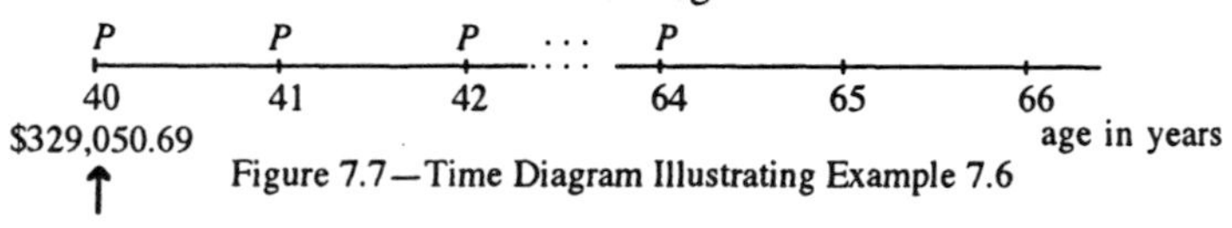

Figure 7.7—Time Diagram Illustrating Example 7.6

We don't show the annual payments of \$60,000 beginning at 65 because we don't need to. Their value is contained in the *NSP* of \$329,050.69, and we are trying to equate their value with Monika's annual premiums. The inputs into Formula 7.4 are $\widehat{S} = \$329{,}050.69$, $c = x = 40$, P (to be determined), $g = female$, $i = .03$, $a = 40$, and $b = 65$. Substituting these into Formula 7.4 gives

$$\$329{,}050.69 = P\frac{N_{40} - N_{65}}{D_{40}} = P(17.1374757957).$$

From this we find that $P = \$19{,}200.65$. This is quite a sizeable annual payment and is partly accounted for by the low interest rate of 3%. ∎

As the next example will show, the difference that the interest rate can make in these situations is substantial.

Example 7.7: Redo Examples 7.5 and 7.6, calculating Monika's net single premiums (*NSP*) and net annual premiums (*NAP*) for the interest rates .02, .03, .04, .05, .06, .07, and .08 in Table IV. Put the results in a table.

Solution: In Example 7.5 with i = .03, we found the net single premium via the formula

$$\hat{S} = \$60{,}000\frac{N_{65}}{D_{40}}.$$

In Example 7.6, we needed to find P, the net annual premium, also assuming i =.03. To do that, we set

$$\hat{S} = \$60{,}000\frac{N_{65}}{D_{40}} = P\frac{N_{40} - N_{65}}{D_{40}}.$$

We solved this equation for P. After canceling the D_{40}'s on both sides of the last equation and dividing both sides by $N_{40} - N_{65}$, we now find

$$P = \$60{,}000\frac{N_{65}}{N_{40} - N_{65}}.$$

Of course, all the N's and D's had numbers substituted into them. Thus, the two things we have to find for each of the interest rates in Table IV are

1) $\hat{S} = \$60{,}000\frac{N_{65}}{D_{40}}$ (the *NSP*).

and

2) $P = \$60{,}000\frac{N_{65}}{N_{40} - N_{65}}$ (the *NAP*).

Here are the results I got in the form of a table:

i	*NSP*	*NAP*
2.0%	\$458,304.50	\$24,153.71
3.0%	**\$329,050.69**	**\$19,200.65**
4.0%	\$238,049.13	\$15,293.04
5.0%	\$173,456.21	\$12,202.62
6.0%	\$127,252.22	\$9,753.09
7.0%	\$93,958.24	\$7,807.64
8.0%	\$69,799.00	\$6,259.71

Table for Example 7.7

The row in bold shows the values we had already found in Examples 7.5 and 7.6. What is informative about the table is the sharp differences in the premiums depending on the assumed interest rates. ∎

Example 7.6 provides a fairly realistic situation. Monika paid periodic premiums into a pension fund to finance her income in retirement.

Example 7.8: Ludwig is 32 and would like to accumulate a pension fund that will provide him with an annual income of \$50,000 beginning at age 55. He will make equal annual deposits beginning immediately into a fund that will earn 7% annually. His last deposit will be at age 54. What must his annual deposits be?

Solution: We need to look at a time diagram for this.

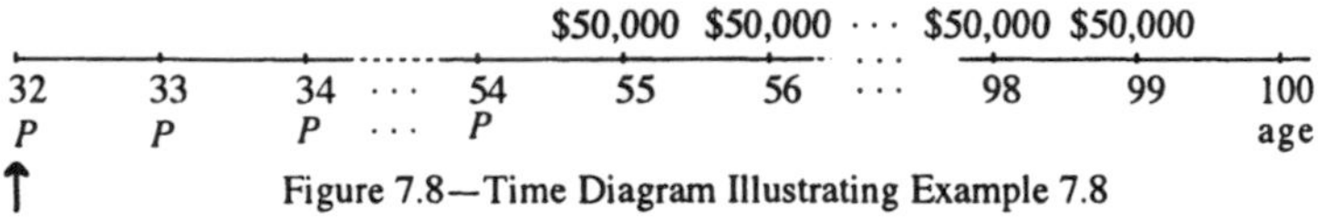

				\$50,000	\$50,000 ···	\$50,000	\$50,000	
32	33	34 ···	54	55	56 ···	98	99	100
P	P	P ···	P					age
↑								

Figure 7.8—Time Diagram Illustrating Example 7.8

As in Example 7.6, we have two life annuities here—from Ludwig's point of view, one is money out (his net annual premium), and one is money in (his pension). The one out is a 23-year temporary life annuity while the one in is a 23-year deferred life annuity. What we need to do is to find the present value of each annuity and set them equal. We can do this in two ways. On the one hand, we can evaluate the pension annuity numerically and then set it equal to the pension fund annuity. Alternately, we can save the calculations until the end by writing down the formulas for both present values, set them equal, and then solve for the unknown quantity. This will be clearer if I show you both ways. We will be using Table III.

Method 1 (Less algebra, more arithmetic): From Formula 7.4 with inputs $c = x = 32$, $P = \$50{,}000$, $g = male$, $i = .07$, $a = 55$, and $b = 100$, the present value with interest and survivorship of the pension is,

$$\widehat{S} = \$50{,}000\frac{N_{55}}{D_{32}} = \$50{,}000\frac{2{,}279{,}119.51}{1{,}095{,}364.93} = \$104{,}034.71.$$

On the other hand, from Formula 7.4 with inputs $c = x = 32$, P (to be determined), $g = male$, $i = .07$, $a = 32$, and b = 55 (remember, one year past the last payment), the present value with interest and survivorship of the 23 annual deposits of P is

$$\widehat{S} = P\frac{N_{32} - N_{55}}{D_{32}} = P\frac{15{,}187{,}786.13 - 2{,}279{,}119.51}{1{,}095{,}364.93} =$$
$$P(11.7848091229).$$

Set this equal to \$104,034.71, and we get P = \$8,827.87 as Ludwig's annual deposit into his pension fund.

Method 2 (More algebra, less arithmetic): This is really the same as Method 1 except that we don't look up any commutation functions until necessary. This saves a little arithmetic but also increases the amount of algebra you have to do. What we do is set the present value of the pension payments that we already know to be $\$50{,}000\frac{N_{55}}{D_{32}}$ equal to the present value of the 23 annual pension deposits, which we know to be $P\frac{N_{32} - N_{55}}{D_{32}}$. Thus, we set

$$\$50{,}000\frac{N_{55}}{D_{32}} = P\frac{N_{32} - N_{55}}{D_{32}}$$

and solve this for P. The D_{32}'s cancel, so the result is

$$P = \$50{,}000\frac{N_{55}}{N_{32} - N_{55}} =$$

$$\$50{,}000\frac{2{,}279{,}119.51}{15{,}187{,}786.13 - 2{,}279{,}119.51} = \$8{,}827.87,$$

just as before. The reason that there is less arithmetic this way is that the D_{32}'s cancel, so they never have to be looked up. ∎

Example 7.9: Rework Example 7.8 with $i = .06$ and $i = .08$.

Solution: This amounts to two quick table look-ups if we use the formula from Method 2. We need to go to Table III, look up N_{32} and N_{55} with $i = .06$ and .08, and substitute into

$$P = \$50{,}000\frac{N_{55}}{N_{32} - N_{55}}.$$

When we do this, we get P = \$10,977.92 when $i = .06$, and P = \$7,121.05 when $i = .08$. The results will look better in a table:

i	P
.06	\$10,977.92
.07	\$8,827.87
.08	\$7,121.05

∎

Example 7.10: (*A forborne annuity*) Bev is starting to deposit \$5,000 at the beginning of every year into an account earning 6% effective. She has just made her first deposit at age 35 and will collect the accumulated value of her deposits at age 65 if she survives. Her last deposit will be at age 64. There will be no payment if she does not survive just as her deposits will stop if she dies before reaching age 65. How much will Bev collect if and when she survives to age 65?

Solution: This is an example of a *forborne annuity*, in which the annuity is valued and collected after the payments have been made. We use Formula 7.4 with inputs $x = 35$, $P = \$5{,}000$, $g = female$, $i = .06$, $a = 35$, $b = 65$, and $c = 65$. Bev will collect

$$\hat{S} = \$5{,}000\frac{N_{35} - N_{65}}{D_{65}} \approx \$479{,}985.71. \qquad ∎$$

Another common type of pension is one whose payments are certain for a fixed period but then contingent upon survival after that.

Example 7.11: Shirley is a 25-year-old woman who has elected to take an annual pension of \$75,000 beginning when she is 65. The first 10 years of payments will be certain—they will be paid to her or her estate if she does not survive to 65 while those beginning at age 75 will be contingent on her survival. To pay for this pension, she and her employer will make annual deposits, contingent upon her survival, into a fund earning 5.5% interest annually. These contributions will begin immediately and end when Shirley is 64. What is the size of the net annual premium put into Shirley's pension fund?

Solution: We'll start out with a time diagram.

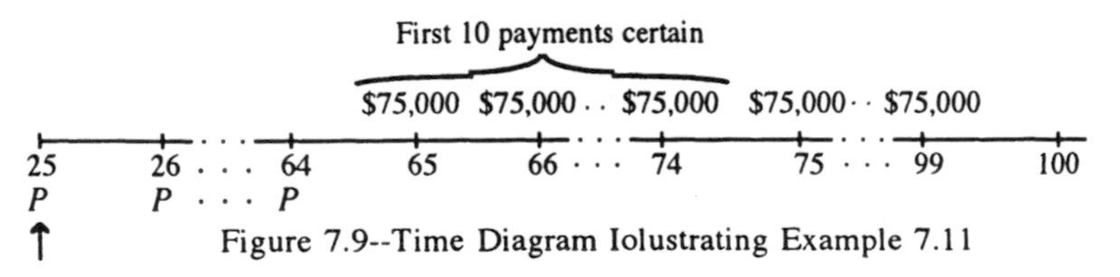

Figure 7.9--Time Diagram Iolustrating Example 7.11

As the time diagram indicates, there are three pieces to this problem: the 40 payments of P (to be determined) into Shirley's pension fund, the 40-year deferred 10-year annuity-certain beginning at age 65, and the 50-year deferred life-annuity beginning at age 75. The annuity certain will be paid regardless of Shirley's status at age 65. All other payments into or out of the fund are contingent upon Shirley's survival. We will find the present value of the Shirley's money out and set this equal to the present value of her money in. Using Formula 7.4 with inputs $c = x = 25$, P (to be determined), $g = female$, $i = .055$, $a = 25$, $b = 65$, and using Table IV, the present value of the money out is

$$\widehat{S}_1 = P\frac{N_{25} - N_{65}}{D_{25}} = P\cdot(16.4633690302).$$

Using Formula 3.2 with inputs $P = \$75{,}000$, $i = .055$, $n = 10$, and $d = -40$, the present value (at age 25) of the 10 certain payments is

$$S = \$75{,}000\frac{1-(1.055)^{-10}}{.055}(1.055)^{-40+1} = \$70{,}056.74.$$

Using Formula 7.4 again, with inputs $c = x = 25$, $P = \$75{,}000$, $g = female$, $i = .055$, $a = 75$, $b = 100$, and Table IV, the present value of the life annuity beginning at age 75 is

$$\widehat{S}_2 = \$75{,}000\frac{N_{75}}{D_{25}} = \$27{,}044.51.$$

We now set the money in, $\widehat{S}_1$, equal to the money out, $S + \widehat{S}_2$, all when Shirley is aged 25:

$$P\cdot(16.4633690302) = \$70{,}056.74 + \$27{,}044.51 = \$97{,}101.25.$$

We find $P = \$5{,}898.02$. These are the annual contributions Shirley and her employer make to her pension fund in order to provide her desired pension. ∎

Example 7.12: Oscar has turned 35 and is depositing \$5,400 into an account at the beginning of every year. The last deposit is to be at age 64. The accumulated value of the account is to provide him a pension for life, beginning at age 65. How much will be in Oscar's account when he turns 65 assuming interest at 4% effective?

Solution: To make using the tables more efficient whenever more than one interest rate appears in the same problem, I will indicate the interest rate we are using by augmenting the commutation functions with that rate. For example, when the interest rate is 7%, I will write $N_{30}(.07)$ instead of just N_{30}, and when it is 2%, I will write $D_{37}(.02)$ instead of D_{37}. Let's look at the time diagram for this:

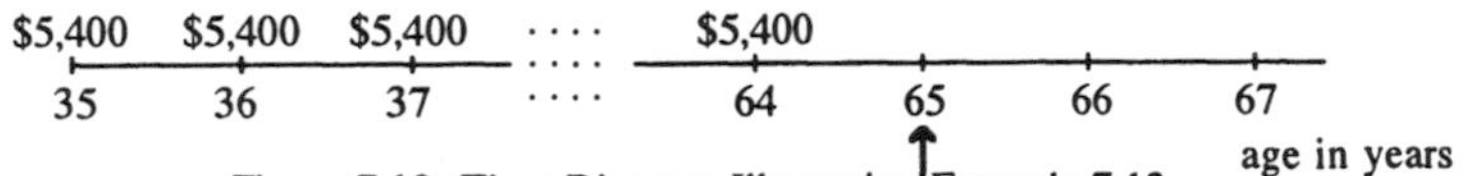

$5,400	$5,400	$5,400	....	$5,400			
35	36	37	....	64	65 ↑	66	67

age in years

Figure 7.10--Time Diagram Illustrating Example 7.12

This is another example of a forborne annuity. At age 65, Oscar's pension fund at 4% would be worth

$$\hat{S} = \$5{,}400\frac{N_{35}(.04) - N_{65}(.04)}{D_{65}(.04)} = \$386{,}753.00.$$

Note that this is the value of Oscar's pension fund with interest and survivorship. If survivorship were not a consideration and payments were certain, his fund would be worth (using Formula 3.2) only $S(\$5400, .04, 30, 30) = \$314{,}973.01$. ∎

Exercises 7.3

7.3.1 What size whole-life annuity can Annika, aged 29, purchase with $90,000 if her first payment is to be at age 50 and interest is at 6.25% effective?

7.3.2 Naomi is 55 and has a pension fund that is about to begin providing her with an annual income of $43,600 for life. How much is in her pension fund if the money earns at the annual rate of 4.75%?

7.3.3 Redo Example 7.8 with $i = 2\%, 3\%, 4\%$ and 5%. Combine this with the results we gathered in Examples 7.8 and 7.9 to make a table and a graph of i versus P for $i = 2\%$ to 8% in steps of 1%.

7.3.4 Sven who is now 72 just inherited $250,000 from his friend Lena. He would like to turn his inheritance into an annual income with equal payments until he dies. If money is worth 4% effective, what will Sven's annual income be if his first payment is in one year when he turns 73?

7.3.5 In Example 7.10, explain why the forborne annuity pays more than the corresponding annuity-certain.

7.3.6 Elena is 40 years old and has arranged for an annuity paying \$20,000 annually for 10 years beginning at age 60 followed by payments of \$30,000 for the rest of her life beginning at age 70, all payments contingent upon her survival. If money earns at the rate of 6% annually, find Elena's net single premium.

7.3.7 (Continuation) In the previous exercise, if Elena makes 20 annual payments beginning immediately, what is the size of those payments, interest remaining at 6% annually?

7.3.8 Calculate the net single premium and net annual premium in Exercises 7.3.6 and 7.3.7 assuming an interest rate of 3% instead of 6%.

7.3.9 Jan is 40 and wants to know the difference between the net single premium for a whole-life annuity of \$50,000 per year if his first payment is right away versus what it would be if his first payment were at age 41. What is the difference?

7.3.10 Rodica is 31 years old and has arranged for a pension of \$80,000 annually, the first payment on her 65th birthday. To finance the pension, she will make annual deposits into a fund earning 5% effective beginning immediately with her last deposit at age 64. What will her annual deposits be?

7.3.11 Martin pays \$5,131.27 annually into a pension fund that will begin making equal annual payments to him when and if he reaches age 60. If Martin began payments at age 25, will continue paying until age 55, and money earns 5.5% annually, what will the size of Martin's pension payments be?

7.3.12 How much should Ingela now aged 35 deposit annually into her pension fund, her first deposit now and her last deposit at age 64 if she wishes to have a pension of \$70,000 per year beginning when she is 65, and money earns at the rate of 4.75% annually?

7.3.13 What is the net single premium for a whole-life annuity of \$30,000 annually, first payment immediately, on the life of a 35-year-old male if money earns *a*) 4% annually? *b*) 5% annually? *c*) 6% annually?

7.3.14 *a*) Would you expect the results in the preceding exercise to be lower or higher for a female aged 35? Why?

b) Redo Exercise 7.3.13 for a 35-year-old female.

7.3.15 Show that the net single premium by (x) for a whole-life annuity of P, first payment immediately, is $P \cdot N_x / D_x$.

*7.4 Life Annuities with Payments Increasing/Decreasing by a Fixed Percentage**

Life annuities whose payments increase or decrease by a fixed percentage can be handled by suitably modifying Formula 7.4. As in the case of the increasing annuities certain of Chapter 4, we can argue that they are important in that they allow the annuitant to anticipate and act on the possible effects of inflation. We'll look at an example first, and then we will work out the general formula.

Example 7.13: Sandy's dad is rich and has just bought her a whole-life annuity whose payments will increase by 3% every year. Sandy, who is 20, will receive her first payment of \$35,000 immediately. If money earns 6%, how much did Sandy's dad pay for her life annuity?

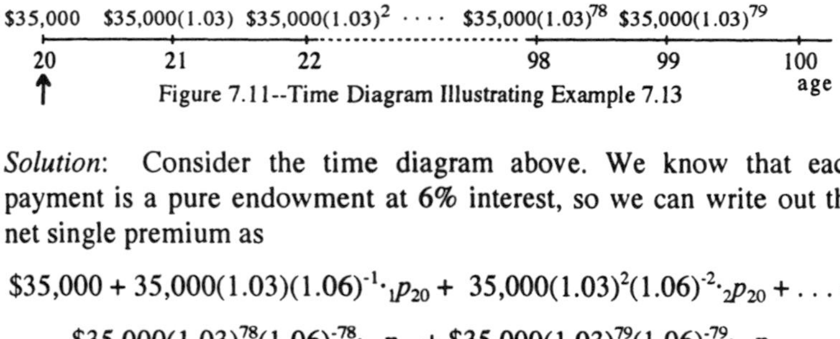

\$35,000 \$35,000(1.03) \$35,000(1.03)^2 .... \$35,000(1.03)^{78} \$35,000(1.03)^{79}

20 21 22 98 99 100 age

↑

Figure 7.11--Time Diagram Illustrating Example 7.13

Solution: Consider the time diagram above. We know that each payment is a pure endowment at 6% interest, so we can write out the net single premium as

$$\$35,000 + 35,000(1.03)(1.06)^{-1} \cdot {}_{1}p_{20} + 35,000(1.03)^2(1.06)^{-2} \cdot {}_{2}p_{20} + \ldots +$$

$$\$35,000(1.03)^{78}(1.06)^{-78} \cdot {}_{78}p_{20} + \$35,000(1.03)^{79}(1.06)^{-79} \cdot {}_{79}p_{20}.$$

What will enable us to modify Formula 7.4 is the fact that we can write

$(1.03)^n(1.06)^{-n}$ in the form $(1 + i^*)^{-n}$, where $i^* = (.06 - .03)/1.03 = .02912621359 \approx .03$. You can now evaluate the net single premium, approximately, by using $i^* = .03$ in place of $i = .06$ and $P = \$35,000$ (I will also evaluate it with $i^* = .02912621359$ just to see how good the approximation is.). The net single premium above is approximately

$$\$35,000 \frac{N_{20}(.03)}{D_{20}(.03)} = \$35,000 \frac{148,660,185.56}{5,437,550.47} = \$956,884.27.$$

The exact net single premium that I calculated with computer help is

$$\$35{,}000\frac{N_{20}(.02912621359)}{D_{20}(.02912621359)} = \$35{,}000\frac{153{,}898{,}521.37}{5{,}530{,}634.83} =$$
$$= \$973{,}929.47. \quad \blacksquare$$

The approximation is an underestimate by about 1.75%. That is good enough for the sake of illustration, but in practice, the exact value would be required. For the examples and exercises in this book, I will contrive things so that the percentage increases or decreases in the examples and exercises are such that you can use our available tables and the results will be very accurate. In fact, in Example 7.13, if I had contrived things so that the annuity payments increased by 2.91262135922% instead of 3%, then we would have had $i^* = 3\%$ to many decimal places.

A General Formula for Annuity Payments Increasing or Decreasing by a Fixed Percentage: 'General' here means that the annuity may start and end at any (whole) ages and that the percentage increase or decrease as well as the interest rates are arbitrary but fixed.

Derivation: Assume that (x) is purchasing an annuity whose first payment of P will be made at age a with subsequent payments increasing or decreasing by a factor of $1 + k$ until there are a total of $b - a$ payments. This means that the last payment will be at age $b - 1$ in the amount of $P(1 + k)^{b-a-1}$. Assume that the effective rate is i throughout, and the annuity will be evaluated when the annuitant is aged c. Here is a time diagram for this.

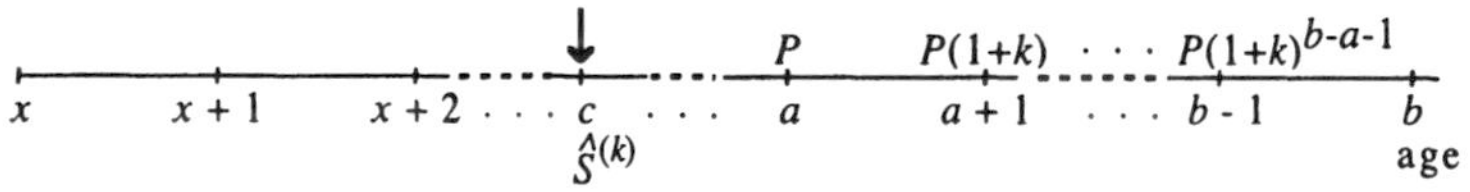

P $P(1+k)$ . . . $P(1+k)^{b-a-1}$

x $x+1$ $x+2$ . . . c . . . a $a+1$ . . . $b-1$ b

$\hat{S}^{(k)}$ age

Figure 7.12--Time Diagram for a Life Annuity with $b - a$ Increasing/Decreasing Payments

The net single premium for this annuity purchased by (x) is a sum of premiums for pure endowments. The way we will evaluate this is first to find the present value of the premium as if it were purchased at age a and then discount or accumulate with interest and survivorship by the factor ${}_{a-c}E_c$. The net single premium at age a is

$$\hat{S}^{(k)} = P + P(1+k)(1+i)^{-1}\cdot{}_1p_a + P(1+k)^2(1+i)^{-2}\cdot{}_2p_a +$$

$$+ P(1+k)^3(1+i)^{-3}\cdot{}_3p_a + \ldots + P(1+k)^{b-a-1}(1+i)^{-(b-a-1)}\cdot{}_{b-a-1}p_a .$$

Next, we set each power of $(1 + k)(1 + i)^{-1}$ equal to $(1 + i^*)^{-1}$ making $i^* = \frac{i-k}{1+k}$ (This is analogous to what we did for increasing/decreasing annuities certain in Chapter 4.). The net single premium at age a can then be written

$$\widehat{S}^{(k)} = P + P(1+i^*)^{-1} \cdot {}_1p_a + P(1+i^*)^{-2} \cdot {}_2p_a +$$
$$+ P(1+i^*)^{-3} \cdot {}_3p_a + \ldots + P(1+i^*)^{-(b-a-1)} \cdot {}_{b-a-1}p_a.$$

This in turn is

$$\widehat{S}^{(k)} = P\left(\frac{D_a(i^*) + D_{a+1}(i^*) + D_{a+2}(i^*) + D_{a+3}(i^*) + \ldots + D_{b-a-1}(i^*)}{D_a(i^*)}\right) =$$
$$P\frac{N_a(i^*) - N_b(i^*)}{D_a(i^*)}.$$

We are almost done. To evaluate this when the annuitant is aged c, we need to accumulate or discount to age c with interest (at rate i) and survivorship. We do this by multiplying by the factor ${}_{a-c}E_c(i) = D_a(i)/D_c(i)$. This can all be summarized as follows:

Final Result: The net single premium for a life annuity consisting of $b - a$ payments to (x), with payments of P, $P(1+k)$, $P(1+k)^2$, ..., $P(1+k)^{b-a-1}$, first payment at age a, last payment at age $b - 1$, at interest rate i annually, evaluated when the annuitant is aged c is

$$\widehat{S}^{(k)} = P\frac{N_a(i^*) - N_b(i^*)}{D_a(i^*)}\frac{D_a(i)}{D_c(i)}. \tag{7.5}$$

The inputs for Formula 7.5 are

x—the age of the annuitant at purchase,
P—the first payment,
g—gender: *male*, *female*,
k—the percentage increase or decrease of subsequent payments,
i—annual effective rate of interest,
i^*—adjusted interest rate $i^* = (i - k)/(1 + k)$,
a—age at the time of first payment,
b—one year past the age of last payment,
c—the age at which the annuity is to be evaluated. ∎

Remarks: In most cases we will have $c = x$. Notice that when $k = 0$, so that $i^* = i$, Formula 7.5 reduces to Formula 7.4, which is as it should be. Also, if $a = c$, the formula simplifies because then the last factor equals one. If $i = k$, then $i^* = 0$. The algebraically inclined should try Exercise 7.4.3 below in which you are asked to verify that when $i^* = 0$ and $a = c$, $\widehat{S}^{(k)}$ reduces to

$$\widehat{S}^{(k)} = P \cdot \left\{ \frac{1}{2} + AAD(a) - a - \frac{l_b}{l_a}\left(\frac{1}{2} + AAD(b) - b \right) \right\} \qquad (7.6)$$

where *AAD* is given in Formulas 6.4 and 6.5.

Remark: If $i^* = 0$ but $a \neq c$, Formula 7.6 can still be used with an additional multiplicative factor of $D_a(i)/D_c(i)$. See Exercise 7.4.6.

Example 7.14: Veronica is a 30-year-old woman and has just won a somewhat unusual lottery. She can receive an increasing life annuity whose first payment is \$50,000 with subsequent payments increasing by 20% annually to a final payment when she is 59, but payments cease if she dies. Assume that interest earned is 20% annually. Find the equivalent cash payment that Veronica could get.

Solution: We have $a = c = 30$, $b = 60$, and $i^* = 0$ so consulting Formula 7.6 and Table II, we get

$$\widehat{S}^{(.2)} = \$50{,}000 \left\{ \frac{1}{2} + 77.65 - 30 - \frac{8{,}603{,}801}{9{,}707{,}590}\left(\frac{1}{2} + 81.25 - 60 \right) \right\} = \$1{,}443{,}700$$

to the nearest thousand dollars. Because the *AAD* column in Table II is rounded to two decimal places, we cannot calculate the payment with any more accuracy. ∎

Example 7.15: Pilar is 60 and just retired early, so her retirement income won't start until she is 65. She has no further premiums to pay. Her first payment at age 65 will be \$44,000 with subsequent payments increasing by 3.902439% annually. (Remember, I'll be choosing the values of k to make i^* come out to be one of our tabled interest rates.) Assume that money earns 6.5% effective.

a) How much is in Pilar's retirement fund now, when she is 60?

b) How much will be in Pilar's retirement fund if she survives to age 65?

Solution:

a) This is a life annuity with payments increasing by k = 3.902439% annually. The inputs for Formula 7.5 are therefore $c = x = 60$, P = \$44,000, g = *female*, k = .03902439, i = .065, i^* = (.065 - .03902439)/1.03902439 = .025, a = 65, and b = 100. Applying these to Formula 7.5, we have the amount in Pilar's fund at age 60.

$$\widehat{S}^{(.039...)} = \$44{,}000\frac{N_{65}(.025)}{D_{65}(.025)}\frac{D_{65}(.065)}{D_{60}(.065)} =$$

$$\$44{,}000\left(\frac{22{,}986{,}404.34}{1{,}634{,}007.68}\right)\left(\frac{135{,}700.31}{196{,}659.03}\right) = \$427{,}106.88.$$

Incidentally, Pilar's first four payments will be \$44,000, \$45,717.07, \$47,501.15, and \$49,354.86. Her payment at age 99 will be $\$44{,}000(1.03902439)^{34} = \$161{,}706.66$. Of course, all of the payments are contingent upon Pilar's survival.

b) We can accumulate the amount found part *a*) with interest at 6.5% and survivorship according to Table II. This amounts to multiplying \$427,106.88 by $D_{60}(.065)/D_{65}(.065)$ = 196,659.03/ 135,700.31 = 1.44921577... yielding a value of \$618,970.03. ∎

Example 7.16: Olle is planning for his pension. He is now 36 and would like to retire at 65 with an annual income of \$50,000, with subsequent payments increasing by 3.846154% each year for life. To pay for this, he will make annual deposits into a pension fund with these deposits increasing at 2.857143% per year until his last contribution at age 64. If money earns at the annual effective rate of 8%, what is the size of Olle's first deposit into his fund?

Solution: What we have here is two increasing life annuities that we set equal to solve for the one unknown quantity. Let's look at a time diagram for this. I'll put the payments in on the bottom and the payments out on the top.

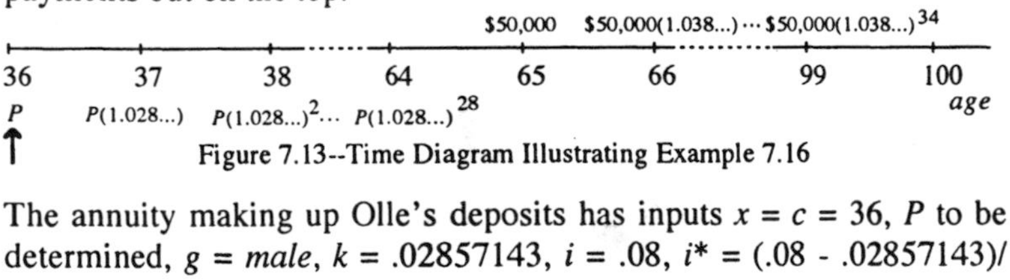

$50,000 $50,000(1.038...) ⋯ $50,000(1.038...)^{34}

36 37 38 64 65 66 99 100 *age*

P $P(1.028...)$ $P(1.028...)^2 \cdots P(1.028...)^{28}$

↑

Figure 7.13--Time Diagram Illustrating Example 7.16

The annuity making up Olle's deposits has inputs $x = c = 36$, P to be determined, g = *male*, k = .02857143, i = .08, i^* = (.08 - .02857143)/

1.02857143 = .05, a = 36, and b = 65 for Formula 7.5. The present value of this annuity is

$$\widehat{S}^{(.02857...)} = P\frac{N_{36}(.05) - N_{65}(.05)}{D_{36}(.05)}\cdot 1 = P\frac{27{,}782{,}022.30 - 3{,}054{,}397.96}{1{,}635{,}356.30}$$

$$= P(15.1206341639).$$

The deferred annuity beginning at age 65 has inputs $x = c = 36$, P = \$50,000, g = *male*, k = .03846154, i = .08, i^* = (.08 - .03846154)/ 1.03846154 = .04, a = 65, and b = 100 for Formula 7.5. The present value of this annuity is

$$\widehat{S}^{(.03846...)} = \$50{,}000\frac{N_{65}(.04)}{D_{65}(.04)}\cdot\frac{D_{65}(.08)}{D_{36}(.08)} =$$

$$\$50{,}000\left(\frac{6{,}086{,}115.74}{572{,}692.53}\right)\left(\frac{49{,}265.68}{593{,}160.14}\right) =$$

$$= \$44{,}132.77.$$

We now set the two annuities equal, $P(15.1206341639)$ = \$44,132.77 and solve for P to get P = \$2,918.71. So, Olle's first few deposits are \$2,918.71, \$3,002.10, and \$3,087.88. His last deposit at age 64 will be \$6,423.26. ∎

Example 7.17: Redo Example 7.16 with the same increasing pension but with level (i.e. equal) deposits into Olle's pension fund.

Solution: Half of the work has already been done in Example 7.16. That is, we know that the present value of the increasing pension that

P	P	......	P					
36	37	......	64	65	66	......	99	100 age
\$44,132.77 ↑								

Figure 7.14--Time Diagram Illustrating Example 7.17

begins at age 65 is \$44,132.77. What is different now is that the 29 contributions into the fund will all be equal. Look at the time diagram above for an illustration of this.

The present value of the level annuity beginning at age 36 and ending at age 64 has inputs x = 36, P to be determined, g = *male*, i = .08, a = 36, and b = 65 for Formula 7.4. This gives

$$\hat{S} = P\frac{N_{36} - N_{65}}{D_{36}} = P\frac{7,280,566.52 - 408,497.75}{593,160.14} = P(11.5855201767).$$

We set this equal to \$44,132.77 and find that P = \$3,809.30. ∎

Exercises 7.4

7.4.1 Gloria, who just turned 60, has started receiving her annual pension payments. Her first payment was \$43,000 and subsequent payments will increase by 3.38983% annually for life. How much must have been in her pension fund just before her first payment if the fund earns 6.75% annually?

7.4.2 What net single premium is Manu going to pay for his 20-payment increasing annuity if he is 30 now, will receive his first payment of \$10,000 at 40, subsequent payments will increase by 2.91262%, and money earns at the annual rate of 6%?

7.4.3 Verify Formula 7.6 when $i^* = 0$ and $a = c$.

7.4.4 Blake who just turned 23 will make annual deposits into his retirement fund together with his employer beginning in one year. Their combined first deposit will be \$6,400 with subsequent deposits increasing by 3.43137255% annually, until the final contribution when Blake is 65. Assuming that Blake's fund will accumulate with interest at 5.5% annually and survivorship, how much will Blake have in his retirement fund when and if he reaches age 65?

7.4.5 The widow of a 22-year-old father of two killed by a drunk driver is suing the driver's insurance company for lost income. Assuming an effective rate of 3%, a presumed income of \$32,000 in his 23rd year, and annual increases of 3% until retirement at age 65, what would the present value of the lost income be today ignoring any retirement income? *Hint*: Refer to Formula 7.6.

7.4.6 Redo Exercise 7.4.5 assuming that if the husband had survived to retirement, he would have an income worth 75% of his most recent salary beginning at age 66 with annual increases of 3%. *Hint*: See the remark following Formula 7.6.

7.4.7 Magnus wishes to begin financing his retirement income today at age 32. He would like to have an income of $80,000 when he retires at 60 and then receive annual increases of 3.398058% for life. If money earns $j_1 = 6.5\%$, what is the present value of Magnus's retirement income?

7.4.8 Chill is planning to have a nest egg of $1,000,000 when she reaches 65. She is now 27 years old and will make her first deposit at age 28. Later deposits will increase by 3.3816425% until her final deposit at age 64. All deposits, as well as her payment at age 65, are contingent upon her survival. If Chill's deposits earn interest at $j_1 = 7\%$, what is the amount of her first deposit at age 28? What will be the amount of her final deposit at age 64?

7.4.9* (💻) Mini wants to put some of her money into an increasing life annuity for herself. She is now 50 and would like to begin receiving her annuity with a payment of $50,000 at age 55, with subsequent payments increasing by 3% annually for life. If money earns interest at the rate of $j_1 = 5.5\%$, what is Mini's net single premium?

7.5 *Summary of Formulas*

Commutation Function D_y $$D_y = (1+i)^{-y} l_y$$

Pure Endowment Symbol $${}_nE_x = \frac{D_{x+n}}{D_x}$$

Commutation Function N_y

$$N_y = D_y + D_{y+1} + D_{y+2} + \ldots + D_{99}$$

NSP for Regular Life-Annuity of P $$\hat{S} = P\frac{N_a - N_b}{D_c}$$

NSP for Increasing/Decreasing Life-Annuity of P

$$\hat{S}^{(k)} = P\frac{N_a(i^*) - N_b(i^*)}{D_a(i^*)}\frac{D_a(i)}{D_c(i)}$$

Miscellaneous Exercises

7M1. Arvind is 28 and has arranged for an annuity of \$75,000 annually, his first payment at age 60. The arrangement includes the stipulation that the first 15 payments of the annuity be certain with subsequent payments contingent upon Arvind's survival for as long as he lives. What net single premium will he pay for this annuity if interest is assumed at:

a) 2.25% effective?

b) 7.75% effective?

c) Discuss the difference between the two answers.

7M2. What size pure endowment will Nelly receive at 65 if she is now 25 years old and will pay \$1,000 at the beginning of every year for 40 years into a fund earning at the annual rate of 6.75%?

7M3. In the preceding exercise, how much would Nelly or her estate receive when she reaches 65 if all of the payments were certain? In other words, what would Nelly or her estate receive even if Nelly died before 65 if her deposits were made for her if she died? Would you expect the amount to be less or more than the amount found in the preceding exercise?

7M4. You have been called upon to estimate the loss of income incurred by the wife of a 35-year-old man killed in a work-related accident. If the man was otherwise expected to live another 39 years, if his income was expected to be \$35,000 one year after his death with annual increases estimated at 2.912621359%, and if money is valued at 6% per annum, what is the present value of the loss to his wife? Assume that he would have worked until his death. (See also, Exercise 4M2.)

7M5.* (💻) Females on earth have survival function given in Table II. Females on planet Q0317B have survival function

$$l_x = 10,000,000(1 - x/100), \; 0 \le x \le 100.$$

Females on planet Q0317C have survival function

$$l_x = 10,000,000\sqrt{1 - x/100}, \; 0 \le x \le 100.$$

Find the net single premium for a whole-life annuity of $10,000 to a 35-year-old female on all three planets if money earns 5% annually. See also Exercise 8M4.

Chapter 8—Life Insurance

8.1 Introduction

When I was in my teens, an insurance salesman came to our house and, since no one else was home, tried to sell me life insurance. I declined, telling him that it sounded to me as though his company was betting that I would not die in a certain amount of time and that I was betting that I would. He said that life insurance was not like that.

Looking back, I know that I was right to decline but for the wrong reason. At the time, I had little income, and no family members depending on me.

Some people have the idea that they should be heavily insured so that their death will make someone rich. A few gladly pay for insurance for which they are the beneficiaries then do away with the insured. We have all likely read of cases like this, and we have certainly seen it as a plot element in movies, plays and television dramas. These are not the reasons that we should buy life insurance.

The real purpose of any type of insurance is to replace a loss. In the case of life insurance, the loss is likely in the form of future income.

8.2 How Life Insurance Works

The way life insurance works is that a large number of people buy insurance policies and pay premiums. The insurance company invests the premium income and pays off *death benefits* as necessary from funds called reserves. Although the companies calculate their premiums so that they are likely to make a profit, for now, we will talk only about net premiums. But, if the company were to collect only net premiums, they would in theory have exactly enough income to pay off all of the death benefits and nothing left over for paying expenses or dividends to stockholders and policyholders. In Chapter 9, we take up the concept of *gross premiums*, where profits and expenses of various kinds are taken into account.

We begin by deriving the net single premiums (*NSP's*) for different kinds of life insurance. These premiums are rarely paid up in lump sums. More commonly, insurance is paid for with periodic payments, which, by the way, always are assumed begin immediately. These periodic payments constitute a life-annuity—an annuity that will cease to be paid if the insured dies, or if the term of the payments expires. On the death of the insured an agreed-upon *benefit* is paid to one or more *beneficiaries*. To determine the amount of the periodic payments, their present value is set equal to the present value of the life insurance and the amount solved for.

In the first two examples, I will introduce the idea of how the net single premium for an insurance policy is calculated and then how we calculate the periodic payments. In these two examples, Hannah is assumed to be very old for the sake of making the calculations brief.

Example 8.1: Hannah is 97 and is buying a whole-life insurance policy in the amount of $200,000 payable to her daughter. A *whole-life* insurance policy of $200,000 (the *benefit*) is a contract between Hannah (the *insured*) and her insurance company (the *insurer*) to pay her daughter (the *beneficiary*) $200,000 at the end of the year of her death. In actual practice, the company pays out the death benefit with as little delay as possible once proof of death is received. We assume that the payment is at the end of the year of death to simplify the mathematics. Find Hannah's net single premium, assuming that money earns 5% effective.

Solution: Hannah is 97, and she will die

—at 97 with probability $q_{97} = {}_{0|}q_{97} = \dfrac{d_{97}}{l_{97}} = \dfrac{80,695}{169,895} = .47496983431$;

—at 98 with probability ${}_{1|}q_{97} = \dfrac{d_{98}}{l_{97}} = \dfrac{58,502}{169,895} = .34434209365$; and

—at 99 with probability ${}_{2|}q_{97} = \dfrac{d_{99}}{l_{97}} = \dfrac{30,698}{169,895} = .18068807204$.

Note that these three probabilities sum to one—they constitute the probability distribution of the ages at death of a 97-year-old woman. Now, assuming that the death benefit is paid at the end of the year of death, at the time of her 97th birthday, that benefit is worth

—$\$200,000(1.05)^{-1} = \$190,476.190476$ if Hannah dies at 97,

—$\$200,000(1.05)^{-2} = \$181,405.895691$ if Hannah dies at 98, and

—$\$200,000(1.05)^{-3} = \$172,767.519706$ if Hannah dies at 99.

The net single premium for Hannah's policy is the weighted average of the three present values $\$200,000(1.05)^{-1}$, $\$200,000(1.05)^{-2}$, and $200,000(1.05)^{-3}$, the weights being the corresponding probabilities above. That is,

$$I = (\$190{,}476.190476)(.47496983431) +$$
$$(\$181{,}405.895691)(.34434209365)$$
$$+ (\$172{,}767.519706)(.18068807204) = \$184{,}153.16$$

where we will be using the symbol I to denote the net single premium for an insurance. I will describe this result one more time. The present value of the payment to Hannah's beneficiaries at the end of the year of her death is

— $200,000 $(1.05)^{-1}$ if she dies at 97 happening with probability ${}_{0|}q_{97}$,

— $200,000 $(1.05)^{-2}$ if she dies at 98 happening with probability ${}_{1|}q_{97}$, and

— $200,000 $(1.05)^{-3}$ if she dies at 99 happening with probability ${}_{2|}q_{97}$,

so her net single premium is the weighted average of these three values

$$I = \$200{,}000\,(1.05)^{-1}\cdot{}_{0|}q_{97} + \$200{,}000\,(1.05)^{-2}\cdot{}_{1|}q_{97} +$$
$$\$200{,}000\,(1.05)^{-3}\cdot{}_{2|}q_{97} = \$184{,}153.16.$$ ∎

Example 8.2: Assume now that instead of paying the net single premium of $184,153.16 that we found in Example 8.1, Hannah will make equal payments at the beginning of each year until her death. Find the size of those payments.

Solution: The payments constitute a whole-life annuity whose present value we found in the previous exercise. Let's look at a time diagram.

From Formula 7.4 and Table IV, with $i = .05$, we have

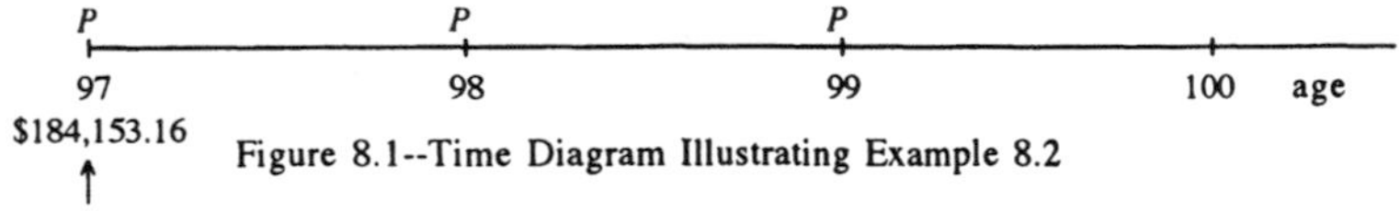

P	*P*	*P*		
97	98	99	100	age
$184,153.16 ↑				

Figure 8.1--Time Diagram Illustrating Example 8.2

$$\$184{,}153.16 = P\frac{N_{97}}{D_{97}} = P\frac{2{,}488.57}{1{,}495.61} = P(1.66392),$$

so $P = \$110{,}674.53$. This seems a very high premium in comparison to the benefit, but remember, Hannah has only about an 18% chance of living to pay the third premium. ▮

Derivation of the Net Single Premium for a Whole-Life Insurance: We now derive the net single premium for a whole-life insurance policy. I'll state the results at the end of the derivation and indicate how to modify the results to accommodate other types of insurance including a more general formula, Formula 8.3.

A whole-life insurance policy with *benefit B* to (x) is a contract between insurance company (the *insurer*) and (x) (the *insured*, now aged x) to pay the amount B to the insured's *beneficiaries* at the end of the year of death of (x). Here is a time diagram illustrating the situation where the net single premium is I.

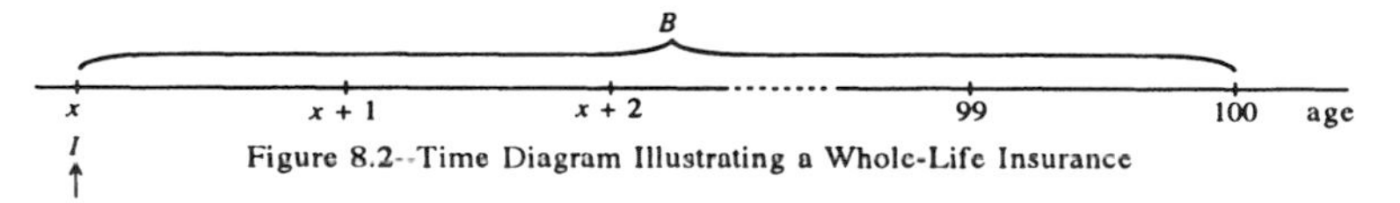

Figure 8.2--Time Diagram Illustrating a Whole-Life Insurance

Let's call the insured Xavier. I am using an *umbrella* to indicate the period of insurance coverage. Here, that period runs from age x to the end of the table.

The benefit, without survivorship, has a different present value depending on when it is paid. Here is a table showing the present value of the benefit and the probability of death according to Xavier's age.

Xavier's Age at Death	*PV of Benefit Without Survivorship*	*Probability of Death*
x	$B(1 + i)^{-1}$	${}_{0\|}q_x$
$x + 1$	$B(1 + i)^{-2}$	${}_{1\|}q_x$
$x + 2$	$B(1 + i)^{-3}$	${}_{2\|}q_x$
....	....	....
98	$B(1 + i)^{-(99 - x)}$	${}_{98-x\|}q_x$
99	$B(1 + i)^{-(100 - x)}$	${}_{99-x\|}q_x$

Table 8.1—Present Values of Benefit and Probabilities of Death Beginning at Age x

Remember, if Xavier dies at age x, the benefit is paid on his $(x + 1)$st birthday, et cetera.

Xavier's net single premium, I, is the average of the present values of the death benefits weighted by the probabilities of death. That is, the net single premium is

$$I = B(1+i)^{-1}({}_{0|}q_x) + B(1+i)^{-2}({}_{1|}q_x) + B(1+i)^{-3}({}_{2|}q_x)\ldots$$
$$+ B(1+i)^{-(99-x)}({}_{98-x|}q_x) + B(1+i)^{-(100-x)}({}_{99-x|}q_x).$$

That is a nice formula, but as it stands it would involve too much work to be practical. Instead, we are going to rewrite it in a way suitable to introducing our third (and last) commutation function after which calculations will become relatively simple. Here is how to get to the next step:

a) Factor out the benefit, B, from the above formula;

b) Rewrite the probabilities ${}_{t|}q_x$ as $\frac{d_{x+t}}{l_x}$, $t = 0, 1, 2, \ldots, 99 - x$;

c) Collect terms to get l_x as a common denominator;

d) Finally, multiply and divide by $(1 + i)^{-x}$, and collect terms again;

After doing these things, this is what the revised version of the net single premium formula:

$$I = B\left(\frac{d_x(1+i)^{-(x+1)} + d_{x+1}(1+i)^{-(x+2)} + \ldots + d_{98}(1+i)^{-99} + d_{99}(1+i)^{-100}}{l_x(1+i)^{-x}}\right).$$

You should recognize the common denominator of this expression as D_x because we used it quite a bit in the last chapter. The numerator contains only terms of the form $d_y(1+i)^{-(y+1)}$.

The Commutation Function M_x: We now define M_x to be the sum of the terms of the above form from age x, up to age 99:

$$M_x = d_x(1+i)^{-(x+1)} + d_{x+1}(1+i)^{-(x+2)} + \ldots + d_{98}(1+i)^{-99} + d_{99}(1+i)^{-100} \quad (8.1)$$

For example, $M_{97} = d_{97}(1+i)^{-98} + d_{98}(1+i)^{-99} + d_{99}(1+i)^{-100}$. This commutation function is tabulated along with the earlier ones in Tables III and IV for values of i running from 2% to 8% in steps of 0.25%. Here is the result of our derivation:

Result: The net single premium for a whole-life insurance of B to (x) with interest at i per annum is $I = I(x, g, B, i)$

$$I = B\frac{M_x}{D_x} \tag{8.2}$$

where g stands for *gender*: *male* (Table III) or *female* (Table IV). ∎

Example 8.3: Wanda is buying a whole-life insurance policy that will pay \$100,000 to her beneficiary when she dies. Find the net single premium Wanda will pay for each of the nine possible cases where her age is 30, 50, and 70 and interest is $i = 3\%$, 5%, and 7%. Put the results in a table.

Solution: As we did in the last chapter, we will write $M_x(i)$ and $D_x(i)$ for our commutation functions when we need to emphasize the interest rate, i, as well as the age. For each of the nine combinations of x and i, the net single premium will be found from Table IV using Formula 8.2. That is

$$I = I(x, \textit{female}, \$100{,}000, i) = \$100{,}000\frac{M_x(i)}{D_x(i)}.$$

You should verify the results in the table below:

		Interest Rate (i)		
		3%	5%	7%
Age (x)	**30**	\$26,365.29	\$12,457.78	\$6,651.71
	50	\$43,474.59	\$26,944.18	\$17,750.76
	70	\$67,177.72	\$53,024.26	\$42,736.41

Net Single Premiums for Females for Example 8.3

Note that the net single premiums decrease as the interest rate increases and that they increase as age at purchase increases. Were these things to be expected? ∎

Example 8.4: Zehman just turned 30 years old and is buying a \$150,000 whole-life insurance policy payable to his beneficiary at his death. To pay for this policy, Zehman will make annual payments at the beginning of every year for the next 25 years. Assume that money earns the effective rate of 8%.

a) Make a time diagram.
b) Find the net single premium.
c) Find the net annual premium that Zehman will be paying.

Solution:

a) Below is a time diagram showing the coverage and the 25 annual payments, *P*.

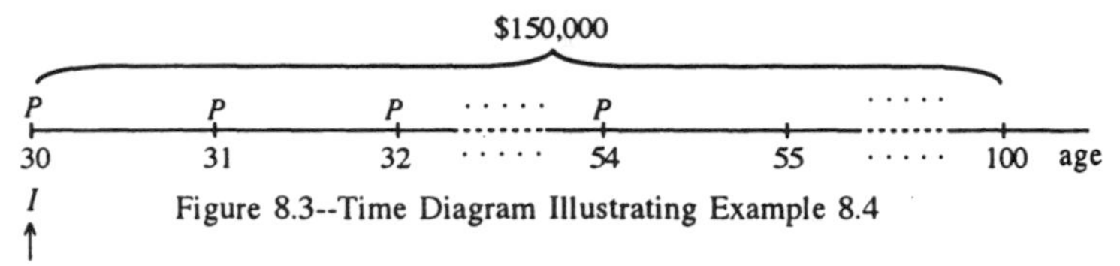

Figure 8.3--Time Diagram Illustrating Example 8.4

b) $I(30, male, \$150{,}000, .08) = \$150{,}000\frac{M_{30}}{D_{30}} = \$150{,}000\frac{62{,}082.59}{952{,}034.65}$

$= \$9{,}781.56$.

c) The 25 payments of *P* that Zehman will make constitute a 25-year temporary life annuity. The inputs for Formula 7.4 are $x = c = 30$, $g = male$, *P* is to be determined, $i = .08$, $a = 30$, amd $b = 55$ giving us

$$\hat{S} = P\frac{N_{30} - N_{55}}{D_{30}} = P\frac{12{,}014{,}352.75 - 1{,}269{,}390.45}{952{,}034.65} =$$

$$P(11.2863143164).$$

This is the present value of the 30-payment annuity, and must equal the present value *I* of the insurance policy. That is,

$$\hat{S} = P(11.2863143164) = \$9{,}781.56 = I$$

from which we get Zehman's net annual premium, $P = \$886.67$. ∎

Exercises 8.2

8.2.1 What is the net single premium for a whole-life insurance policy on the life of a 35-year-old male paying a benefit of $200,000 if money earns *a*) 4% annually? *b*) 5% annually? *c*) 6% annually?

8.2.2 Redo the preceding exercise for a female aged 35. Do you expect results lower or higher than in the preceding problem?

8.2.3 Write a paragraph or two on the implications of the two preceding exercises and Exercises 7.3.13 and 7.3.14.

8.2.4 Use Formula 8.1 to verify that $M_{96} = 9{,}845.91216\ldots$ when $i = .02375$ and $g = male$.

8.2.5 Rework Example 8.3 for males, and compare the results with those of the example.

8.2.6 Now 37 years of age, Dmitrij bought a whole-life insurance policy paying a benefit of 4,365,000 **R**ussian **R**ubles (RUR). Assuming the mortality experience of Tables I and III and an interest rate of $j_1 = 5\%$, what net single premium should Dmitrij pay in Russian Rubles?

8.2.7 Referring to the preceding exercise, what net annual premium in Russian Rubles should Dmitrij pay for life for his whole-life policy?

8.2.8 Who do think pays more for a given life insurance benefit: a man or a woman of the same age? Explain.

8.3 A General Formula for the Net Single Premium of a Life Insurance

Suppose now that we want the value at age c of a life insurance policy to an insured (x), paying a benefit of B at the end of the year of death. The term of the coverage is to begin when the insured is a years old and to cease when the insured is b years old $(a < b)$. Here is a time diagram for the policy.

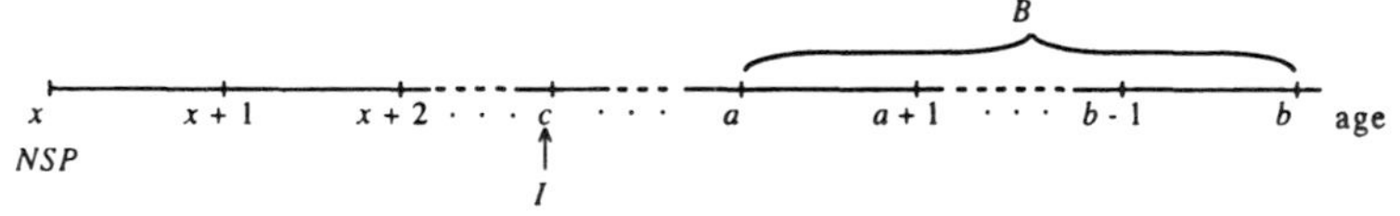

Figure 8.4--Time Diagram Illustrating General Life Insurance Policy

Notice that the value of the net single premium is at age x, but the time diagram shows the value of the insurance at age c which may be different from x. A derivation for the general formula is similar to the derivation of Formula 8.2 (See Exercise 8.3.2). Here are the results:

The General Formula: The value of this policy at age c, paying a benefit of B if (x) dies after reaching age a, but before reaching age b, assuming effective interest rate i, is

$$I = \begin{cases} B\dfrac{M_a - M_b}{D_c} & \text{as a general insurance formula, which reduces to} \\ B\dfrac{M_a}{D_c} & \text{when } b = 100. \end{cases} \tag{8.3}$$

The reduction occurs because $M_{100} = 0$. Note that the inputs for I are x, B, g, i, a, b, and c.

The Names and Descriptions of Some Different Types of Life Insurance: The formulas for the first three insurances are special cases of Formula 8.3 with $c = x$.

Whole-life insurance: The *NSP* for a whole-life insurance to (x) paying benefit B is

$$I = B\frac{M_x}{D_x}.$$

Notice that in Formula 8.3, $a = x$ and $b = 100$.

n-year term insurance: The *NSP* for an n-year term insurance to (x) paying a benefit of B if the insured's death occurs between age x and $x + n$ is

$$I = B\frac{M_x - M_{x+n}}{D_x}. \tag{8.4}$$

In Formula 8.3, we have $a = x$ and $b = x + n$.

m-year-deferred n-year term insurance: The net single premium for an m-year-deferred n-year temporary insurance to (x) paying benefit B is

$$I = B\frac{M_{x+m} - M_{x+m+n}}{D_x}. \tag{8.5}$$

Here we have $a = x + m$ and $b = x + m + n$ in Formula 8.3.

Another common insurance but *not* a special case of Formula 8.3, is *endowment insurance*. It is a combination of an n-year term insurance and a pure endowment.

Endowment insurance: An insurance to (x) paying a benefit of B if (x) dies between ages x and $x + n$ and paying the insured a pure endowment of B if and when the insured survives to age $x + n$ is called an *endowment insurance with benefit and pure endowment B*. The net single premium for this insurance is

$$I_E = B\frac{M_x - M_{x+n}}{D_x} + B\frac{D_{x+n}}{D_x} = B\frac{M_x - M_{x+n} + D_{x+n}}{D_x}. \quad (8.6)$$

Special Endowment Insurance: An insurance to (x) paying a benefit of B if (x) dies between ages x and $x + n$ and paying the insured a pure endowment of E if he or she survives to age $x + n$, is called a *special endowment insurance with benefit B and pure endowment E*. The net single premium for this insurance is

$$I_{SE} = B\frac{M_x - M_{x+n}}{D_x} + E\frac{D_{x+n}}{D_x} = \frac{B(M_x - M_{x+n}) + E \cdot D_{x+n}}{D_x}. \quad (8.7)$$

Let's illustrate these types of insurance with some examples.

Example 8.5: Jacob is 60 and wants to buy a 10-year term policy with a death benefit of $175,000.

a) Find Jacob's net single premium if $i = 6.25\%$ effective.

b) Find Jacob's net annual premium if he pays for his coverage with equal payments at the beginning of each year for the next 10 years.

Solution:

a) We can use Formula 8.3 or 8.4. From Formula 8.4 with $x = 60$, $n = 10$, and $i = .0625$, Jacob's net single premium is

$$I = \$175{,}000\frac{M_{60} - M_{70}}{D_{60}} = \$175{,}000\frac{82{,}008.13 - 48{,}489.63}{212{,}767.00} =$$
$$\$27{,}568.83.$$

b) It might help to look at a time diagram at this point.

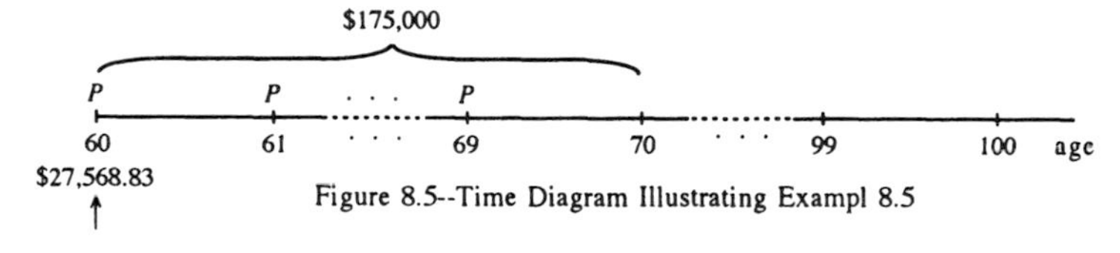

Figure 8.5--Time Diagram Illustrating Exampl 8.5

We found the net single premium of $27,568.83 in part *a*), and this must equal the present value of the 10-year life annuity that Jacob will pay. The present value of this annuity is, from Formula 7.4,

$$\hat{S} = P\frac{N_{60} - N_{70}}{D_{60}} = P\frac{2,222,900.86 - 706,689.04}{212,767.00} = P(7.12616063581).$$

Setting this equal to \$27,568.83, we get P = \$3,868.68. ▮

Example 8.6: Holden is 23 years old and has plans to expand his family beginning in about 5 years. He thinks it would be a good idea to buy a 5-year deferred whole-life policy of \$150,000. He will pay for the policy with equal annual payments at the beginning of each year for the next 10 years. Find Holden's annual payments if money earns at the annual rate of 5.25%.

Solution: Notice that in this problem, we are not even asked for the net single premium. Of course, it plays a role, but we do not have to solve for it. Look at the time diagram.

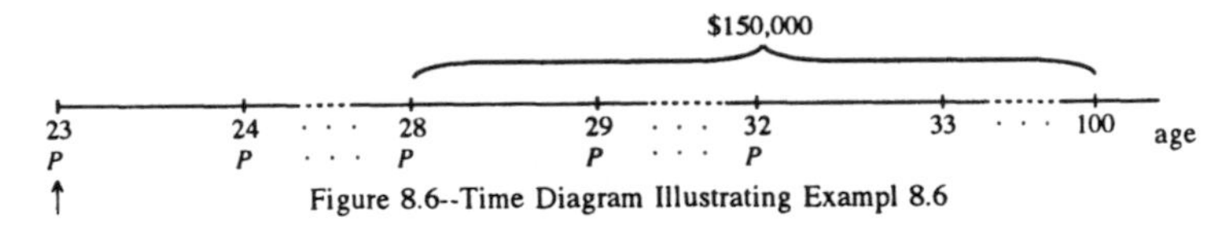

Figure 8.6--Time Diagram Illustrating Exampl 8.6

What we will do is equate the present value of Holden's payments,

$$P\frac{N_{23} - N_{33}}{D_{23}},$$

at 5.25%, to the net single premium for his deferred insurance,

$$\$150,000\frac{M_{28}}{D_{23}},$$

at 5.25%. In other words, we set

$$P\frac{N_{23} - N_{33}}{D_{23}} = \$150,000\frac{M_{28}}{D_{23}}$$

and solve for P. When we do this, the D_{23}'s terms cancel, and we get

$$P = \$150,000\frac{M_{28}}{N_{23} - N_{33}} = \$150,000\frac{294,355.42}{53,574,255.34 - 29,743,377.35}$$

$$= \$1,852.78.$$

I should point out that you could first have calculated Holden's net single premium,

$$\$150{,}000\,\frac{M_{28}}{D_{23}} = \$14{,}769.29$$

and then his annual payment,

$$P\frac{N_{23} - N_{33}}{D_{23}} = P(7.97143378543)$$

and solved for P to get P = \$14,769.29/7.971433... = \$1,852.78, as before. The first way takes less arithmetic but more algebra. The second route takes less algebra but more arithmetic. ∎

Example 8.7: Tom, a 43-year-old divorced man, is thinking of buying an endowment insurance policy providing a benefit of \$300,000 to his children or their estates if he dies before age 60 or a payment of \$300,000 to himself if he survives to age 60. Assume interest at i = 7.5% annually:

a) Find Tom's net single premium.
b) Find Tom's equal annual payment for the insurance at the beginning of each of the next 17 years.

Solution: A time diagram for both parts is below.

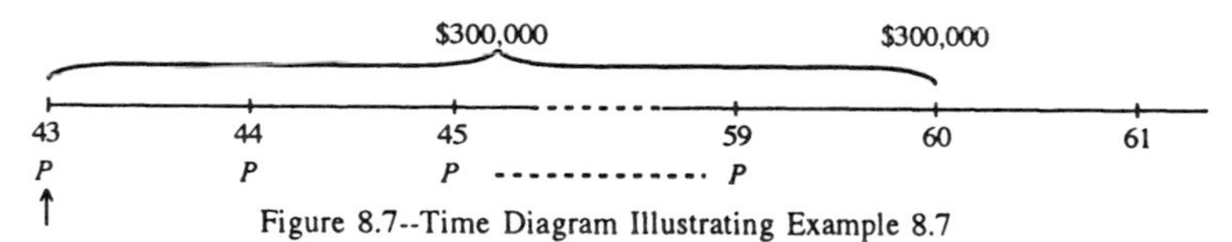

Figure 8.7--Time Diagram Illustrating Example 8.7

a) We use Formula 8.6 with x = 43, B = \$300,000, g = *male*, i = .075, and n = 17. The net single premium is

$$I_E = \$300{,}000\frac{M_{43} - M_{60} + D_{60}}{D_{43}} =$$

$$I_E = \$300{,}000\frac{60{,}923.93 - 34{,}861.93 + 105{,}470.92}{414{,}206.37} = \$95{,}266.22.$$

b) The net annual premium for a 17-payment life annuity for a male aged 43 is

$$P\frac{N_{43} - N_{60}}{D_{43}} = P\frac{5{,}063{,}714.99 - 1{,}012{,}062.17}{414{,}206.37} = P(9.78172503721).$$

This must equal the net single premium of \$95,266.22, so P = \$95,266.22/9.78172... = \$9,739.20. Tom might want to rethink the pure endowment part of this policy since it makes up over 80% of his premium. ∎

Example 8.8: Tom (from Example 8.7) rethought his endowment policy strategy. He will now buy a special endowment policy with the same benefit as before, \$300,000, but a pure endowment of only \$50,000 to himself if he survives to age 60. Find both his net single premium and his net annual premium as in Example 8.7.

Solution: A time diagram for Tom's special endowment policy looks like this:

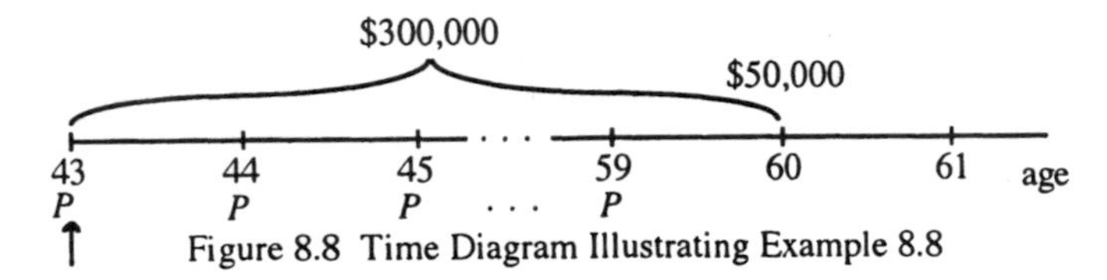

Figure 8.8 Time Diagram Illustrating Example 8.8

From Formula 8.7, Tom's net single premium is now

$$I_{SE} = \frac{\$300,000(60,923.93 - 34,861.93) + (\$50,000)(105,470.92)}{414,206.37}$$

$$= \$31,607.79.$$

The formula for Tom's net annual premium is unchanged from Example 8.7 at $P(9.78172503721)$. Equating this with his net single premium of \$31,607.79, we find $P = \$3,231.31$. ∎

Example 8.9: Leslie is 63 years old. How much should he pay for a one-year term policy of \$100,000 if money is worth $i = 6\%$ per annum?

Solution: From Formula 8.4 with inputs $x = 63$, $B = \$100,000$, g = *male*, $i = .06$, and $n = 1$, the premium for this one-year term insurance policy to Leslie is

$$I = \$100,000\frac{M_{63} - M_{64}}{D_{63}} = \$100,000\frac{86,237.92 - 82,361.93}{195,088.04} =$$

$$\$1,986.79.$$ ∎

Example 8.10: Todd estimates that he needs to protect his family with a term life insurance policy if he dies in the next 20 years. He is now 35 years old. He intends to pay for this insurance with equal annual premiums of \$1,843.13 for the next 15 years. To the nearest \$5.00, what is the death benefit if money is worth 5.75% effective?

Solution: Todd's net annual premium is

$$NAP = \$1{,}843.13\frac{N_{35}-N_{50}}{D_{35}},$$

and this must equal

$$B\frac{M_{35}-M_{55}}{D_{35}},$$

where B is the death benefit. The solution for B is

$$B = \$1{,}843.13\frac{N_{35}-N_{50}}{M_{35}-M_{55}} = \$1{,}843.13\frac{20{,}990{,}186.80-7{,}227{,}074.59}{200{,}006.35-136{,}588.44}$$

$$= \$400{,}000.65 \approx \$400{,}000.$$

∎

Exercises 8.3

8.3.1 What net single premium will Linda aged 36 pay for a 29-year term insurance of \$150,000 with interest at 5% annually?

8.3.2 What net annual premium will Gabriel aged 25 pay for a 35-year endowment insurance of \$200,000 if interest is at 4% annually, and Gabriel's payments begin now and end when he is 49?

8.3.3 Find Adelaida's premiums for a 1-year term insurance of \$1,000,000 if she is 25, and money earns at 2%, 4%, 6%, or 8% annually. Show your results in a table.

8.3.4 Find Pierre's premiums for a 1-year term insurance of \$1,000,000 if he is 25 and money earns at 2%, 4%, 6%, or 8% annually. Put the results in a table. Compare your results with those of the preceding exercise, and discuss.

8.3.5 At the age of 25, Eva bought a 30-year term policy paying for it with premiums of \$1,000 at the beginning of every year, last payment to be made at age 44. If interest is at 5% annually, what benefit is to be paid if she dies before age 55?

8.3.6 Sarafino just turned 35 and is buying a 30-year special endowment policy paying a death benefit of $100,000. His payments at the beginning of every year beginning now and ending when he is 64 are in the amount of $1,200. If interest is assumed to be 6% effective, what is the size of Sarafino's pure endowment payable at age 65?

8.3.7 At age 30, Elliot buys an insurance policy paying his beneficiary $150,000 if he dies before reaching age 60 or paying Elliot a life annuity of $50,000 beginning at age 60 if he survives to that age. What will Elliot's 30 equal annual premiums be beginning at age 30 if interest is at $j_1 = .055$?

8.3.8 Kai is 27 years old and wants to buy a $200,000 whole-life insurance policy payable to her beneficiary at the end of the year of her death. To pay for this policy, Kai will make annual payments at the beginning of every year for the next 33 years. Assume that money earns 4.5% effective.
a) Find the net single premium for this policy.
b) Find the net annual premium that Kai will be paying.

8.3.9 Assuming interest at $i = 5\%$ annually, find the net single premium for a one-year $100,000 term insurance to a man and a woman aged 30, 40, 50, 60, 70, 80, and 90. Comment. *Hint:* See Example 8.9.

8.3.10 Find the net annual premium payable for life for 50-year-old men and women buying $100,000 whole-life insurance policies for interest rates 2%-8% in steps of 1%.

8.3.11 Molly inherited $27,500 from her great-grandmother at the age of 27. With her inheritance, she bought endowment insurance with the endowment portion payable at age 60. If interest is at 4.75%, how much endowment insurance did Molly buy?

*8.4 Policies with Benefits Increasing/Decreasing by a Fixed Percentage**

In Chapters 4 and 7, we learned about annuities whose payments increase or decrease by a fixed percentage. It is also desirable to have insurance benefits that have the possibility of keeping up with inflation by increasing each year. It is also not uncommon to have decreasing insurance. Perhaps the best example of this is *mortgage insurance*, where the benefit decreases more or less with the outstanding balance on the mortgage.

The General Formula: Suppose that a person aged x buys insurance that goes into force at age a and out of force at age b. The benefit is to have value B at age a and change (increase or decrease) by a factor of $1 + k$ for each year the insurance is in force. Thus, the benefit moves in value from B at age a to $B(1 + k)^{b-a}$ at age b. Note that B is used only as a starting value at age a. No one would actually receive a benefit of B, since we assume that benefits are paid at the end of the year of death. The beneficiary of an insured dying at age a receives $B(1 + k)$. In this sense, we call B the *nominal benefit* of the policy.

To repeat, the benefit is to be $B(1 + k)$ if death occurs after age a but before age $a + 1$, $B(1 + k)^2$ if death occurs after age $a + 1$ but before age $a + 2$, $B(1 + k)^3$ if death occurs after age $a + 2$ but before age $a + 3$, ..., $B(1 + k)^{b-a}$ if death occurs after age $b - 1$ but before age b. So, the benefit increases annually by a factor of $1 + k$ if $k > 0$ and decreases by a factor of $1 + k$ if $k < 0$. The benefit stays level at B if $k = 0$. Assuming that the annual interest rate stays constant at i, the value of the policy at age c is

$$I^{(k)} = B\frac{M_a(i^*) - M_b(i^*)}{D_a(i^*)}\frac{D_a(i)}{D_c(i)} \tag{8.8}$$

where the inputs for the formula are

x—the age of the annuitant at purchase,
B—the benefit at age a,
g—gender: *male*, *female*,
k—the percentage increase or decrease of subsequent benefits,
i—annual effective rate of interest,
i^*—adjusted interest rate $i^* = (i - k)/(1 + k)$,
a—age at the time insurance goes into force,
b—age at the time insurance goes out of force, and
c—the age where the annuity is to be valued.

The formula reduces to

$$I^{(k)} = B\frac{M_a(i^*)}{D_a(i^*)}\frac{D_a(i)}{D_c(i)}$$

when $b = 100$, because $M_{100} = 0$.

Derivation: The derivation is similar to that of the net single premium for a whole-life insurance except that the coverage here runs from age a to age b, and each year the benefit B is augmented by a factor of $1 + k$. Factors of $(1 + k)(1 + i)^{-1}$ are combined and equated to $(1 + i^*)^{-1}$ with the result that $i^* = (i - k)/(1 + k)$. The details are left as an exercise. See Exercise 8.4.6 below.

As in Chapter 7, for the examples and exercises, I will manipulate the values of k to produce values of i^* that coincide with interest rates in our tables. In actual practice, it would not be difficult to produce a table with any desired interest rate by using a spreadsheet program. ▮

Example 8.11: Stig-Ove is 30 years old and wants to buy an increasing term insurance with a nominal benefit of \$100,000 with the benefit increasing by 2.6699% each year until he is age 65 at which time the insurance is to go out of force. Assuming that money earns 5.75% annually, find Stig-Ove's

a) net single premium.

b) net annual premium assuming he makes equal payments at the beginning of each year from now until age 54.

Solution:

a) Notice first that the death benefit is

— \$100,000(1.026699) = \$102,669.90 if Stig-Ove dies before age 31,

— $\$100,000(1.026699)^2 = \$105,411.08$ if he dies before age 32,

— $\$100,000(1.026699)^3 = \$108,225.45$ if he dies before age 33,

. . .

— $\$100,000(1.026699)^{35} = \$251,483.71$ if he dies before age 65, and

— \$0.00 if Stig-Ove dies at age 65 or older.

Here is what we need for Formula 8.8:

$a = c = x = 30$, $B = \$100,000$, $g = male$, $k = 2.6699\%$, $i = 5.75\%$, $i^* = (.0575 - .026699)/1.026699 = .03$, and $b = 65$. The *NSP* is

$$I^{(.0266...)} = \$100,000\frac{M_{30}(.03) - M_{65}(.03)}{D_{30}(.03)}\frac{D_{30}(.0575)}{D_{30}(.0575)} =$$

$$\$100{,}000\frac{M_{30}(.03)-M_{65}(.03)}{D_{30}(.03)}=$$

$$\$100{,}000\frac{1{,}177{,}429.40-716{,}385.98}{3{,}946{,}832.33}=\$11{,}681.35.$$

b) Stig-Ove will make payments at the beginning of each year for 25 years. From Formula 7.4, the present value of his 25-payment annuity with P to be determined is

$$P\frac{N_{30}(.0575)-N_{55}(.0575)}{D_{30}(.0575)}=$$

$$P\frac{28{,}993{,}303.14-4{,}803{,}943.12}{1{,}790{,}416.77}=$$

$$P(13.5104632761).$$

Equating $P(13.5104632761)$ with \$11,681.35, we find Stig-Ove's 25 annual payments are \$864.62 each. ∎

Example 8.12: Catarina has decided that due to a declining need her family will have for her income, she will purchase an insurance whose nominal benefit decreases from \$150,000 at her present age of 25 by 1.415094% annually until she reaches age 45 at which time the insurance will go out of force. To pay for this decreasing term insurance, she will make equal payments at the beginning of each year, her last payment at age 44.

a) Find the net single premium for Catarina's insurance.

b) Find the annual payment, assuming that the annual interest rate is 4.5%.

Solution:

a) Here, we have k = -.01415094 so that the factor by which the benefit decreases each year is $1 + k = 0.98584906$. The other values for Formula 8.8 are $a = c = x = 25$, B = \$150,000, g = *female*, $i = .045$, $i^* = (.045 - (-.01415094))/(1 + (-.01415094)) = .06$, and $b = 45$. Formula 8.8 gives us

$$I^{(-.0141...)}=\$150{,}000\frac{M_{25}(.06)-M_{45}(.06)}{D_{25}(.06)}\frac{D_{25}(.045)}{D_{25}(.045)}=$$

$$\$150{,}000\frac{M_{25}(.06)-M_{45}(.06)}{D_{25}(.06)}=$$

$$=\$150{,}000\frac{163{,}194.74-120{,}191.32}{2{,}275{,}771.48}=\$2{,}834.43.$$

This is low enough to pay off the policy with a single payment. Notice, incidentally, that the benefit in the final year of coverage is $\$150{,}000(.98584906)^{19}=\$114{,}416.69$.

b) From Formula 7.4, the 20-payment temporary annuity that Catarina will pay has present value

$$\hat{S}=P\frac{N_{25}(.045)-N_{45}(.045)}{D_{25}(.045)}=$$

$$P\frac{66{,}122{,}795.30-22{,}457{,}821.62}{3{,}249{,}885.21}=P(13.4358510712),$$

and this must equal the net single premium of \$2,834.43. Solving for P, we have P = \$210.96 per year. This is a very modest annual premium, but remember that it does reflect the fact that it is unlikely that the insurance company will have to make a benefit payment. In fact, from Table II, the probability that the company will have to make a payment at all is

$$_{20}q_{25}=1-\frac{l_{45}}{l_{25}}=1-\frac{9{,}409{,}244}{9{,}767{,}317}=.03666\ldots$$ ∎

Example 8.13: Sonny is starting a family at age 25 and thinks that an increasing term insurance until he is aged 55 is right for him because there will be little need to replace his income after that. He elects that the nominal coverage begin at \$100,000 with annual increases of 3%. Sonny will pay for this policy with equal annual payments at the beginning of every year beginning now and ending with his last payment at age 54. If interest earns 6.5%, find Sonny's net single premium and the size of his annual payments.

Solution: This time, I have not contrived the percentage increase in the benefits to make i^* come out to be one of our tabled values. Instead, we will try to use our existing tables to approximate the result. The benefit program is

— \$100,000(1.03) = \$103,000 if Sonny dies before age 26,

— $\$100{,}000(1.03)^2 = \$106{,}090$ if Sonny dies before age 27,

— $\$100{,}000(1.03)^3 = \$109{,}272.70$ if Sonny dies before age 28,

. . . .

— $\$100{,}000(1.03)^{30} = \$242{,}726.25$ if Sonny dies before age 55, and no benefit beginning at age 55.

Using Formula 8.8 with values $x = c = 25$, $B = \$100{,}000$, $g = male$, $k = .03$, $i = .065$, $i^* = (.065 - .03)/1.03 = .03398058252$, $a = 25$, and $b = 55$, we find Sonny's net single premium

$$I^{(.03)} = \$100{,}000\frac{M_{25}(.03398...) - M_{55}(.03398...)}{D_{25}(.03398...)}\frac{D_{25}(.065)}{D_{25}(.065)}$$

or

$$I^{(.03)} = \$100{,}000\frac{M_{25}(.03398...) - M_{55}(.03398...)}{D_{25}(.03398...)}.$$

The value .03398058252 lies part way between our tabled values of .0325 and .035. In fact, .03398058252 is 59.2233% of the way from .0325 to .035. This is because

$$\frac{.03398058252 - .0325}{.035 - .0325} = .592233 = 59.2233\%.$$

We will evaluate $I^{(k)}$ at .0325 and at .035 and then take our approximate net single premium to be 59.2233% of the way from the former to the latter. Now,

$$I^{(k)} = \$100{,}000\frac{M_{25}(.0325) - M_{55}(.0325)}{D_{25}(.0325)} = \$5{,}933.65,$$

and

$$I^{(k)} = \$100{,}000\frac{M_{25}(.035) - M_{55}(.035)}{D_{25}(.035)} = \$5{,}685.43.$$

To determine 59.2233% of the way from \$5,933.65 down to \$5,685.43, note that the difference between the premiums is

$$\$5{,}933.65 - \$5{,}685.43 = \$248.22,$$

and 59.2233% of that is

$$.592233(\$248.22) = \$147.00,$$

so

$$I^{(k)} = \$100,000\frac{M_{25}(.03398...) - M_{55}(.03398...)}{D_{25}(.03398...)} \approx$$

$$\$5,933.65 - \$147.00 = \$5,786.65.$$

Sonny's payments form a 30-year life-annuity of P (to be determined) with present value

$$\hat{S} = P\frac{N_{25}(.065) - N_{55}(.065)}{D_{25}(.065)} =$$

$$P\frac{30,275,640.95 - 3,065,216.60}{2,001,576.07} = P(13.5944992338).$$

Since this must equal \$5,786.65, we have P = \$5,786.65/13.594499... = \$425.66. To see how close we came to the value of the net single premium when i^* = .03398..., I constructed a table like Table III but with interest rate 3.398058252%. The result was that I = \$5,784.98, a difference of only \$1.67. With the correct net single premium, the annual payment becomes P = \$5,784.98/13.59449 = \$425.54, a difference of only 12¢. ∎

Example 8.14: Peter feels that the right insurance for him is a policy with a benefit of \$150,000 if he dies between his current age of 35 and age 45, \$200,000 if he dies between ages 45 and 55, and \$250,000 if he dies between ages 55 and 65. Find Peter's net single premium and his net annual premium if he is to make equal annual payments at the beginning of every year until his last payment at age 64. Assume the interest rate to be 7.25% annually.

Solution: A time diagram will show things clearly.

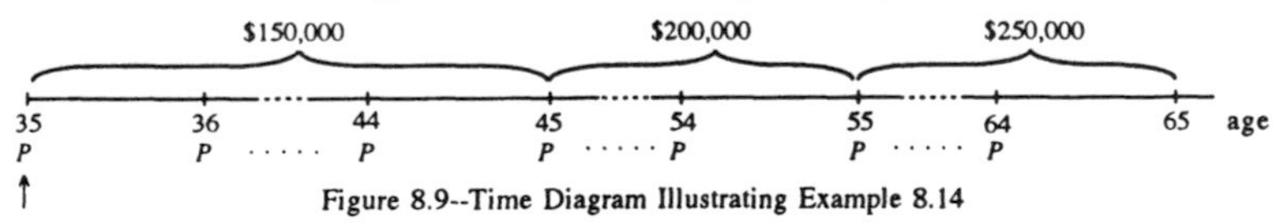

Figure 8.9--Time Diagram Illustrating Example 8.14

Actually, there are several ways to attack this problem. I'll show one way here, and ask you to use another in Exercise 8.4.3 below. We can consider this policy as three separate 10-year term policies—the second is deferred 10 years, and the third is deferred 20 years. Formula 8.3 can be applied three times. With the help of Table III, the net single premium at 7.25% is

$$I = \$150{,}000\frac{M_{35} - M_{45}}{D_{35}} + \$200{,}000\frac{M_{45} - M_{55}}{D_{35}} + \$250{,}000\frac{M_{55} - M_{65}}{D_{35}}$$

which can be rewritten

$$I = \frac{\$150{,}000M_{35} + \$50{,}000M_{45} + \$50{,}000M_{55} - \$250{,}000M_{65}}{D_{35}} =$$

$$\$12{,}669.69.$$

Peter's payments constitute a 30-year life annuity of P (to be determined). From Formula 7.4, the present value of the annuity at 7.25% is

$$\hat{S} = P\frac{N_{35} - N_{65}}{D_{35}} = P(12.4658973745)$$

which, when equated with the net single premium, gives $P = \$1{,}016.35$ annually. ∎

Exercises 8.4

8.4.1 Melinda is planning to buy a whole-life insurance policy that increases by 3.90244% annually. Her equal annual premiums beginning immediately and ending when she is 59, are to be in the amount of \$950. If Melinda is 35 and money earns 6.5% annually, what will Melinda's nominal death benefit be worth when her coverage begins?

8.4.2 Lonny is 40 years old and has decided to buy increasing term insurance that will go out of force when he is aged 65. He chooses a nominal coverage of \$125,000 with annual increases of 3%. He will pay for the policy with equal annual payments at the beginning of every year starting now and ending with his last payment at age 64. If money earns 5.5%, find the approximate value of Lonny's net single premium and that of his net annual premiums. *Hint:* Follow the method of approximation used in Example 8.13.

8.4.3 Recalculate the result in Example 8.14 by consulting the time diagram below:

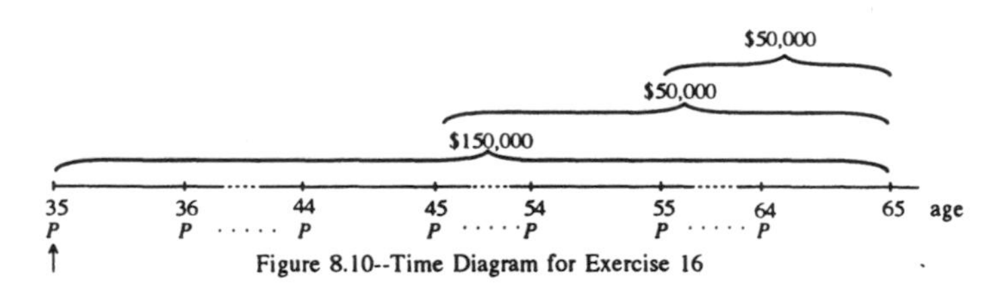

Figure 8.10--Time Diagram for Exercise 16

8.4.4 Syed wants to buy an increasing whole-life insurance policy with a nominal benefit of $50,000 at age 30 and annual increases to the benefit of 4.411764706%.

a) If Syed were to die at age 79, what would his death benefit be?

b) What would Syed's net single premium be with interest at 6.5% annually?

c) What would Syed's equal net annual premium for life be with interest at 6.5% annually?

8.4.5 Lars and Linnea are both 25 years old and are starting a family. They think that since Lars has a good paying job at the moment he should buy life insurance. They decide that there should be a $50,000 benefit plus second benefit with a nominal value of $50,000 with decreases of 2.7907% each year until Lars is 55 at which time the decreasing portion of the insurance should go out of force. The original $50,000 benefit is to continue for Lars's life. Assume interest is at 4.5%.

a) Find Lars's net single premium.

b) Find Lars's equal net annual premiums if his last payment is to be age 54.

8.4.6* Derive Formula 8.8.

8.4.7 Show that when $i^* = 0$ and $a = c$, Formula 8.8 reduces to

$$I^{(k)} = B\frac{l_a - l_b}{l_a},$$

and that when $a \neq c$, the multiplicative factor $D_a(i)/D_c(i)$ is appended.

8.4.8 Adelle purchased a 20-year term life insurance policy with a nominal benefit of $250,000. The benefits will increase by 4% annually. She is now 35 years old and will make equal annual payments at the beginning of each year for 20 years. If money earns $j_1 = 4\%$, find Adelle's net single premium and net annual premium. *Hint*: Refer to the preceding exercise.

8.5 Summary of Formulas

Commutation Function M_y

$$M_y = (1+i)^{-(y+1)}d_y + (1+i)^{-(y+2)}d_{y+1} + \ldots + (1+i)^{-99}d_{98} + (1+i)^{-100}d_{99}$$

NSP for Life Insurance with Benefit B

$$I = B\frac{M_a - M_b}{D_c}$$

Special Cases of Insurance to (x):

Whole-Life Insurance $$I = B\frac{M_x}{D_x}$$

n-Year Term Insurance $$I = B\frac{M_x - M_{x+n}}{D_x}$$

m-Year Deferred, n-Year Term $$I = B\frac{M_{x+m} - M_{x+m+n}}{D_x}$$

Endowment Insurance $$I_E = B\frac{M_x - M_{x+n} + D_{x+n}}{D_x}$$

Special Endowment Insurance $$I_{SE} = \frac{B\cdot(M_x - M_{x+n}) + E\cdot D_{x+n}}{D_x}$$

NSP for Increasing or Decreasing Insurance, Nominal Benefit B

$$I^{(k)} = B\frac{M_a(i^*) - M_b(i^*)}{D_a(i^*)}\frac{D_a(i)}{D_c(i)}$$

Miscellaneous Exercises

8M1. Derive Formula 8.3

8M2. Derive a formula for the *NSP* of a life insurance for a woman now aged 30 with a benefit of \$300,000 if she dies between ages 30 and 40; \$150,000 if she dies between the ages of 40 and 55; and \$75,000 if she dies after age 55. Simplify as much as possible.

8M3. Kosta is 30 years old, and he and his employer will begin making deposits at the beginning of every year for 30 years into a fund providing a 30-year term life insurance followed by a retirement annuity paying Kosta annually for life beginning when he reaches age 60. The deposits, the insurance, and the annuity payments will all increase by 4.368932039% annually. The term insurance is to have a nominal value of $250,000. If the fund earns 7.5% effective annually and the first deposit into Kosta's fund is $8,000, what will his first annuity payment be?

8M4.* (🖳) Females on earth have the survival function given in Table II. Females on planet Q0317B have survival function

$$l_x = 10{,}000{,}000(1 - x/100),\ 0 \le x \le 100.$$

Females on planet Q0317C have survival function

$$l_x = 10{,}000{,}000\sqrt{1 - x/100},\ 0 \le x \le 100.$$

Find the net single premium for a whole-life insurance of $100,000 (or its equivalent) to a 35-year-old female on all three planets if money earns 5% annually.

*Chapter 9—Loose Ends: Reserves & Gross Premiums**

9.1 Introduction

Until now, we have not taken into account the facts that

a) insurance companies need to have funds in reserve to meet obligations as they arise;
b) the net premiums we discussed in Chapters 7 and 8 are neither sufficient to pay the expenses of running the company nor to allow the company to enjoy a profit.

These are the topics of this final chapter.

9.2 Reserves

If a person aged x buys a one-year term policy, then for each dollar of benefit, the cost of insuring that person for the year is $c_x = (1+i)^{-1}q_x = (M_x - M_{x+1})/D_x$. This follows from Equation 8.3. The numbers c_x, c_{x+1}, . . . , and c_{99} are called *natural premiums*. Note that

$$c_{x+t} = (1+i)^{-1}q_{x+t} = (M_{x+t} - M_{x+t+1})/D_{x+t},$$

$t = 0, 1, \ldots, 99 - x$. Generally, except for the younger ages, the sequence of natural premiums increases with age. This is because the force of mortality increases—after the twenties for men and even earlier for women.

If a woman were to buy a life insurance policy at age x, she could elect to renew her policy annually. That is, she could choose to buy a succession of one-year term policies. Assuming that the benefit and interest rate remain constant, her premium at age $x + t$ would be $c_{x+t} = (M_{x+t} - M_{x+t+1})/D_{x+t}$ for each dollar of benefit. So, if the benefit were B, payable at the end of the year of her death, her t^{th} premium would be

$$P_{x+t} = B\cdot c_{x+t} = B\cdot(1 + i)^{-1}q_{x+t} = B\cdot(M_{x+t} - M_{x+t+1})/D_{x+t}.$$

On the other hand, our insured woman could elect to pay level annual premiums. These premiums would have to be more than enough to cover the cost of insurance (as measured by the natural premiums defined above) in the earlier years of the policy. But as the years pass, the natural premiums far exceed the net level premiums. Because of this, a fund must be created to take advantage of the expected surplus in the early years of the policy so that they will be sufficient to cover the cost of paying benefits when they occur in the later years of the policy. This fund is called a *reserve.*

A good way of viewing this reserve fund is to imagine that *all* l_x individuals aged x purchase an insurance policy with certain benefits to the policyholders. Premiums accumulate at annual interest rate i and claims are paid out from the fund as necessary. When the last of the l_x survivors has died or when the policy goes out of force, there would have been *exactly* enough money in the fund to meet obligations. We will begin by illustrating these ideas with a few examples.

Example 9.1: Aaron is 35 years old and buys a whole-life insurance policy paying \$200,000 to his beneficiary at the end of the year of his death. Assume an annual interest rate of 5%.

a) What equal annual premiums would Aaron pay for life?

b) If, instead of level premiums every year, Aaron were to buy one-year term insurance each year of his life beginning at age 35 with the same \$200,000 benefit, what would his one-year premiums be? Make a graph showing the level premiums and one-year term premiums on the same set of axes.

c) Interpret the graph found in part b).

Solution:

a) Aaron's net level premiums, P, satisfy the equation

$$P\frac{N_{35}}{D_{35}} = \$200{,}000\frac{M_{35}}{D_{35}},$$

so

$$P = \$200{,}000\frac{M_{35}}{N_{35}} = \$2{,}141.23.$$

b) When Aaron reaches age 35 + t, his one-year term premium for a benefit of \$200,000 would be

$$P_{35+t} = \$200{,}000 \cdot c_{35+t} = \$200{,}000(1.05)^{-1} q_{35+t} =$$

$$\$200{,}000\frac{M_{35+t} - M_{36+t}}{D_{35+t}},$$

$t = 0, 1, 2, \ldots 64$, where c_{35+t} is the natural premium discussed above. A few selected values of the premiums are:

Age	One-year Premium	Net Level Premium	Excess of Net Level Premium over One-Year
35	$401.91	$2,141.23	$1,739.32
50	$1,278.09	$2,141.23	$863.14
65	$4,841.91	$2,141.23	-$2,700.68
80	$18,826.70	$2,141.23	-$16,685.47
95	$62,849.72	$2,141.23	-$60,708.49

Table 9.1

The one-year premiums reflect the actual cost of insurance for Aaron age by age. Here is the graph of the premiums versus age from 35 on:

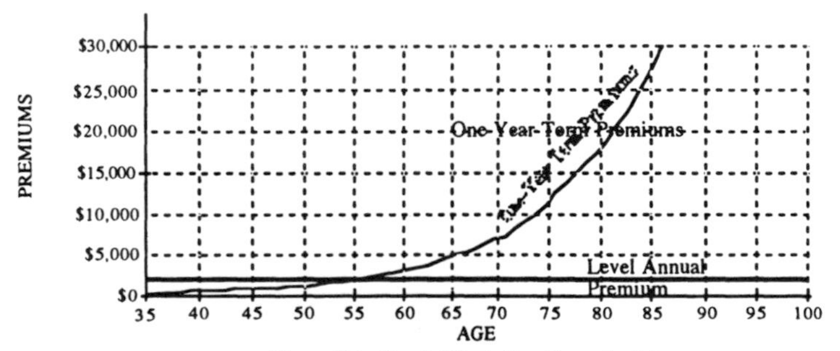

Figure 9.1--Graph Illustrating Example 9.1

c) The one-year term premiums start out significantly lower than the level annual premiums, then exceed them by greater and greater amounts in the later ages. This means that the level annual premiums of $2,141.23, although more than paying the cost of insuring Aaron for the first twenty years or so of his coverage, fall shorter and shorter in the later years. This fact is what makes it necessary for the company to build a reserve fund from the early level premiums so that they will accumulate to be sufficient to pay the benefits should they need to be paid later.

Although the graph does not show it, the actual net cost of one-year term insurance for Aaron at age 99 is $P_{99} = \$190,476.19$. If he chose to pay level premiums beginning at age 35, his premium at age 99 would still be $2,141.23. ∎

We will soon see how to calculate reserves systematically and expeditiously. But first, you might find it instructive to verify the calculations in a particular example.

Example 9.2: At age 30, Martina bought a 10-year special endowment policy paying $100,000 in the event of her death before age 40 or a payment of $50,000 to her if she were to survive to age 40. She paid for the policy with 5 equal annual payments beginning at age 30. Assuming an annual interest rate of $i = 4.75\%$, make a table showing the accumulation of the reserves for Martina's policy.

Solution: Let's look at a time diagram for Martina's policy:

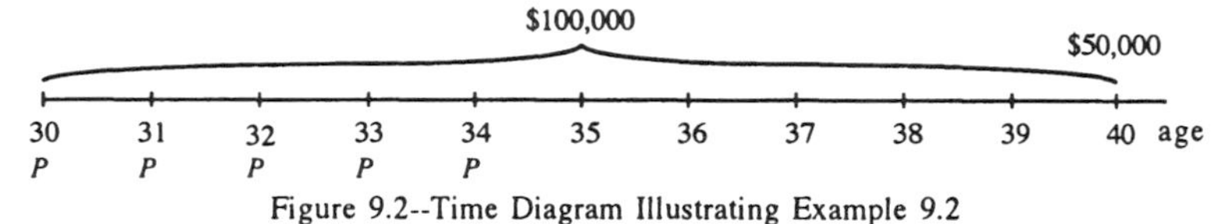

Figure 9.2--Time Diagram Illustrating Example 9.2

Applying Formulas 7.4 and 8.3, we have the equation of value

$$P\frac{N_{30}-N_{35}}{D_{30}} = \$100,000\frac{M_{30}-M_{40}}{D_{30}} + \$50,000\frac{D_{40}}{D_{30}}$$

which, when solved for P, gives $P = \$7,067.94464105 \approx \$7,067.94$ rounded to the nearest cent. The calculations below will not reflect this rounding. To calculate the reserves for Martina's policy, we will assume that all $l_{30} = 9,707,590$ female survivors to age 30 purchase the same policy as she. Then we will see how the premium income placed in a fund is exactly sufficient to satisfy all death claims as well as the claims for the $50,000 endowment. You can follow the calculations in Table 9.2 below.

Yr.	Premiums	Beginning-of-Year Fund	Fund Plus Interest	No. of Deaths	End-of-Year Fund	Number of Survivors	Reserve /Policy in Force
1	$68,612,708,718	$68,612,708,718	$71,871,812,382	13,105	$70,561,312,382	9,694,485	$7,279
2	$68,520,083,303	$139,081,395,685	$145,687,761,980	13,572	$144,330,561,980	9,680,913	$14,909
3	$68,424,157,159	$212,754,719,139	$222,860,568,298	14,037	$221,456,868,298	9,666,876	$22,909
4	$68,324,944,420	$289,781,812,718	$303,546,448,822	14,500	$302,096,448,822	9,652,376	$31,298
5	$68,222,459,222	$370,318,908,044	$387,909,056,176	15,251	$386,383,956,176	9,637,125	$40,093
6		$386,383,956,176	$404,737,194,095	15,901	$403,147,094,095	9,621,224	$41,902
7		$403,147,094,095	$422,296,581,064	16,933	$420,603,281,064	9,604,291	$43,793
8		$420,603,281,064	$440,581,936,915	18,152	$438,766,736,915	9,586,139	$45,771
9		$438,766,736,915	$459,608,156,918	19,556	$457,652,556,918	9,566,583	$47,839
10		$457,652,556,918	$479,391,053,372	21,238	$477,267,253,372	9,545,345	$50,000

Table 9.2--Fund Values for 5-Pay Special Endowment

The 9,707,590 30-year-old women each pay the \$7,067.94 premium resulting a fund of 9,707,590·\$7,067.94 = \$68,612,708,718 at the beginning of the first year. Earning at 4.75% for one year, the fund grows to \$68,612,708,718(1.0475) = \$71,871,812,382 at the end of the year at which time there are d_{30} = 13,105 death claims expected, for \$100,000 each. The fund at the end of the first year is therefore expected to be \$71,871,812,382 - \$100,000·13,105 = \$70,561,312,382. Now l_{31} = 9,694,485 new premiums of \$7,067.94 are added to the fund giving us a fund of \$70,561,312,382 + 9,694,485·\$7,067.94 = \$139,081,395,685 at the beginning of the second year.

The remaining calculations follow the same pattern except that premium income ceases after the fifth payment. Notice that at the end of the tenth year, when and if Martina is 40, the fund is just sufficient to pay her the endowment of \$50,000 at which time the fund will be depleted. The table thus shows that the premium income if allowed to earn at 4.75% annually, and pay claims as they come due would be exactly enough to pay all death claims occurring before age 40 as well as the endowment payments of \$50,000 for each survivor to age 40. ∎

It would be impractical if we always had to calculate reserves as we did in Example 9.2. Fortunately, there are better ways. There are two methods of determining reserves for life insurance and annuity products. They are analogous to the two methods we studied for determining outstanding balances on loans in Chapter 5. As in Chapter 5, these are called the *retrospective method* and the *prospective method*. If the policy goes into force at age x, the tth *retrospective reserve* when the policyholder is $x + t$ years old, is defined as follows:

$$RR_t = \left\{t^{\text{th}} \text{ retrospective reserve}\right\} = \begin{Bmatrix}\text{Accumulated value} \\ \text{at age } x+t \text{ of} \\ \text{past premiums}\end{Bmatrix} - \begin{Bmatrix}\text{Accumulated value} \\ \text{at age } x+t \text{ of past} \\ \text{benefits}\end{Bmatrix}$$

or, in abbreviated form,

$$RR_t = \left\{t^{\text{th}} \text{ retrospective reserve}\right\} = VPP_t - VPB_t. \tag{9.1}$$

The corresponding prospective reserve is

$$PR_t = \left\{t^{\text{th}} \text{ prospective reserve}\right\} = \begin{Bmatrix}\text{Value of future} \\ \text{benefits measured} \\ \text{at age } x+t\end{Bmatrix} - \begin{Bmatrix}\text{Value of future} \\ \text{premiums measured} \\ \text{at age } x+t\end{Bmatrix}$$

or, in abbreviated form,

$$PR_t = \{t^{\text{th}} \text{ prospective reserve}\} = VFB_t - VFP_t \qquad (9.2)$$

So,

VPP stands for *Value of Past Premiums*,

VPB stands for *Value of Past Benefits*,

VFB stands for *Value of Future Benefits*, and

VFP stands for *Value of Future Premiums*.

Formulas 7.4 and 8.3 are often useful in calculating the above reserves.

In the exercises, you will be asked to show that Equations 9.1 and 9.2 are equivalent (i.e., $RR_t = PR_t$) so we will write R_t for their common value.

Example 9.3: Using Formula 9.2, verify the calculations we carried out in Example 9.2 above. In particular, verify the last column of Table 9.2.

Solution: The five premiums for Martina's 10-year policy are all equal to \$7,067.94. For the second five-year period, no premiums are paid. The benefits are \$100,000 at the end of the year of death if death occurs before age 40 or \$50,000 if Martina survives to age 40. At the end of the year when Martina is 30 + t, and thinking prospectively, the value of her future benefits is

$$VFB_t = \$100{,}000\frac{M_{30+t} - M_{40}}{D_{30+t}} + \$50{,}000\frac{D_{40}}{D_{30+t}}.$$

We can calculate these for $1 \le t \le 10$. At age $x + t$, the value of her future premiums is

$$VFP_t = \begin{cases} \$7{,}067.94\dfrac{N_{30+t} - N_{35}}{D_{30+t}}, & 1 \le t \le 4 \\ 0, & 5 \le t \le 10. \end{cases}$$

I calculated the ten reserves using a spreadsheet program. Here they are in a table:

Policy Year	*t-th Reserve*
1	$7,278.50
2	$14,908.78
3	$22,908.84
4	$31,297.63
5	$40,093.28
6	$41,901.85
7	$43,793.27
8	$45,770.95
9	$47,838.66
10	$50,000.00

Table 9.3

Note that the values in Table 9.3 match those in the last column of Table 9.2. In the exercises, you will be asked to repeat this example using Formula 9.1. ∎

Example 9.4: At age 40, Dag bought a whole-life insurance with a death benefit of $150,000. What is the 10th reserve of Dag's policy assuming that he is to make equal annual payments for 20 years beginning at age 40 and that interest is 3.5% annually? Do this by both the retrospective and prospective methods.

Solution: First, we need to know Dag's net annual premiums, P. for those, we have

$$P\frac{N_{40}-N_{60}}{D_{40}} = \$150{,}000\frac{M_{40}}{D_{40}}.$$

These are applications of Formulas 7.4 and 8.3. Canceling the D_{40}'s and solving for P gives

$$P = \$150{,}000\frac{M_{40}}{N_{40}-N_{60}} = \$150{,}000\frac{790{,}757.21}{46{,}654{,}004.48 - 13{,}258{,}529.37} =$$

$$\$3{,}551.786016\ldots$$

From the retrospective point of view (Formula 9.1), the 10th reserve is

$$\$3{,}551.78601620\frac{N_{40}-N_{50}}{D_{50}} - \$150{,}000\frac{M_{40}-M_{50}}{D_{50}} =$$

$$\$44{,}386.8842 - \$7{,}888.2287 = \$36{,}498.6555\ldots$$

Notice that we can write the above reserve as

$$\left(\$3{,}551.78601620\frac{N_{40}-N_{50}}{D_{40}} - \$150{,}000\frac{M_{40}-M_{50}}{D_{40}}\right)\frac{1}{{}_{10}E_{40}}.$$

The expression in parentheses gives the difference between 10 years of future premiums and benefits when Dag is age 40. The multiplicative term, $1/{}_{10}E_{40}$, is the 10-year accumulation factor bringing the value of the difference up to age 50.

On the other hand, the prospective method (Formula 9.2) gives the 10th reserve as

$$\$150{,}000\frac{M_{50}}{D_{50}} - \$3{,}551.78601620\frac{N_{50}-N_{60}}{D_{50}}$$

$$\$65{,}991.1511482 - \$29{,}492.4956485 = \$36{,}498.6555,$$

exactly the same as with the retrospective method. ∎

Example 9.5: At age 60, Ray purchases a six-year temporary annuity paying him \$25,000 annually beginning at age 65. He is to pay for this annuity with five annual payments beginning at age 60. Assuming i = 4% and using Formula 9.1, construct a table showing the reserves for this product.

Solution: It is helpful to see a time diagram for Ray's annuity.

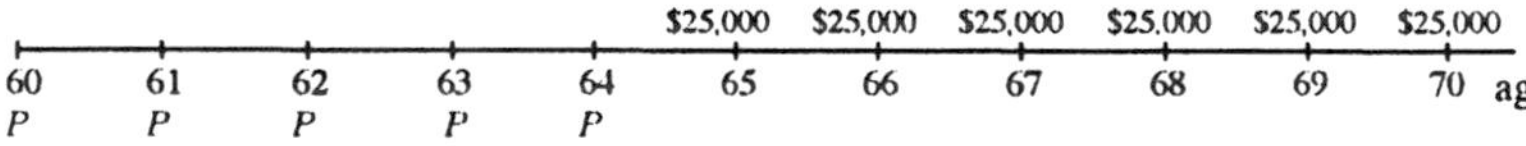

					\$25,000	\$25,000	\$25,000	\$25,000	\$25,000	\$25,000	
60	61	62	63	64	65	66	67	68	69	70	age
P	*P*	*P*	*P*	*P*							

Figure 9.3--Time Diagram Illustrating Example 9.5

First, we'll find P, Ray's equal annual payments. From Formula 7.4, we get

$$P\frac{N_{60}-N_{65}}{D_{60}} = \$25{,}000\frac{N_{65}-N_{71}}{D_{60}}.$$

The solution, from Table III, 4% is P = \$21,195.72479208333 ≈ \$21,195.72. By putting the focal date successively at 61, 62, . . . , 70 and looking backward in time, we see that the tth reserve is

$$R_t = \begin{cases} P\dfrac{N_{60}-N_{60+t}}{D_{60+t}}, \ 1 \le t \le 5 \\ P\dfrac{N_{60}-N_{65}}{D_{60+t}} - \$25{,}000\dfrac{N_{65}-N_{60+t}}{D_{60+t}}, \ 6 \le t \le 10. \end{cases}$$

I used a spreadsheet program to construct the following table of these values:

Policy Year	*t-th Reserve*
1	\$22,403.81
2	\$46,153.04
3	\$71,413.13
4	\$98,385.20
5	\$127,310.11
6	\$109,177.82
7	\$90,052.90
8	\$69,779.10
9	\$48,168.99
10	\$25,000.00

Table 9.4

Notice that the 10th policy year reserve is exactly \$25,000, and that when paid, the fund is depleted. ∎

Exercises 9.2

9.2.1 (🖳) Make a table similar to Table 9.2 showing how reserves accumulate for a five-year term policy paying \$100,000 that is sold to a 55-year-old man. Assume equal annual premiums and an annual interest rate of $i = 4.25\%$.

9.2.2 (🖳) Make a table similar to Table 9.2 showing how reserves accumulate for a five-year endowment policy paying \$100,000 that is sold to a 55-year-old man. Assume equal annual premiums and an annual interest rate of $i = 4.25\%$

9.2.3 (🖳) Verify the values in Table 9.3 using the retrospective method of Formula 9.1.

9.2.4 (🖳) Verify the values in Table 9.4 using the prospective method of Formula 9.2.

9.2.5 Determine the 15th reserve, retrospectively, for a $250,000 whole-life policy issued to a woman aged 30 if she is to make 25 equal annual payments beginning immediately, and annual interest is at $i = 6\%$ effective.

9.2.6 Repeat Exercise 9.2.5 prospectively.

9.2.7 Determine the 10th and 30th reserves retrospectively for a 20-year deferred life annuity of $75,000 (first payment at age 60) issued to a man aged 40 if his equal annual payments begin immediately and continue for 15 years, and annual interest is at 5.5%. Start with a time diagram.

9.2.8 Repeat Exercise 9.2.7 prospectively.

9.2.9 Show that the retrospective and prospective reserves for the tth year of a whole life insurance policy with death benefit B and annual premiums are equivalent.

9.2.10 Justify the claim that Formulas 9.1 and 9.2 are equivalent.

9.3 Gross Premiums

The net premiums we discussed in Chapters 7 and 8 were based on assumptions that the mortality tables and interest rates were correct and that the interest rates were constant. But, even if these assumptions are correct (and they may not be), net premiums are insufficient to meet the obligations of an insurance company because they do not allow for the expenses of running the company, nor do they provide for the earning of profits or paying dividends to policyholders or stockholders. In other words, an insurance company could not carry on its business if it collected only net premiums. In practice, therefore, the premiums actually charged by the company are not net premiums at all. Rather, they are *gross premiums*. Think of gross premiums as consisting of net premiums plus amounts called *loadings*. That is,

$$\textit{Gross Premium} = \textit{Net Premium} + \textit{Loading}. \tag{9.3}$$

The loadings consist of expenses in doing business, profit considerations, and other contingencies like allowing for the possibility that interest and/or mortality assumptions are incorrect. Since different companies have different methods of determining loadings, we will illustrate some of the basic ideas through examples.

A significant portion of the expenses involved in issuing an insurance policy are incurred at the time the policy is issued. Then, administrative costs may continue throughout the life of the policy. Finally, when the policy terminates either by death or maturity, there is often a cost associated with settlement. When all assumptions have been made, and expenses estimated, the gross premiums are calculated according to the following principle:

$$\begin{Bmatrix}\text{Present Value of}\\ \text{Gross Premiums}\end{Bmatrix} = \begin{Bmatrix}\text{Present Value}\\ \text{of Benefits}\end{Bmatrix} + \begin{Bmatrix}\text{Present Value}\\ \text{of Expenses}\end{Bmatrix} \tag{9.4}$$

Example 9.6: Håkan, who is 35, is buying a whole-life policy paying \$100,000 to his beneficiary at the end of the year of his death. He will pay premiums at the beginning of every year for 25 years. Assume expenses on the policy as follows: 65% of the first gross premium, 25% of the second, and 10% of the remaining gross premiums at the times they are paid. There is also an expense of \$1,250 to settle the policy. Find Håkan's *gross annual premium.* Assume an effective rate of i = 3.5%.

Solution: Here is a time diagram for the policy with focal date age 35.

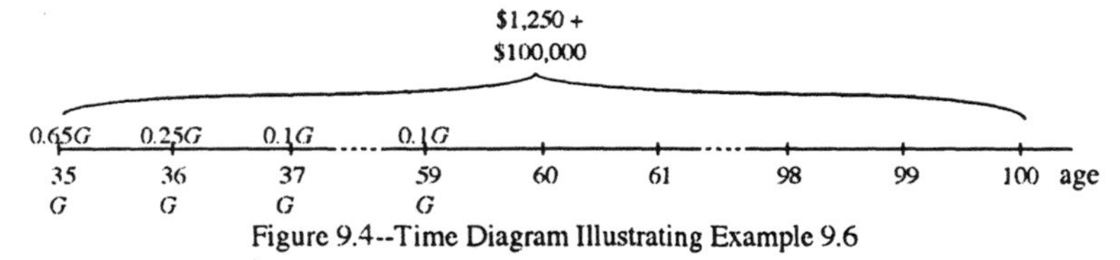

Figure 9.4--Time Diagram Illustrating Example 9.6

With the help of Formula 9.4 and the time diagram, we write the equation of value:

$$G\frac{N_{35}-N_{60}}{D_{35}} = \$100{,}000\frac{M_{35}}{D_{35}} + \$1{,}250\frac{M_{35}}{D_{35}} + 0.65\cdot G +$$

$$0.25\cdot G\cdot\frac{N_{36}-N_{37}}{D_{35}} + 0.1\cdot G\cdot\frac{N_{37}-N_{60}}{D_{35}}.$$

Solving this equation for G gives

$$G = \frac{\$101{,}250\cdot M_{35}}{N_{35}-N_{60}-0.65\cdot D_{35}-0.25\cdot(N_{36}-N_{37})-0.1\cdot(N_{37}-N_{60})} =$$

$$\$2{,}079.67.$$

For the sake of comparison, the net premium is

$$P = \$100{,}000 \cdot \frac{M_{35}}{N_{35} - N_{60}} = \$1{,}761.50. \qquad \blacksquare$$

Example 9.7: At 30 years of age, Elise buys a \$75,000 twenty-five-year endowment policy. The company estimates expenses for her policy as:

i) 60% of the first premium,

ii) 15% of the second premium,

iii) 4% of the third through twenty-fifth premium,

iv) \$800 miscellaneous expenses for the first year, \$300 annually thereafter to age 54,

v) \$100 settlement cost at the end of the year of death or at maturity,

vi) 7% tax on each gross premium, and

vii) 3.75% effective yield on invested income.

a) Find the gross premium for Elise's policy.

b) Find the net premium for Elise's policy.

Solution: *a)* Here is a time diagram with focal date at age 30:

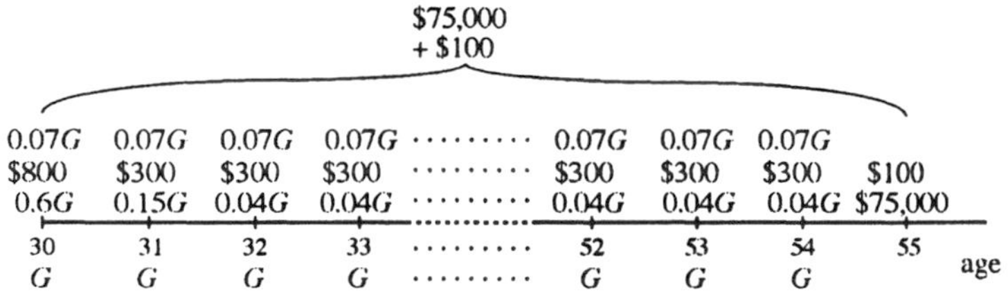

\$75,000
+ \$100

0.07*G*	0.07*G*	0.07*G*	0.07*G*	·········	0.07*G*	0.07*G*	0.07*G*	
\$800	\$300	\$300	\$300	·········	\$300	\$300	\$300	\$100
0.6*G*	0.15*G*	0.04*G*	0.04*G*	·········	0.04*G*	0.04*G*	0.04*G*	\$75,000
30	31	32	33	·········	52	53	54	55 age
G	*G*	*G*	*G*	·········	*G*	*G*	*G*	

Figure 9.5--Time Diagram 1 Illustrating Example 9.7

The quantities in the time diagram above can be rearranged to make writing the equation of value a little simpler as follows:

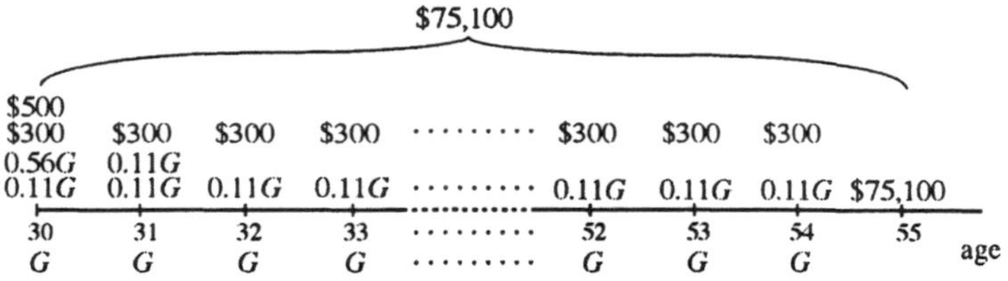

$75,100								
\$500								
\$300	\$300	\$300	\$300	·········	\$300	\$300	\$300	
$0.56G$	$0.11G$							
$0.11G$	$0.11G$	$0.11G$	$0.11G$	·········	$0.11G$	$0.11G$	$0.11G$	\$75,100
30	31	32	33	·········	52	53	54	55 age
G	G	G	G	·········	G	G	G	

Figure 9.6--Time Diagram 2 Illustrating Example 9.7

We can now write the equation of value for this policy as:

$$G\frac{N_{30}-N_{55}}{D_{30}}=\$75{,}100\frac{M_{30}-M_{55}+D_{55}}{D_{30}}+\$500+\$300\frac{N_{30}-N_{55}}{D_{30}}+$$

$$0.56G+0.11\frac{N_{31}-N_{32}}{D_{30}}+0.11\frac{N_{30}-N_{55}}{D_{30}}.$$

Solving this for G, and using Table III with $i=.0375$ gives

$$G=\frac{\$75{,}100(M_{30}-M_{55}+D_{55})+\$300(N_{30}-N_{55})+\$500D_{30}+0.11(N_{31}-N_{32})}{0.89(N_{30}-N_{55})-0.56D_{30}}$$

$$=\$2{,}603.33.$$

b) For comparison, the net single premium, P, for this policy satisfies the equation of value

$$P\frac{N_{30}-N_{55}}{D_{30}}=\$75{,}000\frac{M_{30}-M_{55}+D_{55}}{D_{30}},$$

so, using Table III with $i=.0375$,

$$P=\$75{,}000\frac{M_{30}-M_{55}+D_{55}}{N_{30}-N_{55}}=\$1{,}894.22.$$ ∎

Example 9.8: Algonquin buys a 5-year term insurance at age 40 paying a benefit of \$150,000 at the end of the year of his death if death occurs before age 45. He pays for this immediately with a single premium, G, which is to be returned to his beneficiary if there is a claim (This is an example of a ***return of premium*** policy.). Assuming interest at 4% and expenses at 1% of the insured amount for the first year (\$1,500) in addition to an annual administrative cost of 1% of the gross single premium for the entire term insured (five installments) find the gross single premium. How much of the gross premium is loading?

Solution: Let's begin with a time diagram for this problem:

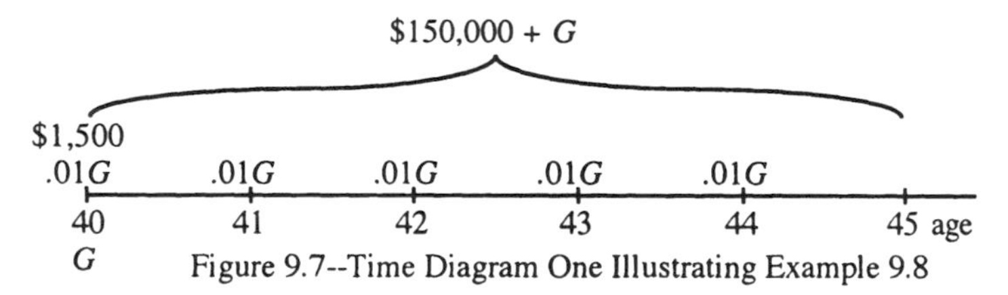

Figure 9.7--Time Diagram One Illustrating Example 9.8

Here is an equation of value for the transaction:

$$G = (\$150{,}000 + G)\frac{M_{40} - M_{45}}{D_{40}} + \$1{,}500 + 0.01G\frac{N_{40} - N_{45}}{D_{40}}.$$

The solution of this equation for G is

$$G = \frac{\$150{,}000(M_{40} - M_{45}) + \$1{,}500D_{40}}{D_{40} - (M_{40} - M_{45}) - 0.01(N_{40} - N_{45})} = \$4{,}117.34.$$

A time diagram and an equation of value for the net version of this return of premium policy follow:

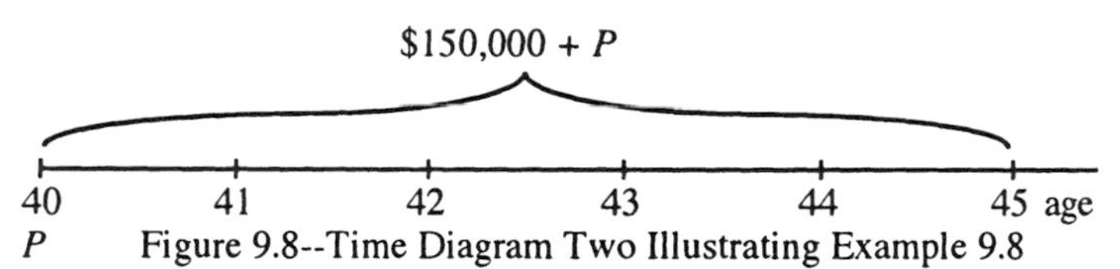

Figure 9.8--Time Diagram Two Illustrating Example 9.8

$$P = (\$150{,}000 + P)\frac{M_{40} - M_{45}}{D_{40}}.$$

From this, we find

$$P = \frac{\$150{,}000(M_{40} - M_{45})}{D_{40} - (M_{40} - M_{45})} = \$2{,}400.87.$$

The premium loading is therefore \$4,117.34 - \$2,400.87 = \$1,716.47. ▮

Example 9.9: Julie chooses a special five-year term insurance policy at age 55. Her beneficiary is to receive \$50,000 and the return of any gross premiums she pays, without interest, at the end of the year of her death if she dies before age 60. In addition, expenses include agent's commissions in the amount of 15% of the first gross premium and 4%

of the remaining four gross premiums and a \$500 charge at issue. Determine the amount of Julie's five gross annual premiums if $i = 3.5\%$.

Solution: In the time diagram below,

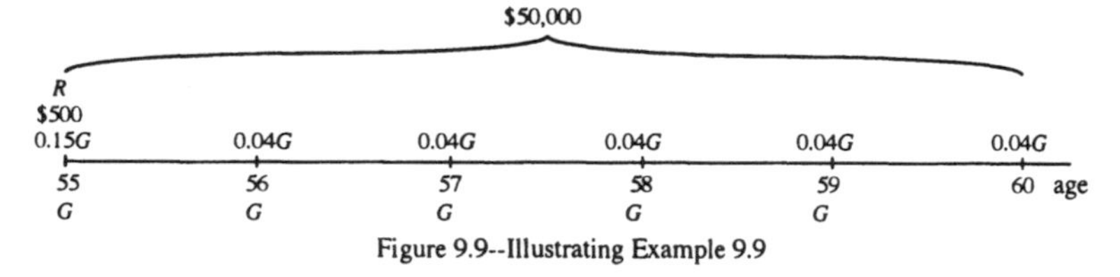

Figure 9.9--Illustrating Example 9.9

R denotes the present value of the return of premium portion of the benefit. Since the gross premiums are returned without interest, we can write R as

$$R = G\frac{M_{55} - M_{56}}{D_{55}} + 2G\frac{M_{56} - M_{57}}{D_{55}} + 3G\frac{M_{57} - M_{58}}{D_{55}} +$$

$$4G\frac{M_{58} - M_{59}}{D_{55}} + 5G\frac{M_{59} - M_{60}}{D_{55}} = G\frac{M_{55} + ... + M_{59} - 5M_{60}}{D_{55}}$$

$$= (0.10763060839)G.$$

That is, G is returned if death occurs between ages 55 and 56, 2G is returned if death occurs between ages 56 and 57, and so on. We can now write the equation of value for Julie's policy as

$$G\frac{N_{55} - N_{60}}{D_{55}} = \$50,000\frac{M_{55} - M_{60}}{D_{55}} + R + \$500 + 0.11G + 0.04G\frac{N_{55} - N_{60}}{D_{55}}.$$

Replacing R by $(0.10763060839)G$, we get

$$G = \frac{\$50,000(M_{55} - M_{60}) + \$500D_{55}}{0.96(N_{55} - N_{60}) - 0.21763060839D_{55}} = \$541.37. \qquad ■$$

Some insurance companies share profits with their policyholders in the form of dividends. In such cases, a simpler form for determining a gross premium, G, may be used because precision in choosing G is not as critical as otherwise. As long as G is sufficient to cover errors in estimating mortality, interest, and expenses, dividends may be suitably adjusted. This *alternative formula* may be written as

$$G = (P + c)(1 + k). \qquad (9.5)$$

In the formula, we first augment P by an amount c and then scale by the factor $1 + k$. The constants c and k will likely differ from one type of insurance to another.

Example 9.10: At age 45, Rosalie decided to buy a whole life insurance policy paying a benefit of \$100,000 plus the return of the gross single premium to her beneficiary at the end of the year of her death. Determine P and G using Formula 9.5 with $c = \$980$, $k = 5\%$, and interest at $i = 2.5\%$.

Solution: Since the net single premium, P, pays for a benefit of $\$100,000 + G$, we have

$$P = (\$100,000 + G)\frac{M_{45}}{D_{45}} = (\$100,000 + G)(0.01972807063) =$$

$$\$1,972.80706316 + (0.01972807063)G.$$

Formula 9.5 gives $G = (P + \$980)(1.05) = 1.05P + \$1,029$. We have two equations in two unknowns solving of which gives $P = \$2,035.27$ and $G = \$3,166.03$. ∎

Example 9.11: Lisa is 33 years old and planning an insurance and retirement strategy. She and her employer will contribute a percentage of her salary at the end of each year until her retirement at age 55. They estimate that the first combined contribution in one year should be \$7,200 with each succeeding contribution increasing by 3.163017032% until the final contribution at age 54. She is to have term insurance with a nominal premium of \$150,000 with annual increases of 3.667481663% in the event of her death before age 55. The amount of her first pension payment at age 55 is to be determined, but subsequent payments should increase by 3.414634146% until her death. The pension provider's expenses are estimated as follows:

a) a one time administrative charge of 50% of the first fund contribution,

b) a fixed charge of \$250 annually thereafter, for administrative costs, until Lisa's death,

c) 4% premium tax on all contributions to the fund, and

d) a settlement fee of 5% of the insured amount if there is a death benefit.

In addition, 6% effective is assumed earned on invested income. Find the amount of Lisa's first pension payment. If $7,200 is the limit on the first contribution, should Lisa reconsider the amount of her term insurance?

Solution: To cut down on notation, let G = $7,200, the amount of the first gross premium, let j = 3.163017032%, and k = 3.414634146%. Then we will have $i_1^* = (0.06 - 0.03163\ldots)/1.03163\ldots = 0.0275$ and $i_2^* = (0.06 - 0.03414\ldots)/1.03414\ldots = 0.025$ as inputs for Formula 7.5. Also, let l = 3.667481663%. Then, we will need $i_3^* = (0.06 - 0.03667\ldots)/1.03667\ldots = 0.0225$ as an input for Formula 8.8. Let's look at time diagram:

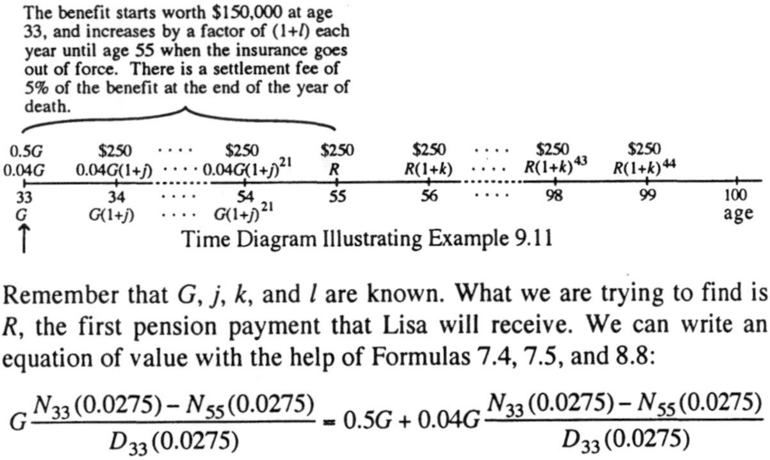

The benefit starts worth $150,000 at age 33, and increases by a factor of $(1+l)$ each year until age 55 when the insurance goes out of force. There is a settlement fee of 5% of the benefit at the end of the year of death.

0.5G	$250	....	$250	$250	$250	....	$250	$250	
0.04G	0.04$G(1+j)$	....	0.04$G(1+j)^{21}$	R	$R(1+k)$	....	$R(1+k)^{43}$	$R(1+k)^{44}$	
33	34	....	54	55	56	....	98	99	100
G	$G(1+j)$	....	$G(1+j)^{21}$						age

Time Diagram Illustrating Example 9.11

Remember that G, j, k, and l are known. What we are trying to find is R, the first pension payment that Lisa will receive. We can write an equation of value with the help of Formulas 7.4, 7.5, and 8.8:

$$G\frac{N_{33}(0.0275)-N_{55}(0.0275)}{D_{33}(0.0275)} = 0.5G + 0.04G\frac{N_{33}(0.0275)-N_{55}(0.0275)}{D_{33}(0.0275)}$$

$$+\ \$250\frac{N_{34}(0.06)}{D_{33}(0.06)} + \$150{,}000(1.05)\frac{M_{33}(0.0225)-M_{55}(0.0225)}{D_{33}(0.0225)}$$

$$+\ R\frac{N_{55}(0.025)}{D_{55}(0.025)}\frac{D_{55}(0.06)}{D_{33}(0.06)}.$$

The solution for R in terms of G = $7,200, $i = 0.06$, $i_1^* = 2.75\%$, $i_2^* = 2.5\%$, and $i_3^* = 2.25\%$ is

$$R = \frac{0.96G\dfrac{N_{33}(i_1^*)-N_{55}(i_1^*)}{D_{33}(i_1^*)} - 0.5G - \$250\dfrac{N_{34}(i)}{D_{33}(i)} - \$157{,}500\dfrac{M_{33}(i_3^*)-M_{55}(i_3^*)}{D_{33}(i_3^*)}}{\dfrac{N_{55}(i_2^*)D_{55}(i)}{D_{55}(i_2^*)D_{33}(i)}}$$

= \$20,367.95.

This is a pretty low first pension payment. Lisa should probably shift some of the benefit over to the pension payments. Remember, though, that later pension payments increase by 3.414634146% annually so that if Lisa is alive at 99, her payment will be \$89,239.13. ∎

Exercises 9.3

9.3.1 At age 37, Ellis buys a \$300,000 twenty-five year term policy for which he will make equal annual payments for twenty-five years. The company estimates expenses as follows: 50% of the first gross premium, 25% of the second, and 8% of the remaining premiums. There is also an expense of 1.5% of the insured amount for settling the policy if there is a claim. Assume an annual rate of $i = 3\%$. Find the net and gross annual premiums.

9.3.2 Dawn bought an \$85,000 whole life policy at age 25. She will pay equal annual premiums for 30 years. Interest is $i = 2.75\%$. The company's estimated expenses are:
- *i*) 55% of first gross premium for issuing and administrative expenses,
- *ii*) 7% of the remaining gross premiums for administrative expenses, and
- *iii*) \$250 at issue for miscellaneous expenses.

a) Make a time diagram, and find the gross annual premium.
b) Make a time diagram, and find the net annual premium.

9.3.3 Devean buys a ten-year term policy in the amount of \$1,000,000 at age 25. He is to make ten equal annual payments beginning immediately. The actuarial staff estimates expenses as follows:
- *i*) agent's commissions 40% of the first gross premium,
- *ii*) tax of 3.5% on each gross premium,
- *iii*) a settlement charge of 2.5% of the insured amount, and
- *iv*) an administrative charge of \$550 at issue.

Assuming interest at $i = 4\%$, calculate Devean's gross annual premiums.

9.3.4 Rigoberta buys a \$90,000 whole life policy at age 41 which returns her gross single premium to her beneficiary upon her death. Expenses are estimated as: 10% commission on the gross single premium plus a $450 settlement expense. If interest is at $i = 2.25\%$, find Rigoberta's gross single premium.

9.3.5 At 33, Yvonne will pay annual premiums for a $250,000 twenty-seven year endowment policy with expenses as follows: administrative expenses of $500 the first year, 20% commission on her first gross premium, 4% tax on all of her gross premiums, and a settlement charge of 1.5% of the benefit. Assuming annual interest at $i = 3.75\%$, find Yvonne's gross annual premiums.

9.3.6 Mauricio inherited $500,000 at age 30 and would like to use it to purchase a life annuity that begins when he reaches 50 years of age. His insurance company estimates expenses for this policy at 2% of the gross single premium for commission, an administrative fee of $450 at issue, and administrative expenses of $200 annually when the annuity begins at age 50. Assuming annual interest is calculated at $i = 5\%$, what will be the size of Mauricio's annuity payments?

9.3.7 Sophia buys a 15-year term insurance with a benefit of $175,000. Sophia is 45 and pays for the policy with a single premium. Expenses were estimated to be 1% of the insured amount at issue and 0.5% of the insured amount for each of the remaining 14 years. In addition, half of the gross single premium is to be returned to the beneficiary in the event of a claim. Assuming annual interest at $i = 4\%$, find Sophia's gross single premium.

9.3.8 Liam is 40 years old and is buying a 20-year term insurance with a benefit of $50,000 to be followed by a life annuity paying $40,000 annually, the first payment at age 60. Expenses are 10% of the first gross premium as commission, 1% of the amount insured for the 20 payments, and 1% of the annuity amount until the death of the annuitant. If annual interest is $i = 5.5\%$ and Liam will pay 20 equal premium payments beginning immediately, calculate his gross annual premiums.

9.3.9 Using Formula 9.5, derive a formula in terms of commutation functions, c and k, for the gross single premium for a whole life insurance purchased at age x, providing for the return of the premium at the end of the year of death, in addition to a death benefit of B.

9.3.10 Derive a formula in terms of commutation functions for the gross single premium paying for a whole life annuity of R to (x), first payment immediately. Estimated expenses include an annual charge of c and a one-time charge at issue of $p\%$ of the gross premium.

9.3.11 Based on his present salary and his projections for the future, Edsger is planning his retirement strategy. He and his employer will make annual contributions to his pension fund beginning now, and their contributions will increase annually by 2.926829268% until his the final contribution at age 59. His pension payments will begin when Edsger is 60, thirty-two years from now. His first pension payment will be $70,000 with subsequent payments increasing by 3.178484108% annually. Expenses in administering Edsger's pension fund are estimated as follows:

a) agent commission of 50% of the first gross premium,
b) 2% of all subsequent gross premiums,
c) 3.4% premium tax on all gross premiums, and
d) fixed charges of $200 per year until Edsger's death.

In addition, the company's invested income will earn a constant 5.5% effective. Find Edsger's gross annual premium as well as his net annual premium.

9.4 Summary of Formulas

---

t^{th} Retrospective Reserve $\qquad RR_t = VPP_t - VPB_t$

t^{th} Prospective Reserve $\qquad PR_t = VFB_t - VFP_t$

Gross Premium (definition)

$$Gross\ Premium = Net\ Premium + Loading$$

Gross Premium (principle)

$$\left\{\begin{matrix}\text{Present Value of}\\ \text{Gross Premiums}\end{matrix}\right\} = \left\{\begin{matrix}\text{Present Value}\\ \text{of Benefits}\end{matrix}\right\} + \left\{\begin{matrix}\text{Present Value}\\ \text{of Expenses}\end{matrix}\right\}$$

Gross Premium (Alternative Form) $G = (P + c)(1 + k)$

Tables[†]

[†]Tables I—IV are based on "Report of the Special Committee to Recommend New Mortality Tables for Valuation," in the *Transactions of the Society of Actuaries*, Volume 33 (1981), pages 617-669.

Table I 1980 Commissioners Standard Ordinary Mortality Table - Males

Age	Number Surviving	Number Dying	Probability of Dying	Expected Life	Age	Number Surviving	Number Dying	Probability of Dying	Expected Life
x	l_x	d_x	q_x	$AAD(x)$	x	l_x	d_x	q_x	$AAD(x)$
0	10,000,000	41,800	0.00418	70.83	50	8,966,618	60,166	0.00671	75.36
1	9,958,200	10,655	0.00107	71.13	51	8,906,452	65,017	0.00730	75.52
2	9,947,545	9,848	0.00099	71.20	52	8,841,435	70,378	0.00796	75.70
3	9,937,697	9,739	0.00098	71.27	53	8,771,057	76,396	0.00871	75.89
4	9,927,958	9,432	0.00095	71.34	54	8,694,661	83,121	0.00956	76.08
5	9,918,526	8,927	0.00090	71.40	55	8,611,540	90,163	0.01047	76.29
6	9,909,599	8,522	0.00086	71.46	56	8,521,377	97,655	0.01146	76.51
7	9,901,077	7,921	0.00080	71.52	57	8,423,722	105,212	0.01249	76.74
8	9,893,156	7,519	0.00076	71.57	58	8,318,510	113,049	0.01359	76.99
9	9,885,637	7,315	0.00074	71.62	59	8,205,461	121,195	0.01477	77.24
10	9,878,322	7,211	0.00073	71.66	60	8,084,266	129,995	0.01608	77.51
11	9,871,111	7,601	0.00077	71.71	61	7,954,271	139,518	0.01754	77.79
12	9,863,510	8,384	0.00085	71.75	62	7,814,753	149,965	0.01919	78.08
13	9,855,126	9,757	0.00099	71.80	63	7,664,788	161,420	0.02106	78.38
14	9,845,369	11,322	0.00115	71.86	64	7,503,368	173,628	0.02314	78.70
15	9,834,047	13,079	0.00133	71.93	65	7,329,740	186,322	0.02542	79.04
16	9,820,968	14,830	0.00151	72.00	66	7,143,418	198,944	0.02785	79.39
17	9,806,138	16,376	0.00167	72.09	67	6,944,474	211,390	0.03044	79.76
18	9,789,762	17,426	0.00178	72.18	68	6,733,084	223,471	0.03319	80.14
19	9,772,336	18,177	0.00186	72.27	69	6,509,613	235,453	0.03617	80.54
20	9,754,159	18,533	0.00190	72.37	70	6,274,160	247,892	0.03951	80.96
21	9,735,626	18,595	0.00191	72.47	71	6,026,268	260,937	0.04330	81.39
22	9,717,031	18,365	0.00189	72.57	72	5,765,331	274,718	0.04765	81.84
23	9,698,666	18,040	0.00186	72.66	73	5,490,613	289,026	0.05264	82.30
24	9,680,626	17,619	0.00182	72.75	74	5,201,587	302,680	0.05819	82.79
25	9,663,007	17,104	0.00177	72.84	75	4,898,907	314,461	0.06419	83.31
26	9,645,903	16,687	0.00173	72.93	76	4,584,446	323,341	0.07053	83.84
27	9,629,216	16,466	0.00171	73.01	77	4,261,105	328,616	0.07712	84.40
28	9,612,750	16,342	0.00170	73.09	78	3,932,489	329,936	0.08390	84.97
29	9,596,408	16,410	0.00171	73.16	79	3,602,553	328,012	0.09105	85.57
30	9,579,998	16,573	0.00173	73.24	80	3,274,541	323,656	0.09884	86.18
31	9,563,425	17,023	0.00178	73.31	81	2,950,885	317,161	0.10748	86.80
32	9,546,402	17,470	0.00183	73.38	82	2,633,724	308,804	0.11725	87.44
33	9,528,932	18,200	0.00191	73.46	83	2,324,920	298,194	0.12826	88.09
34	9,510,732	19,021	0.00200	73.54	84	2,026,726	284,248	0.14025	88.77
35	9,491,711	20,028	0.00211	73.61	85	1,742,478	266,512	0.15295	89.46
36	9,471,683	21,217	0.00224	73.69	86	1,475,966	245,143	0.16609	90.18
37	9,450,466	22,681	0.00240	73.78	87	1,230,823	220,994	0.17955	90.91
38	9,427,785	24,324	0.00258	73.87	88	1,009,829	195,170	0.19327	91.66
39	9,403,461	26,236	0.00279	73.96	89	814,659	168,871	0.20729	92.41
40	9,377,225	28,319	0.00302	74.05	90	645,788	143,216	0.22177	93.18
41	9,348,906	30,758	0.00329	74.16	91	502,572	119,100	0.23698	93.94
42	9,318,148	33,173	0.00356	74.26	92	383,472	97,191	0.25345	94.70
43	9,284,975	35,933	0.00387	74.38	93	286,281	77,900	0.27211	95.44
44	9,249,042	38,753	0.00419	74.50	94	208,381	61,660	0.29590	96.17
45	9,210,289	41,907	0.00455	74.62	95	146,721	48,412	0.32996	96.87
46	9,168,382	45,108	0.00492	74.76	96	98,309	37,805	0.38455	97.54
47	9,123,274	48,536	0.00532	74.90	97	60,504	29,054	0.48020	98.20
48	9,074,738	52,089	0.00574	75.04	98	31,450	20,693	0.65798	98.84
49	9,022,649	56,031	0.00621	75.20	99	10,757	10,757	1.00000	99.50

Table II — 1980 Commissioners Standard Ordinary Mortality Table - Females

Age	Number Surviving	Number Dying	Probability of Dying	Expected Life	Age	Number Surviving	Number Dying	Probability of Dying	Expected Life
x	l_x	d_x	q_x	$AAD(x)$	x	l_x	d_x	q_x	$AAD(x)$
0	10,000,000	28,900	0.00289	75.83	50	9,219,130	45,727	0.00496	79.53
1	9,971,100	8,675	0.00087	76.04	51	9,173,403	48,711	0.00531	79.67
2	9,962,425	8,070	0.00081	76.11	52	9,124,692	52,011	0.00570	79.82
3	9,954,355	7,864	0.00079	76.17	53	9,072,681	55,797	0.00615	79.98
4	9,946,491	7,659	0.00077	76.23	54	9,016,884	59,602	0.00661	80.14
5	9,938,832	7,554	0.00076	76.28	55	8,957,282	63,507	0.00709	80.31
6	9,931,278	7,250	0.00073	76.34	56	8,893,775	67,326	0.00757	80.49
7	9,924,028	7,145	0.00072	76.39	57	8,826,449	70,876	0.00803	80.67
8	9,916,883	6,942	0.00070	76.44	58	8,755,573	74,160	0.00847	80.86
9	9,909,941	6,838	0.00069	76.48	59	8,681,413	77,612	0.00894	81.05
10	9,903,103	6,734	0.00068	76.53	60	8,603,801	81,478	0.00947	81.25
11	9,896,369	6,828	0.00069	76.58	61	8,522,323	86,331	0.01013	81.44
12	9,889,541	7,120	0.00072	76.62	62	8,435,992	92,458	0.01096	81.65
13	9,882,421	7,412	0.00075	76.67	63	8,343,534	100,289	0.01202	81.86
14	9,875,009	7,900	0.00080	76.71	64	8,243,245	109,223	0.01325	82.08
15	9,867,109	8,387	0.00085	76.76	65	8,134,022	118,675	0.01459	82.32
16	9,858,722	8,873	0.00090	76.82	66	8,015,347	128,246	0.01600	82.57
17	9,849,849	9,357	0.00095	76.87	67	7,887,101	137,472	0.01743	82.83
18	9,840,492	9,644	0.00098	76.93	68	7,749,629	146,003	0.01884	83.10
19	9,830,848	10,027	0.00102	76.98	69	7,603,626	154,810	0.02036	83.38
20	9,820,821	10,312	0.00105	77.04	70	7,448,816	164,693	0.02211	83.67
21	9,810,509	10,497	0.00107	77.10	71	7,284,123	176,494	0.02423	83.97
22	9,800,012	10,682	0.00109	77.16	72	7,107,629	190,982	0.02687	84.28
23	9,789,330	10,866	0.00111	77.22	73	6,916,647	208,260	0.03011	84.60
24	9,778,464	11,147	0.00114	77.28	74	6,708,387	227,616	0.03393	84.95
25	9,767,317	11,330	0.00116	77.34	75	6,480,771	247,825	0.03824	85.32
26	9,755,987	11,610	0.00119	77.40	76	6,232,946	267,830	0.04297	85.71
27	9,744,377	11,888	0.00122	77.46	77	5,965,116	286,564	0.04804	86.12
28	9,732,489	12,263	0.00126	77.52	78	5,678,552	303,519	0.05345	86.55
29	9,720,226	12,636	0.00130	77.59	79	5,375,033	319,008	0.05935	87.01
30	9,707,590	13,105	0.00135	77.65	80	5,056,025	333,647	0.06599	87.48
31	9,694,485	13,572	0.00140	77.71	81	4,722,378	347,567	0.07360	87.98
32	9,680,913	14,037	0.00145	77.78	82	4,374,811	360,484	0.08240	88.49
33	9,666,876	14,500	0.00150	77.84	83	4,014,327	371,446	0.09253	89.03
34	9,652,376	15,251	0.00158	77.91	84	3,642,881	378,167	0.10381	89.59
35	9,637,125	15,901	0.00165	77.98	85	3,264,714	379,033	0.11610	90.18
36	9,621,224	16,933	0.00176	78.05	86	2,885,681	373,090	0.12929	90.80
37	9,604,291	18,152	0.00189	78.12	87	2,512,591	360,105	0.14332	91.43
38	9,586,139	19,556	0.00204	78.20	88	2,152,486	340,480	0.15818	92.09
39	9,566,583	21,238	0.00222	78.28	89	1,812,006	315,180	0.17394	92.77
40	9,545,345	23,100	0.00242	78.36	90	1,496,826	285,520	0.19075	93.45
41	9,522,245	25,139	0.00264	78.46	91	1,211,306	253,005	0.20887	94.15
42	9,497,106	27,257	0.00287	78.55	92	958,301	219,269	0.22881	94.85
43	9,469,849	29,262	0.00309	78.66	93	739,032	185,874	0.25151	95.55
44	9,440,587	31,343	0.00332	78.77	94	553,158	154,503	0.27931	96.24
45	9,409,244	33,497	0.00356	78.88	95	398,655	126,501	0.31732	96.91
46	9,375,747	35,628	0.00380	79.00	96	272,154	102,259	0.37574	97.56
47	9,340,119	37,827	0.00405	79.12	97	169,895	80,695	0.47497	98.21
48	9,302,292	40,279	0.00433	79.25	98	89,200	58,502	0.65585	98.84
49	9,262,013	42,883	0.00463	79.39	99	30,698	30,698	1.00000	99.50

Table III Commutation Columns at 2% based on 1980 CSO MortalityTable-Males

Age	D_x	N_x	M_x	Age	D_x	N_x	M_x
0	10,000,000.00	376,668,378.37	2,614,345.52	50	3,331,348.60	65,769,040.84	2,041,759.56
1	9,762,941.18	366,668,378.37	2,573,365.13	51	3,244,112.99	62,437,692.24	2,019,844.51
2	9,561,269.70	356,905,437.20	2,563,123.88	52	3,157,285.29	59,193,579.25	1,996,626.88
3	9,364,513.84	347,344,167.49	2,553,843.89	53	3,070,738.47	56,036,293.96	1,971,987.61
4	9,171,898.59	337,979,653.65	2,544,846.56	54	2,984,306.18	52,965,555.49	1,945,765.88
5	8,983,514.59	328,807,755.07	2,536,303.70	55	2,897,819.81	49,981,249.30	1,917,795.31
6	8,799,440.32	319,824,240.48	2,528,376.78	56	2,811,254.48	47,083,429.50	1,888,049.98
7	8,619,483.36	311,024,800.16	2,520,957.87	57	2,724,546.57	44,272,175.01	1,856,464.71
8	8,443,713.39	302,405,316.80	2,514,197.37	58	2,637,761.84	41,547,628.45	1,823,102.46
9	8,271,858.82	293,961,603.41	2,507,905.81	59	2,550,896.58	38,909,866.61	1,787,958.02
10	8,103,664.66	285,689,744.59	2,501,904.96	60	2,463,940.91	36,358,970.03	1,751,019.93
11	7,938,969.73	277,586,079.94	2,496,105.42	61	2,376,785.04	33,895,029.11	1,712,176.63
12	7,777,310.32	269,647,110.20	2,490,112.08	62	2,289,310.01	31,518,244.07	1,671,305.22
13	7,618,332.94	261,869,799.88	2,483,630.98	63	2,201,351.16	29,228,934.06	1,628,234.80
14	7,461,559.27	254,251,466.95	2,476,236.39	64	2,112,736.10	27,027,582.91	1,582,783.50
15	7,306,841.78	246,789,907.67	2,467,823.98	65	2,023,379.78	24,914,846.80	1,534,853.37
16	7,154,043.03	239,483,065.90	2,458,296.64	66	1,933,279.86	22,891,467.02	1,484,427.57
17	7,003,176.64	232,329,022.87	2,447,705.61	67	1,842,586.34	20,958,187.16	1,431,641.50
18	6,854,393.64	225,325,846.23	2,436,239.80	68	1,751,468.59	19,115,600.82	1,376,652.89
19	6,708,032.03	218,471,452.58	2,424,278.05	69	1,660,134.67	17,364,132.23	1,319,661.49
20	6,564,269.39	211,763,420.56	2,412,045.45	70	1,568,713.26	15,703,997.56	1,260,791.73
21	6,423,330.60	205,199,151.17	2,399,817.83	71	1,477,189.62	14,135,284.31	1,200,027.18
22	6,285,354.97	198,775,820.58	2,387,789.86	72	1,385,517.07	12,658,094.69	1,137,319.13
23	6,150,466.44	192,490,465.61	2,376,143.58	73	1,293,624.69	11,272,577.62	1,072,593.75
24	6,018,653.20	186,339,999.17	2,364,927.73	74	1,201,498.29	9,978,952.93	1,005,832.54
25	5,889,901.07	180,321,345.97	2,354,188.40	75	1,109,395.28	8,777,454.65	937,288.32
26	5,764,191.82	174,431,444.91	2,343,967.41	76	1,017,826.63	7,668,059.37	867,472.52
27	5,641,392.17	168,667,253.09	2,334,191.13	77	927,489.52	6,650,232.74	797,092.80
28	5,521,318.98	163,025,860.92	2,324,733.48	78	839,178.06	5,722,743.22	726,967.41
29	5,403,855.45	157,504,541.94	2,315,531.09	79	753,697.04	4,883,565.17	657,940.86
30	5,288,838.01	152,100,686.49	2,306,471.61	80	671,640.21	4,129,868.13	590,662.41
31	5,176,165.24	146,811,848.48	2,297,501.54	81	593,387.47	3,458,227.91	525,579.08
32	5,065,638.83	141,635,683.24	2,288,468.57	82	519,225.70	2,864,840.44	463,052.35
33	4,957,224.18	136,570,044.41	2,279,380.17	83	449,359.32	2,345,614.75	403,366.88
34	4,850,741.20	131,612,820.23	2,270,097.66	84	384,043.67	1,896,255.42	346,862.19
35	4,746,117.60	126,762,079.04	2,260,586.64	85	323,707.46	1,512,211.75	294,056.25
36	4,643,238.28	122,015,961.44	2,250,768.45	86	268,820.00	1,188,504.30	245,516.00
37	4,541,997.27	117,372,723.16	2,240,571.33	87	219,776.20	919,684.29	201,743.17
38	4,442,251.50	112,830,725.89	2,229,884.33	88	176,779.83	699,908.09	163,056.15
39	4,343,912.10	108,388,474.39	2,218,647.90	89	139,817.19	523,128.26	129,559.77
40	4,246,855.32	104,044,562.28	2,206,765.87	90	108,661.20	383,311.07	101,145.30
41	4,151,009.73	99,797,706.96	2,194,191.94	91	82,905.37	274,649.86	77,520.08
42	4,056,228.29	95,646,697.23	2,180,802.86	92	62,018.01	191,744.49	58,258.32
43	3,962,537.21	91,590,468.94	2,166,645.66	93	45,391.71	129,726.48	42,848.05
44	3,869,806.00	87,627,931.73	2,151,611.26	94	32,392.31	84,334.77	30,738.69
45	3,778,031.10	83,758,125.73	2,135,714.91	95	22,360.21	51,942.46	21,341.73
46	3,687,099.00	79,980,094.63	2,118,861.85	96	14,688.48	29,582.24	14,108.43
47	3,597,018.28	76,292,995.63	2,101,077.19	97	8,862.73	14,893.76	8,570.69
48	3,507,727.52	72,695,977.35	2,082,316.20	98	4,516.52	6,031.04	4,398.26
49	3,419,208.99	69,188,249.82	2,062,576.64	99	1,514.52	1,514.52	1,484.82

Table III	Commutation Columns at 2.25% based on 1980 CSO Mortality Table-Males						
Age	D_x	N_x	M_x	*Age*	D_x	N_x	M_x
0	10,000,000.00	352,574,339.21	2,241,640.46	50	2,947,561.22	56,429,128.37	1,705,844.46
1	9,739,070.90	342,574,339.21	2,200,760.26	51	2,863,357.54	53,481,567.15	1,686,501.54
2	9,514,572.49	332,835,268.30	2,190,569.03	52	2,779,907.15	50,618,209.61	1,666,059.02
3	9,295,993.28	323,320,695.82	2,181,356.94	53	2,697,094.41	47,838,302.46	1,644,417.83
4	9,082,526.31	314,024,702.54	2,172,447.28	54	2,614,770.36	45,141,208.04	1,621,443.04
5	8,874,227.39	304,942,176.23	2,164,008.35	55	2,532,785.47	42,526,437.68	1,596,995.89
6	8,671,139.65	296,067,948.84	2,156,197.01	56	2,451,117.02	39,993,652.22	1,571,061.10
7	8,473,039.31	287,396,809.19	2,148,904.14	57	2,369,708.77	37,542,535.20	1,543,589.41
8	8,279,961.62	278,923,769.88	2,142,274.75	58	2,288,617.30	35,172,826.43	1,514,643.12
9	8,091,607.51	270,643,808.26	2,136,120.29	59	2,207,838.50	32,884,209.13	1,484,225.09
10	7,907,696.85	262,552,200.75	2,130,264.56	60	2,127,362.96	30,676,370.64	1,452,332.80
11	7,728,043.39	254,644,503.90	2,124,619.10	61	2,047,095.32	28,549,007.67	1,418,877.55
12	7,552,168.81	246,916,460.50	2,118,799.27	62	1,966,933.25	26,501,912.35	1,383,761.58
13	7,379,706.07	239,364,291.69	2,112,521.17	63	1,886,736.26	24,534,979.10	1,346,846.75
14	7,210,170.99	231,984,585.62	2,105,375.71	64	1,806,358.63	22,648,242.84	1,307,986.54
15	7,043,402.86	224,774,414.63	2,097,266.60	65	1,725,730.54	20,841,884.21	1,267,107.17
16	6,879,252.16	217,731,011.77	2,088,105.21	66	1,644,853.27	19,116,153.66	1,224,204.41
17	6,717,715.65	210,851,759.60	2,077,945.88	67	1,563,857.36	17,471,300.39	1,179,403.32
18	6,558,921.51	204,134,043.95	2,066,974.33	68	1,482,888.50	15,907,443.03	1,132,846.97
19	6,403,175.04	197,575,122.44	2,055,556.21	69	1,402,123.67	14,424,554.53	1,084,712.93
20	6,250,625.75	191,171,947.41	2,043,908.09	70	1,321,671.19	13,022,430.86	1,035,114.03
21	6,101,466.51	184,921,321.65	2,032,293.17	71	1,241,517.82	11,700,759.68	984,043.89
22	5,955,807.07	178,819,855.15	2,020,895.84	72	1,161,623.65	10,459,241.86	931,469.18
23	5,813,741.53	172,864,048.07	2,009,887.17	73	1,081,928.89	9,297,618.21	877,335.82
24	5,675,234.90	167,050,306.54	1,999,311.28	74	1,002,421.64	8,215,689.32	821,636.30
25	5,540,250.19	161,375,071.64	1,989,209.49	75	923,316.18	7,213,267.68	764,589.01
26	5,408,746.87	155,834,821.45	1,979,618.77	76	845,035.19	6,289,951.51	706,625.50
27	5,280,576.98	150,426,074.59	1,970,467.76	77	768,151.44	5,444,916.32	648,336.66
28	5,155,547.36	145,145,497.60	1,961,636.65	78	693,312.16	4,676,764.87	590,400.46
29	5,033,528.37	139,989,950.24	1,953,064.91	79	621,166.98	3,983,452.72	533,511.54
30	4,914,348.13	134,956,421.88	1,944,646.91	80	552,185.63	3,362,285.74	478,198.90
31	4,797,893.89	130,042,073.75	1,936,332.37	81	486,657.74	2,810,100.11	424,821.79
32	4,683,964.39	125,244,179.86	1,927,979.99	82	424,793.92	2,323,442.37	373,666.83
33	4,572,511.19	120,560,215.47	1,919,596.92	83	366,735.31	1,898,648.45	324,955.76
34	4,463,352.39	115,987,704.27	1,911,055.72	84	312,662.96	1,531,913.14	278,953.38
35	4,356,406.75	111,524,351.88	1,902,325.66	85	262,896.85	1,219,250.18	236,067.39
36	4,251,554.53	107,167,945.12	1,893,335.69	86	217,786.58	956,353.32	196,742.13
37	4,148,685.44	102,916,390.59	1,884,021.59	87	177,618.03	738,566.74	161,365.95
38	4,047,656.38	98,767,705.15	1,874,283.89	88	142,520.05	560,948.71	130,176.44
39	3,948,374.85	94,720,048.78	1,864,070.60	89	112,445.14	418,428.66	103,237.66
40	3,850,717.60	90,771,673.92	1,853,296.90	90	87,174.90	305,983.52	80,441.77
41	3,754,609.80	86,920,956.33	1,841,923.72	91	66,349.32	218,808.62	61,534.46
42	3,659,909.14	83,166,346.53	1,829,842.83	92	49,511.78	152,459.30	46,156.93
43	3,566,630.52	79,506,437.39	1,817,100.11	93	36,149.65	102,947.52	33,884.30
44	3,474,648.02	75,939,806.87	1,803,600.93	94	25,733.94	66,797.88	24,264.06
45	3,383,950.54	72,465,158.85	1,789,362.70	95	17,720.55	41,063.93	16,816.94
46	3,294,428.85	69,081,208.31	1,774,304.46	96	11,612.21	23,343.38	11,098.54
47	3,206,083.54	65,786,779.46	1,758,452.70	97	6,989.44	11,731.17	6,731.30
48	3,118,852.92	62,580,695.92	1,741,771.59	98	3,553.17	4,741.73	3,448.83
49	3,032,714.62	59,461,843.00	1,724,263.31	99	1,188.56	1,188.56	1,162.41

Table III Commutation Columns at 2.5% based on 1980 CSO MortalityTable-Males

Age	D_x	N_x	M_x	Age	D_x	N_x	M_x
0	10,000,000.00	330,909,957.77	1,929,025.42	50	2,608,767.64	48,459,688.29	1,426,824.02
1	9,715,317.07	320,909,957.77	1,888,244.93	51	2,528,061.28	45,850,920.65	1,409,746.14
2	9,468,216.54	311,194,640.70	1,878,103.35	52	2,448,396.55	43,322,859.37	1,391,741.44
3	9,228,139.58	301,726,424.16	1,868,958.50	53	2,369,665.62	40,874,462.83	1,372,727.50
4	8,994,239.95	292,498,284.58	1,860,135.45	54	2,291,732.49	38,504,797.21	1,352,591.10
5	8,766,531.73	283,504,044.63	1,851,798.94	55	2,214,461.97	36,213,064.71	1,331,216.49
6	8,545,016.16	274,737,512.90	1,844,101.21	56	2,137,830.74	33,998,602.74	1,308,596.53
7	8,329,431.87	266,192,496.74	1,836,931.95	57	2,061,786.53	31,860,772.00	1,284,694.53
8	8,119,773.86	257,863,064.87	1,830,430.82	58	1,986,375.50	29,798,985.47	1,259,570.98
9	7,915,709.92	249,743,291.01	1,824,410.14	59	1,911,590.78	27,812,609.97	1,233,234.44
10	7,716,929.36	241,827,581.09	1,818,695.67	60	1,837,420.98	25,901,019.19	1,205,688.81
11	7,523,215.74	234,110,651.73	1,813,199.85	61	1,763,780.73	24,063,598.20	1,176,863.70
12	7,334,070.91	226,587,435.99	1,807,548.08	62	1,690,579.51	22,299,817.47	1,146,681.53
13	7,149,109.21	219,253,365.08	1,801,466.15	63	1,617,694.94	20,609,237.96	1,115,030.60
14	6,967,835.39	212,104,255.87	1,794,560.86	64	1,545,001.34	18,991,543.02	1,081,792.98
15	6,790,070.74	205,136,420.48	1,786,743.41	65	1,472,439.02	17,446,541.67	1,046,913.62
16	6,615,648.92	198,346,349.74	1,777,933.07	66	1,400,009.39	15,974,102.65	1,010,397.13
17	6,444,545.42	191,730,700.83	1,768,186.87	67	1,327,823.57	14,574,093.26	972,357.89
18	6,276,861.65	185,286,155.40	1,757,687.13	68	1,256,004.47	13,246,269.68	932,924.72
19	6,112,867.02	179,009,293.75	1,746,786.69	69	1,184,700.19	11,990,265.21	892,254.70
20	5,952,679.81	172,896,426.73	1,735,693.79	70	1,113,999.54	10,805,565.02	850,449.18
21	5,796,458.20	166,943,746.92	1,724,659.50	71	1,043,888.23	9,691,565.48	807,508.58
22	5,644,280.00	161,147,288.71	1,713,858.32	72	974,329.69	8,647,677.25	763,410.74
23	5,496,207.23	155,503,008.72	1,703,450.92	73	905,271.11	7,673,347.56	718,116.29
24	5,352,179.53	150,006,801.48	1,693,477.05	74	836,700.11	6,768,076.45	671,625.08
25	5,212,135.04	144,654,621.95	1,683,973.53	75	768,792.77	5,931,376.34	624,125.05
26	5,076,009.08	139,442,486.91	1,674,972.81	76	701,896.52	5,162,583.57	575,979.85
27	4,943,636.88	134,366,477.83	1,666,405.71	77	636,479.76	4,460,687.04	527,682.52
28	4,814,812.91	129,422,840.96	1,658,158.26	78	573,067.81	3,824,207.28	479,794.46
29	4,689,392.75	124,608,028.04	1,650,172.55	79	512,182.83	3,251,139.47	432,886.74
30	4,567,193.97	119,918,635.29	1,642,349.21	80	454,193.80	2,738,956.64	387,389.98
31	4,448,090.65	115,351,441.32	1,634,640.86	81	399,318.28	2,284,762.84	343,592.35
32	4,331,876.10	110,903,350.67	1,626,916.32	82	347,706.89	1,885,444.57	301,720.44
33	4,218,486.56	106,571,474.58	1,619,182.30	83	299,451.98	1,537,737.67	261,946.18
34	4,107,735.97	102,352,988.02	1,611,321.63	84	254,677.36	1,238,285.70	224,475.27
35	3,999,532.39	98,245,252.05	1,603,306.73	85	213,618.44	983,608.33	189,628.00
36	3,893,749.43	94,245,719.66	1,595,073.34	86	176,532.20	769,989.89	157,751.96
37	3,790,270.49	90,351,970.23	1,586,563.90	87	143,621.45	593,457.69	129,146.87
38	3,688,950.14	86,561,699.74	1,577,689.17	88	114,960.25	449,836.24	103,988.63
39	3,589,690.27	82,872,749.60	1,568,403.69	89	90,479.84	334,875.99	82,312.14
40	3,492,365.76	79,283,059.34	1,558,632.60	90	69,974.87	244,396.15	64,013.99
41	3,396,896.48	75,790,693.58	1,548,342.98	91	53,128.38	174,421.28	48,874.20
42	3,303,142.10	72,393,797.10	1,537,439.73	92	39,549.23	121,292.90	36,590.87
43	3,211,105.15	69,090,655.00	1,525,967.22	93	28,805.34	81,743.67	26,811.59
44	3,120,661.58	65,879,549.85	1,513,843.29	94	20,455.72	52,938.33	19,164.54
45	3,031,791.39	62,758,888.27	1,501,086.80	95	14,051.58	32,482.61	13,259.32
46	2,944,387.00	59,727,096.88	1,487,628.54	96	9,185.49	18,431.03	8,735.95
47	2,858,439.76	56,782,709.88	1,473,495.62	97	5,515.30	9,245.55	5,289.80
48	2,773,885.67	53,924,270.11	1,458,659.57	98	2,796.93	3,730.25	2,705.95
49	2,690,696.16	51,150,384.45	1,443,125.81	99	933.32	933.32	910.55

Table III Commutation Columns at 2.75% based on 1980 CSO Mortality Table - Males

Age	D_x	N_x	M_x	Age	D_x	N_x	M_x
0	10,000,000.00	311,377,633.01	1,666,292.06	50	2,309,601.97	41,653,062.53	1,194,799.81
1	9,691,678.83	301,377,633.01	1,625,610.80	51	2,232,705.15	39,343,460.56	1,179,717.15
2	9,422,198.54	291,685,954.18	1,615,518.50	52	2,157,086.55	37,110,755.42	1,163,854.65
3	9,160,944.65	282,263,755.64	1,606,440.25	53	2,082,643.40	34,953,668.87	1,147,143.75
4	8,907,023.72	273,102,810.99	1,597,702.75	54	2,009,249.20	32,871,025.47	1,129,489.39
5	8,660,400.64	264,195,787.26	1,589,467.16	55	1,936,779.34	30,861,776.27	1,110,795.06
6	8,421,027.73	255,535,386.62	1,581,881.13	56	1,865,208.00	28,924,996.93	1,091,059.66
7	8,188,599.38	247,114,358.89	1,574,833.08	57	1,794,484.39	27,059,788.94	1,070,256.46
8	7,963,064.12	238,925,759.51	1,568,457.42	58	1,724,643.64	25,265,304.55	1,048,443.28
9	7,744,050.64	230,962,695.39	1,562,567.31	59	1,655,674.59	23,540,660.91	1,025,632.57
10	7,531,212.00	223,218,644.75	1,556,990.37	60	1,587,562.24	21,884,986.32	1,001,832.68
11	7,324,296.21	215,687,432.75	1,551,639.85	61	1,520,227.97	20,297,424.07	976,987.91
12	7,122,779.87	208,363,136.54	1,546,150.91	62	1,453,589.45	18,777,196.10	951,036.76
13	6,926,253.52	201,240,356.67	1,540,258.58	63	1,387,537.80	17,323,606.65	923,888.96
14	6,734,205.58	194,314,103.15	1,533,584.81	64	1,321,962.37	15,936,068.85	895,449.57
15	6,546,434.42	187,579,897.57	1,526,047.86	65	1,256,809.87	14,614,106.49	865,678.07
16	6,362,752.16	181,033,463.15	1,517,574.32	66	1,192,079.58	13,357,296.61	834,585.02
17	6,183,108.70	174,670,710.99	1,508,223.49	67	1,127,863.93	12,165,217.04	802,274.18
18	6,007,574.76	168,487,602.29	1,498,174.21	68	1,064,264.45	11,037,353.10	768,860.84
19	5,836,380.67	162,480,027.54	1,487,766.79	69	1,001,402.94	9,973,088.65	734,483.29
20	5,669,610.44	156,643,646.87	1,477,201.40	70	939,350.02	8,971,685.71	699,231.92
21	5,507,385.04	150,974,036.42	1,466,717.40	71	878,088.87	8,032,335.69	663,111.52
22	5,349,747.89	145,466,651.39	1,456,479.85	72	817,584.12	7,154,246.81	626,107.93
23	5,196,726.98	140,116,903.49	1,446,639.54	73	757,787.09	6,336,662.70	588,192.71
24	5,048,234.36	134,920,176.52	1,437,232.07	74	698,683.37	5,578,875.60	549,370.40
25	4,904,181.45	129,871,942.15	1,428,290.06	75	640,415.60	4,880,192.23	509,802.18
26	4,764,477.67	124,967,760.70	1,419,841.74	76	583,267.45	4,239,776.63	469,794.11
27	4,628,939.49	120,203,283.03	1,411,820.00	77	527,620.04	3,656,509.18	429,757.27
28	4,497,346.95	115,574,343.53	1,404,116.34	78	473,897.84	3,128,889.14	390,156.29
29	4,369,538.98	111,076,996.59	1,396,675.33	79	422,518.54	2,654,991.30	351,460.37
30	4,245,320.69	106,707,457.61	1,389,403.33	80	373,769.61	2,232,472.76	314,019.73
31	4,124,551.30	102,462,136.92	1,382,255.66	81	327,811.37	1,858,703.15	278,065.06
32	4,007,016.60	98,337,585.62	1,375,110.41	82	284,747.66	1,530,891.78	243,774.88
33	3,892,636.22	94,330,569.03	1,367,973.79	83	244,633.58	1,246,144.13	211,281.79
34	3,781,217.90	90,437,932.80	1,360,737.95	84	207,549.30	1,001,510.54	180,744.88
35	3,672,657.57	86,656,714.90	1,353,378.09	85	173,664.76	793,961.24	152,415.19
36	3,566,820.51	82,984,057.33	1,345,836.00	86	143,165.68	620,296.48	126,564.07
37	3,463,582.16	79,417,236.83	1,338,060.00	87	116,192.03	477,130.80	103,422.11
38	3,362,792.80	75,953,654.67	1,329,969.93	88	92,778.37	360,938.77	83,118.21
39	3,264,347.14	72,590,861.87	1,321,526.02	89	72,843.86	268,160.39	65,666.82
40	3,168,116.29	69,326,514.73	1,312,662.13	90	56,198.56	195,316.54	50,971.11
41	3,074,013.29	66,158,398.44	1,303,350.56	91	42,564.91	139,117.97	38,841.56
42	2,981,897.57	63,084,385.15	1,293,507.70	92	31,608.60	96,553.06	29,024.45
43	2,891,758.53	60,102,487.58	1,283,176.14	93	22,965.84	64,944.46	21,227.66
44	2,803,471.90	57,210,729.05	1,272,284.50	94	16,269.19	41,978.63	15,145.68
45	2,717,007.79	54,407,257.15	1,260,852.49	95	11,148.55	25,709.43	10,460.46
46	2,632,258.24	51,690,249.36	1,248,820.91	96	7,270.05	14,560.88	6,880.35
47	2,549,204.54	49,057,991.12	1,236,216.94	97	4,354.58	7,290.83	4,159.45
48	2,467,778.80	46,508,786.58	1,223,018.09	98	2,202.93	2,936.25	2,124.35
49	2,387,945.25	44,041,007.78	1,209,232.15	99	733.31	733.31	713.69

Table III Commutation Columns at 3% based on 1980 CSO Mortality Table-Males

Age	D_x	N_x	M_x	Age	D_x	N_x	M_x
0	10,000,000.00	293,720,655.74	1,445,029.44	50	2,045,349.05	35,834,056.97	1,001,638.65
1	9,668,155.34	283,720,655.74	1,404,446.92	51	1,972,451.22	33,788,707.92	988,314.10
2	9,376,515.22	274,052,500.40	1,394,403.56	52	1,901,021.70	31,816,256.70	974,334.61
3	9,094,400.52	264,675,985.17	1,385,391.25	53	1,830,960.71	29,915,235.00	959,643.18
4	8,820,862.10	255,581,584.65	1,376,738.27	54	1,762,148.56	28,084,274.29	944,159.99
5	8,555,807.66	246,760,722.55	1,368,602.14	55	1,694,468.36	26,322,125.73	927,804.51
6	8,299,133.15	238,204,914.90	1,361,125.92	56	1,627,890.53	24,627,657.37	910,580.12
7	8,050,481.66	229,905,781.75	1,354,196.76	57	1,562,363.98	22,999,766.84	892,467.86
8	7,809,748.70	221,855,300.08	1,347,943.85	58	1,497,912.72	21,437,402.86	873,522.35
9	7,576,517.61	214,045,551.38	1,342,181.16	59	1,434,520.40	19,939,490.14	853,758.55
10	7,350,399.29	206,469,033.78	1,336,738.11	60	1,372,167.45	18,504,969.74	833,187.75
11	7,131,100.61	199,118,634.49	1,331,528.73	61	1,310,779.61	17,132,802.29	811,765.95
12	6,918,067.46	191,987,533.88	1,326,197.54	62	1,250,280.12	15,822,022.68	789,444.50
13	6,710,861.26	185,069,466.41	1,320,488.45	63	1,190,570.15	14,571,742.57	766,150.47
14	6,508,948.75	178,358,605.16	1,314,037.92	64	1,131,550.30	13,381,172.42	741,807.42
15	6,312,100.56	171,849,656.41	1,306,770.76	65	1,073,171.09	12,249,622.11	716,385.98
16	6,120,102.57	165,537,555.85	1,298,620.36	66	1,015,428.23	11,176,451.02	689,900.53
17	5,932,874.76	159,417,453.28	1,289,647.97	67	958,396.68	10,161,022.79	662,444.56
18	5,750,453.41	153,484,578.52	1,280,028.79	68	902,158.31	9,202,626.11	634,120.65
19	5,573,026.67	147,734,125.11	1,270,090.99	69	846,811.34	8,300,467.81	605,050.14
20	5,400,641.34	142,161,098.44	1,260,026.82	70	792,409.84	7,453,656.47	575,313.05
21	5,233,378.71	136,760,457.10	1,250,064.42	71	738,933.72	6,661,246.62	544,916.83
22	5,071,245.61	131,527,078.39	1,240,359.83	72	686,347.52	5,922,312.90	513,852.97
23	4,914,234.03	126,455,832.78	1,231,054.44	73	634,604.91	5,235,965.38	482,101.07
24	4,762,226.52	121,541,598.75	1,222,179.95	74	583,688.64	4,601,360.47	449,668.43
25	4,615,105.96	116,779,372.23	1,213,765.02	75	533,712.46	4,017,671.83	416,692.89
26	4,472,754.36	112,164,266.27	1,205,833.99	76	484,906.25	3,483,959.37	383,431.71
27	4,334,967.66	107,691,511.91	1,198,321.68	77	437,578.46	2,999,053.12	350,227.40
28	4,201,509.56	103,356,544.25	1,191,124.77	78	392,070.34	2,561,474.66	317,464.28
29	4,072,200.83	99,155,034.69	1,184,190.11	79	348,714.20	2,169,404.32	285,527.66
30	3,946,832.33	95,082,833.87	1,177,429.40	80	307,731.86	1,820,690.13	254,702.05
31	3,825,247.06	91,136,001.54	1,170,800.42	81	269,238.45	1,512,958.27	225,171.71
32	3,707,221.44	87,310,754.47	1,164,189.76	82	233,301.66	1,243,719.82	197,076.81
33	3,592,657.47	83,603,533.03	1,157,603.11	83	199,948.60	1,010,418.16	170,518.94
34	3,481,354.94	80,010,875.56	1,150,941.09	84	169,226.42	810,469.56	145,620.51
35	3,373,196.51	76,529,520.62	1,144,181.34	85	141,254.80	641,243.14	122,577.81
36	3,268,037.75	73,156,324.12	1,137,271.03	86	116,164.93	499,988.34	101,602.16
37	3,165,744.86	69,888,286.36	1,130,163.70	87	94,049.62	383,823.41	82,870.30
38	3,066,162.24	66,722,541.51	1,122,787.25	88	74,915.57	289,773.79	66,475.55
39	2,969,176.16	63,656,379.26	1,115,106.86	89	58,676.32	214,858.23	52,418.31
40	2,874,652.47	60,687,203.11	1,107,064.03	90	45,158.52	156,181.91	40,609.54
41	2,782,496.20	57,812,550.64	1,098,635.50	91	34,120.14	111,023.38	30,886.45
42	2,692,564.82	55,030,054.44	1,089,747.70	92	25,276.04	76,903.24	23,036.14
43	2,604,834.15	52,337,489.62	1,080,441.25	93	18,320.22	51,627.20	16,816.51
44	2,519,178.06	49,732,655.47	1,070,654.11	94	12,946.70	33,306.98	11,976.59
45	2,435,556.15	47,213,477.42	1,060,406.32	95	8,850.26	20,360.29	8,257.24
46	2,353,858.56	44,777,921.27	1,049,647.26	96	5,757.31	11,510.03	5,422.07
47	2,274,056.01	42,424,062.71	1,038,403.70	97	3,440.12	5,752.72	3,272.56
48	2,196,075.71	40,150,006.70	1,026,658.04	98	1,736.09	2,312.60	1,668.73
49	2,119,874.02	37,953,930.98	1,014,419.72	99	576.51	576.51	559.72

Table III Commutation Columns at 3.25% based on 1980 CSO Mortality Table - Males

Age	D_x	N_x	M_x	Age	D_x	N_x	M_x
0	10,000,000.00	277,717,133.47	1,258,298.46	50	1,811,864.23	30,854,710.90	840,650.57
1	9,644,745.76	267,717,133.47	1,217,814.20	51	1,743,057.26	29,042,846.67	828,875.65
2	9,331,163.34	258,072,387.70	1,207,819.42	52	1,675,867.28	27,299,789.41	816,551.87
3	9,028,499.33	248,741,224.36	1,198,872.41	53	1,610,195.97	25,623,922.13	803,631.84
4	8,735,739.80	239,712,725.04	1,190,302.94	54	1,545,928.47	24,013,726.16	790,048.47
5	8,452,726.84	230,976,985.23	1,182,264.84	55	1,482,953.40	22,467,797.69	775,734.59
6	8,179,292.11	222,524,258.39	1,174,896.57	56	1,421,236.66	20,984,844.28	760,696.77
7	7,915,019.98	214,344,966.28	1,168,084.00	57	1,360,725.70	19,563,607.62	744,922.07
8	7,659,746.11	206,429,946.30	1,161,951.19	58	1,301,433.69	18,202,881.92	728,461.62
9	7,413,001.98	198,770,200.19	1,156,312.87	59	1,243,338.63	16,901,448.24	711,331.78
10	7,174,350.25	191,357,198.21	1,151,000.19	60	1,186,415.95	15,658,109.61	693,545.67
11	6,943,450.95	184,182,847.95	1,145,927.89	61	1,130,594.07	14,471,693.66	675,068.60
12	6,719,713.63	177,239,397.00	1,140,749.56	62	1,075,799.94	13,341,099.59	655,862.18
13	6,502,665.24	170,519,683.37	1,135,217.58	63	1,021,942.23	12,265,299.65	635,867.42
14	6,291,745.59	164,017,018.13	1,128,982.31	64	968,929.96	11,243,357.42	615,022.83
15	6,086,692.68	157,725,272.54	1,121,974.66	65	916,715.66	10,274,427.46	593,307.53
16	5,887,261.55	151,638,579.86	1,114,134.34	66	865,290.79	9,357,711.81	570,738.12
17	5,693,338.10	145,751,318.31	1,105,524.20	67	814,714.25	8,492,421.01	547,398.34
18	5,504,920.45	140,057,980.21	1,096,315.75	68	765,050.20	7,677,706.76	523,379.04
19	5,322,151.64	134,553,059.75	1,086,825.31	69	716,375.97	6,912,656.56	498,786.29
20	5,145,038.44	129,230,908.11	1,077,237.46	70	668,730.86	6,196,280.60	473,690.55
21	4,973,620.16	124,085,869.67	1,067,769.54	71	622,091.34	5,527,549.74	448,100.67
22	4,807,864.96	119,112,249.51	1,058,568.97	72	576,421.14	4,905,458.39	422,012.04
23	4,647,727.06	114,304,384.55	1,049,768.22	73	531,675.23	4,329,037.25	395,410.14
24	4,493,057.68	109,656,657.49	1,041,395.34	74	487,833.26	3,797,362.02	368,303.70
25	4,343,709.63	105,163,599.81	1,033,475.26	75	444,984.28	3,309,528.76	340,810.25
26	4,199,536.13	100,819,890.17	1,026,028.69	76	403,313.05	2,864,544.48	313,145.79
27	4,060,311.00	96,620,354.04	1,018,992.35	77	363,067.68	2,461,231.43	285,595.50
28	3,925,780.00	92,560,043.04	1,012,267.75	78	324,521.00	2,098,163.75	258,477.11
29	3,795,744.35	88,634,263.04	1,005,803.87	79	287,935.77	1,773,642.74	232,106.82
30	3,669,979.25	84,838,518.69	999,517.40	80	253,481.11	1,485,706.98	206,715.52
31	3,548,310.25	81,168,539.44	993,368.33	81	221,236.82	1,232,225.86	182,450.04
32	3,430,502.88	77,620,229.19	987,251.11	82	191,242.90	1,010,989.05	159,420.00
33	3,316,440.71	74,189,726.31	981,170.87	83	163,505.74	819,746.15	137,702.59
34	3,205,914.18	70,873,285.60	975,035.94	84	138,047.95	656,240.41	117,391.48
35	3,098,791.78	67,667,371.42	968,826.09	85	114,950.85	518,192.46	98,639.70
36	2,994,918.32	64,568,579.64	962,493.30	86	94,304.23	403,241.61	81,611.39
37	2,894,149.70	61,573,661.32	955,995.72	87	76,165.86	308,937.38	66,441.44
38	2,796,323.27	58,679,511.61	949,268.43	88	60,523.29	232,771.52	53,196.34
39	2,701,315.90	55,883,188.34	942,280.92	89	47,289.04	172,248.23	41,867.18
40	2,608,987.05	53,181,872.44	934,981.38	90	36,306.51	124,959.19	32,373.17
41	2,519,232.90	50,572,885.39	927,350.31	91	27,365.46	88,652.68	24,574.94
42	2,431,907.60	48,053,652.49	919,322.90	92	20,223.12	61,287.22	18,293.98
43	2,346,973.28	45,621,744.89	910,937.72	93	14,622.34	41,064.10	13,329.77
44	2,264,300.68	43,274,771.61	902,140.80	94	10,308.43	26,441.76	9,476.12
45	2,183,838.62	41,010,470.93	892,952.13	95	7,029.70	16,133.33	6,521.87
46	2,105,474.20	38,826,632.30	883,328.39	96	4,561.92	9,103.63	4,275.36
47	2,029,167.43	36,721,158.10	873,295.63	97	2,719.24	4,541.71	2,576.28
48	1,954,839.92	34,691,990.67	862,840.21	98	1,368.97	1,822.47	1,311.61
49	1,882,439.84	32,737,150.75	851,972.63	99	453.50	453.50	439.22

Table III Commutation Columns at 3.5% based on 1980 CSO Mortality Table-Males

Age	D_x	N_x	M_x	Age	D_x	N_x	M_x
0	10,000,000.00	263,174,871.74	1,100,366.66	50	1,605,503.21	26,589,933.97	706,326.70
1	9,621,449.28	253,174,871.74	1,059,980.18	51	1,540,802.21	24,984,430.76	695,918.08
2	9,286,139.70	243,553,422.47	1,050,033.63	52	1,477,830.31	23,443,628.56	685,050.60
3	8,963,233.32	234,267,282.77	1,041,151.29	53	1,416,489.61	21,965,798.25	673,684.84
4	8,651,641.84	225,304,049.45	1,032,664.32	54	1,356,668.57	20,549,308.63	661,764.42
5	8,351,132.75	216,652,407.61	1,024,722.83	55	1,298,259.73	19,192,640.06	649,233.25
6	8,061,465.17	208,301,274.86	1,017,460.71	56	1,241,224.08	17,894,380.33	636,100.10
7	7,782,157.02	200,239,809.69	1,010,762.49	57	1,185,506.90	16,653,156.26	622,356.69
8	7,512,976.99	192,457,652.67	1,004,747.19	58	1,131,111.08	15,467,649.35	608,050.47
9	7,253,398.05	184,944,675.68	999,230.27	59	1,078,008.90	14,336,538.28	593,198.43
10	7,002,928.31	177,691,277.63	994,044.53	60	1,026,170.69	13,258,529.37	577,814.63
11	6,761,175.17	170,688,349.32	989,105.39	61	975,526.44	12,232,358.68	561,871.80
12	6,527,506.18	163,927,174.15	984,075.17	62	926,005.51	11,256,832.24	545,339.68
13	6,301,408.49	157,399,667.97	978,714.41	63	877,522.20	10,330,826.73	528,170.57
14	6,082,289.69	151,098,259.47	972,686.71	64	829,991.91	9,453,304.54	510,314.95
15	5,869,850.40	145,015,969.79	965,928.71	65	783,368.01	8,623,312.63	491,758.41
16	5,663,810.30	139,146,119.39	958,385.98	66	737,637.48	7,839,944.61	472,518.58
17	5,464,017.16	133,482,309.09	950,122.65	67	692,844.73	7,102,307.13	452,670.10
18	5,270,427.43	128,018,291.93	941,306.44	68	649,038.18	6,409,462.40	432,293.08
19	5,083,136.18	122,747,864.50	932,242.21	69	606,276.92	5,760,424.22	411,479.96
20	4,902,107.54	117,664,728.33	923,107.07	70	564,587.30	5,154,147.30	390,292.46
21	4,727,336.71	112,762,620.78	914,107.98	71	523,942.47	4,589,560.00	368,739.96
22	4,558,751.22	108,035,284.07	905,384.13	72	484,305.12	4,065,617.52	346,820.47
23	4,396,265.96	103,476,532.85	897,059.54	73	445,630.90	3,581,312.40	324,523.72
24	4,239,699.22	99,080,266.89	889,158.79	74	407,896.50	3,135,681.50	301,858.96
25	4,088,872.32	94,840,567.67	881,703.36	75	371,170.08	2,727,785.00	278,926.14
26	3,943,608.52	90,751,695.35	874,710.60	76	335,598.71	2,356,614.92	255,906.41
27	3,803,658.20	86,808,086.84	868,119.04	77	301,380.61	2,021,016.21	233,037.06
28	3,668,747.76	83,004,428.63	861,834.72	78	268,732.52	1,719,635.61	210,580.60
29	3,538,657.74	79,335,680.87	855,808.63	79	237,860.73	1,450,903.08	188,796.37
30	3,413,146.46	75,797,023.13	849,962.10	80	208,892.31	1,213,042.35	167,871.55
31	3,292,021.12	72,383,876.67	844,257.17	81	181,879.58	1,004,150.04	147,922.82
32	3,175,035.06	69,091,855.55	838,595.50	82	156,841.71	822,270.46	129,035.47
33	3,062,052.87	65,916,820.49	832,981.64	83	133,770.08	665,428.75	111,267.66
34	2,952,854.52	62,854,767.62	827,330.98	84	112,669.32	531,658.67	94,690.52
35	2,847,293.68	59,901,913.10	821,625.12	85	93,591.75	418,989.35	79,423.03
36	2,745,203.61	57,054,619.42	815,820.35	86	76,596.03	325,397.60	65,592.25
37	2,646,429.21	54,309,415.81	809,878.92	87	61,714.21	248,801.57	53,300.63
38	2,550,799.82	51,662,986.60	803,742.30	88	48,921.19	187,087.36	42,594.57
39	2,458,182.29	49,112,186.78	797,383.71	89	38,131.57	138,166.17	33,459.29
40	2,368,428.86	46,654,004.48	790,757.21	90	29,205.09	100,034.59	25,822.27
41	2,281,426.34	44,285,575.62	783,846.49	91	21,959.70	70,829.51	19,564.50
42	2,197,024.56	42,004,149.28	776,594.40	92	16,189.05	48,869.80	14,536.45
43	2,115,172.04	39,807,124.71	769,037.39	93	11,677.23	32,680.75	10,572.09
44	2,035,735.55	37,691,952.67	761,128.45	94	8,212.31	21,003.51	7,502.04
45	1,958,653.07	35,656,217.12	752,887.27	95	5,586.75	12,791.21	5,154.19
46	1,883,807.88	33,697,564.05	744,276.73	96	3,616.76	7,204.46	3,373.13
47	1,811,149.41	31,813,756.17	735,321.90	97	2,150.65	3,587.70	2,029.33
48	1,740,593.29	30,002,606.77	726,012.39	98	1,080.11	1,437.05	1,031.51
49	1,672,079.50	28,262,013.48	716,359.24	99	356.94	356.94	344.87

Table III	Commutation Columns at 3.75% based on 1980 CSO Mortality Table - Males						
Age	D_x	N_x	M_x	Age	D_x	N_x	M_x
0	10,000,000.00	249,927,054.19	966,492.02	50	1,423,060.61	22,933,862.54	594,125.82
1	9,598,265.06	239,927,054.19	926,202.86	51	1,362,421.09	21,510,801.92	584,922.22
2	9,241,441.14	230,328,789.13	916,304.18	52	1,303,590.78	20,148,380.83	575,336.05
3	8,898,594.87	221,087,347.99	907,485.90	53	1,246,471.49	18,844,790.06	565,334.50
4	8,568,553.44	212,188,753.12	899,080.44	54	1,190,953.93	17,598,318.57	554,870.13
5	8,251,000.42	203,620,199.68	891,234.17	55	1,136,933.41	16,407,364.64	543,896.13
6	7,945,613.73	195,369,199.26	884,076.41	56	1,084,365.97	15,270,431.23	532,422.67
7	7,651,836.83	187,423,585.53	877,490.36	57	1,033,194.34	14,186,065.27	520,444.99
8	7,369,364.10	179,771,748.70	871,590.05	58	983,411.84	13,152,870.93	508,006.86
9	7,097,603.11	172,402,384.60	866,191.62	59	934,985.27	12,169,459.09	495,125.30
10	6,836,001.11	165,304,781.49	861,129.49	60	887,880.00	11,234,473.82	481,814.68
11	6,584,106.94	158,468,780.37	856,319.70	61	842,026.88	10,346,593.83	468,053.61
12	6,341,240.50	151,884,673.43	851,433.03	62	797,356.84	9,504,566.95	453,818.27
13	6,106,843.79	145,543,432.93	846,237.78	63	753,788.50	8,707,210.11	439,070.06
14	5,880,286.99	139,436,589.14	840,410.27	64	711,242.18	7,953,421.61	423,769.11
15	5,661,228.69	133,556,302.15	833,892.47	65	669,671.35	7,242,179.44	407,905.83
16	5,449,348.84	127,895,073.46	826,635.34	66	629,058.61	6,572,508.08	391,498.07
17	5,244,453.14	122,445,724.63	818,704.06	67	589,435.51	5,943,449.48	374,612.03
18	5,046,453.05	117,201,271.49	810,262.51	68	550,836.70	5,354,013.97	357,318.12
19	4,855,393.01	112,154,818.44	801,604.39	69	513,305.48	4,803,177.27	339,696.66
20	4,671,192.05	107,299,425.43	792,899.56	70	476,857.06	4,289,871.79	321,801.45
21	4,493,799.26	102,628,233.39	784,345.05	71	441,461.63	3,813,014.74	303,641.82
22	4,323,099.88	98,134,434.12	776,072.14	72	407,080.84	3,371,553.11	285,217.47
23	4,158,968.01	93,811,334.24	768,196.89	73	373,670.78	2,964,472.27	266,521.18
24	4,001,187.59	89,652,366.23	760,740.61	74	341,205.54	2,590,801.49	247,562.11
25	3,849,547.29	85,651,178.64	753,721.56	75	309,735.72	2,249,595.96	228,425.02
26	3,703,839.43	81,801,631.35	747,153.96	76	279,377.13	1,939,860.24	209,261.70
27	3,563,789.82	78,097,791.92	740,978.07	77	250,286.90	1,660,483.11	190,269.44
28	3,429,104.32	74,534,002.10	735,104.24	78	222,635.95	1,410,196.21	171,665.00
29	3,299,541.90	71,104,897.78	729,485.36	79	196,584.85	1,187,560.25	153,660.99
30	3,174,843.02	67,805,355.88	724,047.03	80	172,227.30	990,975.40	136,408.92
31	3,054,795.83	64,630,512.86	718,753.20	81	149,594.54	818,748.10	120,001.24
32	2,939,140.49	61,575,717.03	713,512.17	82	128,690.24	669,153.56	104,503.97
33	2,827,722.26	58,636,576.53	708,327.92	83	109,495.25	540,463.31	89,960.43
34	2,720,309.77	55,808,854.28	703,122.26	84	92,001.35	430,968.07	76,424.19
35	2,616,741.48	53,088,544.51	697,878.42	85	76,239.20	338,966.72	63,987.39
36	2,516,838.57	50,471,803.03	692,556.53	86	62,244.26	262,727.52	52,748.08
37	2,420,434.45	47,954,964.46	687,122.48	87	50,029.99	200,483.26	42,783.61
38	2,327,349.81	45,534,530.02	681,523.43	88	39,563.49	150,453.27	34,125.42
39	2,237,441.13	43,207,180.20	675,735.82	89	30,763.41	110,889.78	26,755.35
40	2,150,552.86	40,969,739.07	669,718.92	90	23,505.01	80,126.37	20,608.88
41	2,066,562.16	38,819,186.21	663,459.04	91	17,631.15	56,621.36	15,584.60
42	1,985,313.88	36,752,624.06	656,905.78	92	12,966.66	38,990.20	11,557.37
43	1,906,743.20	34,767,310.18	650,093.44	93	9,330.37	26,023.55	8,389.76
44	1,830,712.36	32,860,566.98	642,981.03	94	6,546.00	16,693.18	5,942.64
45	1,757,148.70	31,029,854.61	635,587.69	95	4,442.45	10,147.18	4,075.68
46	1,685,931.22	29,272,705.91	627,881.61	96	2,869.03	5,704.73	2,662.84
47	1,616,999.05	27,586,774.69	619,886.71	97	1,701.91	2,835.70	1,599.42
48	1,550,261.77	25,969,775.64	611,595.18	98	852.68	1,133.79	811.70
49	1,485,651.34	24,419,513.87	603,018.31	99	281.11	281.11	270.94

Table III Commutation Columns at 4% based on 1980 CSO MortalityTable-Males

Age	D_x	N_x	M_x	Age	D_x	N_x	M_x
0	10,000,000.00	237,828,590.75	852,746.51	50	1,261,716.27	19,796,813.66	500,300.36
1	9,575,192.31	227,828,590.75	812,554.20	51	1,205,048.23	18,535,097.39	492,159.86
2	9,197,064.53	218,253,398.44	802,703.06	52	1,150,241.72	17,330,049.17	483,701.37
3	8,834,576.45	209,056,333.91	793,948.22	53	1,097,197.86	16,179,807.45	474,897.57
4	8,486,460.11	200,221,757.46	785,623.28	54	1,045,808.90	15,082,609.59	465,708.53
5	8,152,305.38	191,735,297.35	777,870.86	55	995,972.07	14,036,800.70	456,095.12
6	7,831,700.03	183,582,991.98	770,815.73	56	947,638.68	13,040,828.62	446,068.35
7	7,524,004.78	175,751,291.94	764,339.71	57	900,748.79	12,093,189.94	435,626.10
8	7,228,832.19	168,227,287.16	758,551.91	58	855,286.99	11,192,441.16	424,808.48
9	6,945,517.43	160,998,454.97	753,269.16	59	811,214.99	10,337,154.17	413,632.14
10	6,673,440.39	154,052,937.55	748,327.41	60	768,493.57	9,525,939.18	402,111.29
11	6,412,085.48	147,379,497.15	743,643.28	61	727,054.03	8,757,445.61	390,229.20
12	6,160,719.24	140,967,411.68	738,895.72	62	686,828.36	8,030,391.58	377,967.15
13	5,918,733.29	134,806,692.43	733,860.51	63	647,738.59	7,343,563.22	365,293.85
14	5,685,455.28	128,887,959.14	728,226.08	64	609,708.90	6,695,824.63	352,177.18
15	5,460,497.22	123,202,503.86	721,939.38	65	572,692.53	6,086,115.74	338,611.15
16	5,243,495.11	117,742,006.64	714,956.39	66	536,667.96	5,513,423.21	324,613.22
17	5,034,208.89	112,498,511.54	707,343.07	67	501,655.55	4,976,755.25	310,241.89
18	4,832,501.82	107,464,302.64	699,259.41	68	467,678.02	4,475,099.70	295,558.80
19	4,638,365.25	102,631,800.82	690,988.29	69	434,765.18	4,007,421.68	280,633.58
20	4,451,670.84	97,993,435.57	682,692.55	70	402,922.80	3,572,656.49	265,512.93
21	4,272,319.83	93,541,764.73	674,559.64	71	372,118.58	3,169,733.70	250,205.74
22	4,100,153.57	89,269,444.91	666,713.38	72	342,313.33	2,797,615.12	234,712.75
23	3,935,004.19	85,169,291.33	659,262.22	73	313,463.56	2,455,301.79	219,028.88
24	3,776,620.09	81,234,287.14	652,224.43	74	285,541.19	2,141,838.22	203,162.79
25	3,624,756.29	77,457,667.05	645,615.25	75	258,582.27	1,856,297.04	187,186.23
26	3,479,173.35	73,832,910.77	639,446.02	76	232,676.80	1,597,714.76	171,226.23
27	3,339,571.67	70,353,737.41	633,658.69	77	207,948.17	1,365,037.97	155,446.71
28	3,205,635.56	67,014,165.75	628,167.65	78	184,530.03	1,157,089.80	140,026.58
29	3,077,101.80	63,808,530.18	622,927.56	79	162,546.11	972,559.77	125,139.96
30	2,953,692.22	60,731,428.38	617,868.05	80	142,063.75	810,013.66	110,909.38
31	2,835,175.44	57,777,736.16	612,954.82	81	123,098.23	667,949.90	97,407.85
32	2,721,277.69	54,942,560.72	608,102.28	82	105,641.96	544,851.68	84,686.12
33	2,611,824.74	52,221,283.03	603,313.85	83	89,668.70	439,209.72	72,776.01
34	2,506,573.29	49,609,458.29	598,517.20	84	75,161.34	349,541.02	61,717.46
35	2,405,346.41	47,102,885.00	593,696.99	85	62,134.59	274,379.68	51,581.53
36	2,307,952.89	44,697,538.59	588,816.79	86	50,606.83	212,245.08	42,443.56
37	2,214,214.40	42,389,585.70	583,845.71	87	40,578.42	161,638.25	34,361.56
38	2,123,942.60	40,175,371.31	578,736.02	88	32,012.09	121,059.83	27,355.94
39	2,036,983.42	38,051,428.70	573,466.94	89	24,831.83	89,047.75	21,406.91
40	1,953,173.24	36,014,445.28	568,002.26	90	18,927.33	64,215.92	16,457.49
41	1,872,379.52	34,061,272.04	562,330.59	91	14,163.30	45,288.59	12,421.43
42	1,794,441.70	32,188,892.52	556,407.37	92	10,391.22	31,125.29	9,194.09
43	1,719,282.13	30,394,450.82	550,264.79	93	7,459.20	20,734.07	6,661.73
44	1,646,758.15	28,675,168.69	543,867.05	94	5,220.65	13,274.88	4,710.07
45	1,576,786.85	27,028,410.54	537,232.60	95	3,534.48	8,054.23	3,224.70
46	1,509,242.73	25,451,623.69	530,334.13	96	2,277.16	4,519.76	2,103.32
47	1,444,055.12	23,942,380.96	523,194.32	97	1,347.57	2,242.60	1,261.31
48	1,381,127.62	22,498,325.84	515,807.39	98	673.52	895.03	639.10
49	1,320,384.56	21,117,198.22	508,184.63	99	221.51	221.51	212.99

Table III	Commutation Columns at 4.25% based on 1980 CSO Mortality Table - Males						
Age	D_x	N_x	M_x	Age	D_x	N_x	M_x
0	10,000,000.00	226,753,025.70	755,871.85	50	1,118,988.15	17,102,735.95	421,754.31
1	9,552,230.22	216,753,025.70	715,775.93	51	1,066,167.61	15,983,747.80	414,551.99
2	9,153,006.80	207,200,795.49	705,971.98	52	1,015,237.03	14,917,580.19	407,086.28
3	8,771,170.64	198,047,788.68	697,279.97	53	966,096.61	13,902,343.16	399,334.42
4	8,405,347.57	189,276,618.04	689,034.60	54	918,639.71	12,936,246.55	391,262.76
5	8,055,023.61	180,871,270.48	681,374.70	55	872,765.00	12,017,606.83	382,838.58
6	7,719,687.12	172,816,246.86	674,420.47	56	828,419.31	11,144,841.83	374,073.24
7	7,398,607.57	165,096,559.74	668,052.38	57	785,540.17	10,316,422.52	364,966.59
8	7,091,307.99	157,697,952.17	662,374.69	58	744,104.36	9,530,882.35	355,555.20
9	6,797,044.08	150,606,644.18	657,204.87	59	704,069.01	8,786,777.99	345,855.04
10	6,515,121.84	143,809,600.10	652,380.35	60	665,390.78	8,082,708.97	335,879.86
11	6,244,955.32	137,294,478.26	647,818.32	61	628,001.24	7,417,318.20	325,616.56
12	5,985,752.08	131,049,522.94	643,205.59	62	591,833.18	6,789,316.96	315,050.47
13	5,736,848.14	125,063,770.85	638,325.11	63	556,811.42	6,197,483.78	304,156.21
14	5,497,523.66	119,326,922.71	632,876.93	64	522,863.32	5,640,672.35	292,907.85
15	5,267,339.67	113,829,399.06	626,812.61	65	489,941.73	5,117,809.04	281,302.03
16	5,045,884.18	108,562,059.39	620,092.79	66	458,021.50	4,627,867.30	269,355.45
17	4,832,867.84	103,516,175.21	612,783.95	67	427,113.30	4,169,845.81	257,119.58
18	4,628,102.71	98,683,307.37	605,042.21	68	397,229.69	3,742,732.51	244,648.27
19	4,431,524.77	94,055,204.67	597,139.93	69	368,389.11	3,345,502.82	232,001.70
20	4,242,956.29	89,623,679.89	589,233.13	70	340,589.41	2,977,113.71	219,220.26
21	4,062,249.05	85,380,723.61	581,500.12	71	313,796.38	2,636,524.30	206,312.17
22	3,889,199.20	81,318,474.56	574,057.55	72	287,970.28	2,322,727.93	193,278.73
23	3,723,595.87	77,429,275.36	567,006.70	73	263,068.10	2,034,757.65	180,116.35
24	3,565,150.88	73,705,679.49	560,362.99	74	239,060.13	1,771,689.55	166,832.98
25	3,413,584.85	70,140,528.61	554,138.84	75	215,970.49	1,532,629.42	153,489.20
26	3,268,626.03	66,726,943.75	548,342.96	76	193,867.95	1,316,658.93	140,191.21
27	3,129,948.63	63,458,317.72	542,918.89	77	172,848.39	1,122,790.97	127,075.14
28	2,997,214.78	60,328,369.08	537,784.87	78	153,015.19	949,942.58	114,288.52
29	2,870,138.53	57,331,154.30	532,897.22	79	134,462.55	796,927.39	101,973.91
30	2,748,422.59	54,461,015.78	528,189.33	80	117,237.17	662,464.84	90,230.21
31	2,631,815.76	51,712,593.19	523,628.51	81	101,342.39	545,227.67	79,114.88
32	2,520,029.83	49,080,777.43	519,134.83	82	86,762.70	443,885.28	68,666.65
33	2,412,871.13	46,560,747.60	514,711.16	83	73,467.41	357,122.58	58,908.46
34	2,310,084.04	44,147,876.47	510,290.52	84	61,433.56	283,655.17	49,869.68
35	2,211,476.25	41,837,792.42	505,858.81	85	50,664.28	222,221.61	41,604.89
36	2,116,844.05	39,626,316.18	501,382.71	86	41,165.64	171,557.33	34,171.70
37	2,025,997.33	37,509,472.13	496,834.20	87	32,928.96	130,391.69	27,613.24
38	1,938,738.57	35,483,474.80	492,170.06	88	25,915.18	97,462.72	21,941.88
39	1,854,903.18	33,544,736.23	487,371.97	89	20,054.24	71,547.54	17,137.43
40	1,774,319.36	31,689,833.05	482,407.70	90	15,249.10	51,493.30	13,149.85
41	1,696,845.04	29,915,513.69	477,267.74	91	11,383.52	36,244.20	9,905.94
42	1,622,314.06	28,218,668.65	471,912.70	92	8,331.74	24,860.68	7,318.24
43	1,550,636.50	26,596,354.59	466,372.64	93	5,966.49	16,528.94	5,292.64
44	1,481,664.76	25,045,718.09	460,616.30	94	4,165.89	10,562.46	3,735.29
45	1,415,306.15	23,564,053.34	454,661.29	95	2,813.63	6,396.56	2,552.85
46	1,351,430.67	22,148,747.19	448,484.15	96	1,808.39	3,582.94	1,662.32
47	1,289,958.47	20,797,316.51	442,106.23	97	1,067.59	1,774.55	995.25
48	1,230,787.40	19,507,358.05	435,523.40	98	532.31	706.96	503.49
49	1,173,834.70	18,276,570.65	428,746.69	99	174.65	174.65	167.53

Table III	Commutation Columns at 4.5% based on 1980 CSO MortalityTable-Males						
Age	D_x	N_x	M_x	Age	D_x	N_x	M_x
0	10,000,000.00	216,589,915.19	673,161.55	50	992,691.14	14,787,073.81	355,927.20
1	9,529,377.99	206,589,915.19	633,161.55	51	943,569.55	13,794,382.67	349,553.07
2	9,109,264.90	197,060,537.20	623,404.44	52	896,345.94	12,850,813.11	342,961.64
3	8,708,370.13	187,951,272.30	614,774.68	53	850,919.62	11,954,467.17	336,133.96
4	8,325,201.80	179,242,902.17	606,607.93	54	807,184.79	11,103,547.55	329,041.59
5	7,959,131.57	170,917,700.37	599,039.21	55	765,041.24	10,296,362.76	321,657.20
6	7,609,538.84	162,958,568.80	592,184.20	56	724,431.81	9,531,321.52	313,992.13
7	7,275,593.14	155,349,029.96	585,921.99	57	685,291.70	8,806,889.70	306,047.64
8	6,956,720.16	148,073,436.82	580,352.07	58	647,590.84	8,121,598.00	297,856.95
9	6,652,088.91	141,116,716.66	575,292.50	59	611,282.34	7,474,007.16	289,435.14
10	6,360,924.99	134,464,627.76	570,582.17	60	576,319.31	6,862,724.82	280,795.27
11	6,082,566.15	128,103,702.77	566,138.76	61	542,633.58	6,286,405.52	271,927.12
12	5,816,155.43	122,021,136.62	561,656.72	62	510,158.64	5,743,771.94	262,819.18
13	5,560,968.12	116,204,981.19	556,925.87	63	478,821.73	5,233,613.30	253,450.82
14	5,316,232.08	110,644,013.07	551,657.35	64	448,552.89	4,754,791.57	243,801.10
15	5,081,453.12	105,327,780.99	545,807.04	65	419,304.66	4,306,238.68	233,868.55
16	4,856,167.40	100,246,327.88	539,339.88	66	391,048.74	3,886,934.02	223,668.81
17	4,640,032.93	95,390,160.48	532,322.67	67	363,787.60	3,495,885.28	213,247.09
18	4,432,807.84	90,750,127.55	524,907.62	68	337,525.26	3,132,097.67	202,650.24
19	4,234,370.67	86,317,319.70	517,356.90	69	312,270.62	2,794,572.41	191,930.18
20	4,044,492.38	82,082,949.04	509,819.94	70	288,015.10	2,482,301.79	181,121.72
21	3,862,973.98	78,038,456.66	502,466.27	71	264,723.09	2,194,286.69	170,232.27
22	3,689,565.28	74,175,482.68	495,405.74	72	242,354.64	1,929,563.60	159,263.38
23	3,524,011.55	70,485,917.40	488,732.81	73	220,867.41	1,687,208.96	148,212.48
24	3,365,987.29	66,961,905.85	482,460.24	74	200,230.57	1,466,341.56	137,086.67
25	3,215,178.08	63,595,918.57	476,597.86	75	180,458.53	1,266,110.99	125,937.00
26	3,071,279.48	60,380,740.48	471,151.90	76	161,602.77	1,085,652.46	114,852.18
27	2,933,939.04	57,309,461.00	466,067.52	77	143,736.77	924,049.70	103,945.15
28	2,802,796.17	54,375,521.95	461,266.52	78	126,939.52	780,312.93	93,337.53
29	2,677,541.93	51,572,725.78	456,706.85	79	111,281.62	653,373.41	83,145.92
30	2,557,859.61	48,895,183.85	452,325.38	80	96,793.72	542,091.79	73,450.06
31	2,443,478.11	46,337,324.24	448,090.94	81	83,470.45	445,298.06	64,294.94
32	2,334,094.44	43,893,846.13	443,928.82	82	71,290.96	361,827.61	55,709.86
33	2,229,495.72	41,559,751.69	439,841.34	83	60,222.10	290,536.66	47,710.95
34	2,129,413.82	39,330,255.97	435,766.44	84	50,237.34	230,314.56	40,319.49
35	2,033,641.24	37,200,842.14	431,691.10	85	41,331.64	180,077.22	33,577.12
36	1,941,961.87	35,167,200.90	427,584.80	86	33,502.36	138,745.58	27,527.67
37	1,854,173.96	33,225,239.03	423,422.04	87	26,734.89	105,243.22	22,202.88
38	1,770,070.78	31,371,065.07	419,163.67	88	20,990.09	78,508.33	17,609.35
39	1,689,477.45	29,600,994.29	414,793.49	89	16,204.14	57,518.24	13,727.28
40	1,612,214.11	27,911,516.84	410,282.77	90	12,292.04	41,314.10	10,512.96
41	1,538,129.44	26,299,302.72	405,623.58	91	9,154.10	29,022.07	7,904.35
42	1,467,051.65	24,761,173.28	400,781.03	92	6,683.98	19,867.96	5,828.42
43	1,398,879.32	23,294,121.63	395,783.17	93	4,775.05	13,183.99	4,207.31
44	1,333,459.94	21,895,242.31	390,602.61	94	3,326.04	8,408.94	2,963.93
45	1,270,691.68	20,561,782.37	385,256.08	95	2,241.01	5,082.91	2,022.13
46	1,210,440.20	19,291,090.69	379,723.38	96	1,436.91	2,841.89	1,314.53
47	1,152,617.12	18,080,650.49	374,024.52	97	846.26	1,404.98	785.76
48	1,097,115.00	16,928,033.37	368,156.63	98	420.94	558.72	396.88
49	1,043,844.56	15,830,918.36	362,130.37	99	137.78	137.78	131.84

Table III	Commutation Columns at 4.75% based on 1980 CSO Mortality Table - Males						
Age	D_x	N_x	M_x	*Age*	D_x	N_x	M_x
0	10,000,000.00	207,242,599.41	602,364.23	50	880,900.97	12,794,976.29	300,699.18
1	9,506,634.84	197,242,599.41	562,459.69	51	835,312.76	11,914,075.33	295,056.37
2	9,065,835.81	187,735,964.57	552,749.11	52	791,613.36	11,078,762.56	289,235.10
3	8,646,167.73	178,670,128.76	544,180.98	53	749,701.29	10,287,149.21	283,219.58
4	8,246,009.00	170,023,961.04	536,091.92	54	709,471.49	9,537,447.92	276,985.78
5	7,864,606.14	161,777,952.03	528,613.09	55	670,824.76	8,827,976.43	270,510.79
6	7,501,219.79	153,913,345.89	521,855.66	56	633,700.44	8,157,151.68	263,805.73
7	7,154,910.68	146,412,126.10	515,697.32	57	598,031.72	7,523,451.24	256,872.84
8	6,824,999.19	139,257,215.42	510,232.86	58	563,782.65	6,925,419.51	249,742.15
9	6,510,560.43	132,432,216.23	505,280.94	59	530,902.93	6,361,636.86	242,427.75
10	6,210,733.04	125,921,655.80	500,681.82	60	499,342.69	5,830,733.93	234,941.87
11	5,924,772.62	119,710,922.76	496,353.69	61	469,034.14	5,331,391.24	227,276.54
12	5,651,752.17	113,786,150.14	491,998.34	62	439,911.48	4,862,357.11	219,422.73
13	5,390,881.31	108,134,397.97	487,412.19	63	411,904.14	4,422,445.63	211,363.64
14	5,141,330.88	102,743,516.67	482,317.00	64	384,944.59	4,010,541.49	203,082.33
15	4,902,547.44	97,602,185.78	476,672.67	65	358,985.17	3,625,596.90	194,578.63
16	4,674,011.64	92,699,638.34	470,448.09	66	333,995.01	3,266,611.73	185,867.03
17	4,455,325.75	88,025,626.70	463,710.22	67	309,969.70	2,932,616.72	176,987.08
18	4,246,191.38	83,570,300.95	456,607.33	68	286,906.17	2,622,647.02	167,979.45
19	4,046,427.74	79,324,109.57	449,391.75	69	264,805.49	2,335,740.86	158,888.84
20	3,855,752.94	75,277,681.82	442,206.50	70	243,653.90	2,070,935.37	149,745.14
21	3,673,915.96	71,421,928.89	435,212.74	71	223,414.93	1,827,281.46	140,554.91
22	3,500,619.38	67,748,012.93	428,513.78	72	204,048.76	1,603,866.53	131,319.73
23	3,335,563.98	64,247,393.56	422,197.69	73	185,513.93	1,399,817.77	122,037.71
24	3,178,386.32	60,911,829.57	416,274.72	74	167,778.97	1,214,303.84	112,715.07
25	3,028,736.58	57,733,443.25	410,752.28	75	150,850.52	1,046,524.87	103,394.74
26	2,886,277.39	54,704,706.67	405,634.37	76	134,766.04	895,674.34	94,150.73
27	2,750,629.36	51,818,429.28	400,867.65	77	119,580.90	760,908.30	85,076.70
28	2,621,408.85	49,067,799.92	396,377.35	78	105,354.49	641,327.41	76,272.82
29	2,498,283.89	46,446,391.06	392,122.95	79	92,138.66	535,972.92	67,834.40
30	2,380,918.17	43,948,107.18	388,044.57	80	79,951.74	443,834.26	59,825.60
31	2,269,020.79	41,567,189.01	384,112.46	81	68,782.15	363,882.53	52,281.51
32	2,162,273.90	39,298,168.22	380,256.73	82	58,605.67	295,100.38	45,224.03
33	2,060,445.75	37,135,894.32	376,479.18	83	49,388.22	236,494.71	38,664.12
34	1,963,255.71	35,075,448.57	372,722.24	84	41,101.38	187,106.48	32,616.84
35	1,870,481.42	33,112,192.86	368,973.87	85	33,734.53	146,005.11	27,113.77
36	1,781,894.62	31,241,711.44	365,206.03	86	27,279.07	112,270.58	22,188.05
37	1,697,282.19	29,459,816.83	361,395.51	87	21,716.75	84,991.51	17,862.72
38	1,616,428.38	27,762,534.64	357,506.76	88	17,009.56	63,274.75	14,140.30
39	1,539,148.40	26,146,106.26	353,525.44	89	13,099.87	46,265.20	11,001.93
40	1,465,254.52	24,606,957.86	349,425.89	90	9,913.50	33,165.32	8,409.59
41	1,394,586.63	23,141,703.34	345,201.51	91	7,365.15	23,251.82	6,310.77
42	1,326,967.47	21,747,116.71	340,821.36	92	5,364.91	15,886.67	4,644.52
43	1,262,284.87	20,420,149.24	336,311.52	93	3,823.56	10,521.76	3,346.44
44	1,200,381.68	19,157,864.37	331,647.98	94	2,656.92	6,698.20	2,353.19
45	1,141,147.63	17,957,482.69	326,846.51	95	1,785.91	4,041.28	1,602.65
46	1,084,444.29	16,816,335.05	321,889.71	96	1,142.37	2,255.37	1,040.10
47	1,030,175.53	15,731,890.77	316,796.24	97	671.19	1,113.00	620.72
48	978,229.10	14,701,715.23	311,564.21	98	333.06	441.82	313.03
49	928,509.84	13,723,486.14	306,203.79	99	108.75	108.75	103.82

Table III Commutation Columns at 5% based on 1980 CSO MortalityTable-Males

Age	D_x	N_x	M_x	Age	D_x	N_x	M_x
0	10,000,000.00	198,626,306.79	541,604.44	50	781,922.51	11,079,793.33	254,313.30
1	9,484,000.00	188,626,306.79	501,794.91	51	739,691.25	10,297,870.82	249,316.45
2	9,022,716.55	179,142,306.79	492,130.52	52	699,325.25	9,558,179.58	244,173.84
3	8,584,556.31	170,119,590.23	483,623.44	53	660,722.48	8,858,854.33	238,872.27
4	8,167,755.62	161,535,033.92	475,611.14	54	623,778.65	8,198,131.85	233,391.42
5	7,771,424.65	153,367,278.31	468,220.92	55	588,395.54	7,574,353.21	227,712.06
6	7,394,695.35	145,595,853.65	461,559.46	56	554,509.55	6,985,957.66	221,844.90
7	7,036,510.57	138,201,158.30	455,503.03	57	522,052.26	6,431,448.11	215,792.82
8	6,696,077.40	131,164,647.73	450,141.79	58	490,982.71	5,909,395.85	209,582.91
9	6,372,369.75	124,468,570.34	445,294.97	59	461,247.83	5,418,413.14	203,228.16
10	6,064,432.80	118,096,200.58	440,804.20	60	432,795.41	4,957,165.31	196,739.92
11	5,771,434.16	112,031,767.79	436,588.08	61	405,558.16	4,524,369.90	190,111.97
12	5,492,371.44	106,260,333.62	432,355.56	62	379,471.11	4,118,811.74	183,337.21
13	5,226,383.73	100,767,962.18	427,909.34	63	354,465.77	3,739,340.63	176,401.93
14	4,972,580.37	95,541,578.45	422,981.39	64	330,476.90	3,384,874.86	169,292.38
15	4,730,344.75	90,568,998.08	417,535.32	65	307,456.82	3,054,397.96	162,009.30
16	4,499,098.60	85,838,653.33	411,543.68	66	285,372.63	2,746,941.14	154,565.91
17	4,278,385.53	81,339,554.73	405,073.40	67	264,214.30	2,461,568.51	146,996.75
18	4,067,848.32	77,061,169.21	398,268.83	68	243,972.96	2,197,354.21	139,337.05
19	3,867,245.19	72,993,320.89	391,372.77	69	224,643.33	1,953,381.25	131,625.18
20	3,676,239.94	69,126,075.70	384,522.05	70	206,207.59	1,728,737.91	123,886.74
21	3,494,528.62	65,449,835.75	377,869.77	71	188,628.89	1,522,530.32	116,127.45
22	3,321,765.79	61,955,307.14	371,513.07	72	171,867.88	1,333,901.43	108,348.76
23	3,157,607.35	58,633,541.34	365,533.96	73	155,884.17	1,162,033.55	100,549.23
24	3,001,651.47	55,475,933.99	359,940.33	74	140,646.11	1,006,149.39	92,734.24
25	2,853,512.75	52,474,282.51	354,737.39	75	126,154.21	865,503.28	84,939.77
26	2,712,820.85	49,620,769.76	349,927.05	76	112,434.64	739,349.06	77,227.54
27	2,579,169.32	46,907,948.91	345,457.47	77	99,528.21	626,914.42	69,675.15
28	2,452,151.36	44,328,779.60	341,257.09	78	87,478.67	527,386.21	62,365.04
29	2,331,412.02	41,876,628.23	337,286.87	79	76,323.06	439,907.53	55,375.08
30	2,216,595.50	39,545,216.22	333,489.96	80	66,070.33	363,584.48	48,756.79
31	2,107,391.31	37,328,620.72	329,837.94	81	56,704.70	297,514.14	42,537.36
32	2,003,466.79	35,221,229.41	326,265.39	82	48,200.08	240,809.44	36,732.96
33	1,904,571.84	33,217,762.62	322,773.62	83	40,522.50	192,609.36	31,350.62
34	1,810,413.48	31,313,190.78	319,309.16	84	33,642.94	152,086.87	26,400.71
35	1,720,755.00	29,502,777.29	315,860.84	85	27,547.16	118,443.93	21,906.98
36	1,635,356.30	27,782,022.30	312,402.86	86	22,222.69	90,896.77	17,894.27
37	1,553,993.36	26,146,666.00	308,914.03	87	17,649.26	68,674.08	14,379.07
38	1,476,441.71	24,592,672.64	305,362.06	88	13,790.80	51,024.81	11,361.05
39	1,402,507.09	23,116,230.93	301,734.18	89	10,595.67	37,234.01	8,822.62
40	1,331,994.32	21,713,723.85	298,007.47	90	7,999.32	26,638.35	6,730.83
41	1,264,734.98	20,381,729.52	294,176.43	91	5,928.87	18,639.03	5,041.30
42	1,200,546.66	19,116,994.54	290,213.58	92	4,308.42	12,710.15	3,703.18
43	1,139,307.30	17,916,447.88	286,143.11	93	3,063.29	8,401.73	2,663.20
44	1,080,855.39	16,777,140.59	281,943.93	94	2,123.56	5,338.45	1,869.35
45	1,025,073.01	15,696,285.20	277,630.86	95	1,424.00	3,214.89	1,270.91
46	971,818.01	14,671,212.19	273,188.86	96	908.70	1,790.89	823.42
47	920,987.34	13,699,394.18	268,635.24	97	532.63	882.19	490.62
48	872,464.45	12,778,406.83	263,968.89	98	263.68	349.57	247.03
49	826,149.05	11,905,942.38	259,199.41	99	85.89	85.89	81.80

Table III	Commutation Columns at 5.25% based on 1980 CSO Mortality Table - Males						
Age	D_x	N_x	M_x	Age	D_x	N_x	M_x
0	10,000,000.00	190,666,538.02	489,317.58	50	694,262.08	9,601,811.87	215,311.84
1	9,461,472.68	180,666,538.02	449,602.62	51	655,205.30	8,907,549.79	210,885.71
2	8,979,904.20	171,205,065.33	439,984.07	52	617,978.44	8,252,344.49	206,341.30
3	8,523,528.89	162,225,161.14	431,537.48	53	582,479.16	7,634,366.05	201,667.56
4	8,090,428.30	153,701,632.25	423,601.03	54	548,604.05	7,051,886.89	196,847.22
5	7,679,564.88	145,611,203.95	416,298.17	55	516,255.96	6,503,282.84	191,864.18
6	7,289,931.60	137,931,639.08	409,731.08	56	485,368.88	5,987,026.88	186,728.58
7	6,920,344.37	130,641,707.47	403,774.64	57	455,873.21	5,501,658.00	181,443.71
8	6,569,888.84	123,721,363.10	398,514.43	58	427,723.86	5,045,784.79	176,033.89
9	6,237,430.49	117,151,474.26	393,770.25	59	400,865.63	4,618,060.93	170,511.05
10	5,921,914.51	110,914,043.78	389,385.01	60	375,244.49	4,217,195.30	164,885.58
11	5,622,414.84	104,992,129.27	385,277.75	61	350,793.88	3,841,950.81	159,152.63
12	5,337,848.40	99,369,714.43	381,164.31	62	327,449.84	3,491,156.93	153,306.62
13	5,067,279.07	94,031,866.03	376,853.45	63	305,145.92	3,163,707.09	147,336.30
14	4,809,750.35	88,964,586.96	372,086.87	64	283,819.06	2,858,561.18	141,230.50
15	4,564,578.83	84,154,836.61	366,831.64	65	263,421.84	2,574,742.12	134,990.52
16	4,331,124.06	79,590,257.78	361,063.70	66	243,919.86	2,311,320.28	128,628.35
17	4,108,868.33	75,259,133.72	354,849.78	67	225,298.53	2,067,400.42	122,174.04
18	3,897,393.46	71,150,265.39	348,330.35	68	207,544.36	1,842,101.89	115,658.04
19	3,696,395.26	67,252,871.93	341,738.94	69	190,646.99	1,634,557.53	109,113.24
20	3,505,481.99	63,556,476.67	335,206.43	70	174,585.54	1,443,910.54	102,561.50
21	3,324,296.00	60,050,994.67	328,878.21	71	159,323.20	1,269,325.00	96,007.70
22	3,152,443.34	56,726,698.67	322,845.54	72	144,821.39	1,110,001.80	89,453.13
23	2,989,534.71	53,574,255.34	317,184.68	73	131,041.00	965,180.41	82,896.61
24	2,835,129.71	50,584,720.63	311,901.37	74	117,950.59	834,139.40	76,342.69
25	2,688,807.32	47,749,590.91	306,998.75	75	105,545.90	716,188.81	69,821.52
26	2,550,164.37	45,060,783.60	302,476.82	76	93,844.09	610,642.91	63,384.47
27	2,418,767.40	42,510,619.23	298,285.21	77	82,874.36	516,798.82	57,095.80
28	2,294,186.51	40,091,851.83	294,355.42	78	72,668.03	433,924.46	51,023.34
29	2,176,044.00	37,797,665.32	290,649.77	79	63,250.52	361,256.44	45,230.61
30	2,063,964.79	35,621,621.31	287,114.32	80	54,623.82	298,005.91	39,758.92
31	1,957,619.20	33,557,656.53	283,721.85	81	46,769.40	243,382.09	34,629.20
32	1,856,659.97	31,600,037.32	280,411.08	82	39,660.45	196,612.69	29,853.17
33	1,760,819.26	29,743,377.35	277,182.86	83	33,263.91	156,952.24	25,434.94
34	1,669,792.06	27,982,558.10	273,987.50	84	27,551.06	123,688.32	21,381.33
35	1,583,327.84	26,312,766.04	270,814.57	85	22,505.49	96,137.27	17,710.04
36	1,501,175.24	24,729,438.20	267,640.32	86	18,112.37	73,631.78	14,439.53
37	1,423,099.80	23,228,262.96	264,445.35	87	14,350.68	55,519.41	11,581.30
38	1,348,868.77	21,805,163.16	261,200.30	88	11,186.72	41,168.72	9,133.17
39	1,278,278.99	20,456,294.39	257,893.76	89	8,574.50	29,982.01	7,078.96
40	1,211,128.31	19,178,015.40	254,505.22	90	6,458.04	21,407.51	5,390.20
41	1,147,240.60	17,966,887.09	251,030.08	91	4,775.15	14,949.47	4,029.45
42	1,086,428.66	16,819,646.49	247,443.92	92	3,461.78	10,174.33	2,954.28
43	1,028,561.45	15,733,217.83	243,769.12	93	2,455.48	6,712.54	2,120.65
44	973,473.54	14,704,656.37	239,987.12	94	1,698.17	4,257.06	1,485.82
45	921,040.13	13,731,182.83	236,111.77	95	1,136.04	2,558.89	1,008.40
46	871,115.80	12,810,142.69	232,130.06	96	723.22	1,422.86	652.25
47	823,591.41	11,939,026.89	228,058.00	97	422.90	699.64	388.00
48	778,346.68	11,115,435.48	223,895.03	98	208.86	276.73	195.06
49	735,276.93	10,337,088.80	219,650.18	99	67.87	67.87	64.49

Age	Table III D_x	Commutation Columns at 5.5% based on 1980 CSO MortalityTable-Males N_x	M_x	Age	D_x	N_x	M_x
0	10,000,000.00	183,297,686.35	444,196.45	50	616,603.08	8,327,192.61	182,484.03
1	9,439,052.13	173,297,686.35	404,575.59	51	580,536.18	7,710,589.53	178,562.32
2	8,937,395.84	163,858,634.22	395,002.58	52	546,254.29	7,130,053.35	174,545.34
3	8,463,078.56	154,921,238.38	386,615.89	53	513,655.06	6,583,799.07	170,423.83
4	8,014,013.92	146,458,159.82	378,754.41	54	482,636.13	6,070,144.01	166,183.13
5	7,589,004.98	138,444,145.90	371,537.66	55	453,101.55	5,587,507.87	161,809.67
6	7,186,895.38	130,855,140.92	365,063.39	56	424,983.47	5,134,406.33	157,313.00
7	6,806,364.77	123,668,245.54	359,205.05	57	398,211.53	4,709,422.85	152,696.59
8	6,446,369.28	116,861,880.76	354,043.74	58	372,737.33	4,311,211.33	147,982.23
9	6,105,658.68	110,415,511.49	349,399.79	59	348,504.08	3,938,474.00	143,180.79
10	5,783,071.77	104,309,852.81	345,117.36	60	325,456.55	3,589,969.92	138,301.72
11	5,477,583.16	98,526,781.04	341,115.90	61	303,529.11	3,264,513.37	133,341.21
12	5,188,023.97	93,049,197.88	337,117.92	62	282,658.96	2,960,984.26	128,294.85
13	4,913,378.33	87,861,173.91	332,937.98	63	262,781.74	2,678,325.30	123,153.41
14	4,652,619.78	82,947,795.58	328,327.12	64	243,836.56	2,415,543.56	117,907.75
15	4,404,994.65	78,295,175.80	323,255.63	65	225,776.48	2,171,706.99	112,559.53
16	4,169,797.28	73,890,181.15	317,702.53	66	208,566.10	1,945,930.52	107,119.49
17	3,946,446.20	69,720,383.87	311,734.25	67	192,187.24	1,737,364.41	101,613.74
18	3,734,460.42	65,773,937.67	305,487.36	68	176,622.80	1,545,177.17	96,068.54
19	3,533,472.03	62,039,477.25	299,186.48	69	161,858.48	1,368,554.36	90,512.04
20	3,343,032.80	58,506,005.22	292,956.70	70	147,871.14	1,206,695.88	84,962.82
21	3,162,730.81	55,162,972.42	286,936.04	71	134,624.41	1,058,824.75	79,425.01
22	2,992,123.23	52,000,241.61	281,210.16	72	122,080.74	924,200.34	73,899.68
23	2,830,775.52	49,008,118.38	275,849.92	73	110,202.46	802,119.60	68,385.80
24	2,678,208.66	46,177,342.85	270,859.04	74	98,958.67	691,917.14	62,887.16
25	2,533,966.12	43,499,134.19	266,238.74	75	88,341.49	592,958.47	57,428.96
26	2,397,612.20	40,965,168.07	261,987.32	76	78,360.99	504,616.98	52,053.95
27	2,268,686.67	38,567,555.87	258,055.79	77	69,037.15	426,255.99	46,815.27
28	2,146,736.69	36,298,869.20	254,378.58	78	60,391.48	357,218.84	41,768.70
29	2,031,362.24	34,152,132.52	250,919.31	79	52,440.41	296,827.36	36,966.00
30	1,922,169.27	32,120,770.28	247,626.74	80	45,180.77	244,386.95	32,440.22
31	1,818,809.47	30,198,601.01	244,474.82	81	38,592.51	199,206.18	28,207.36
32	1,720,921.30	28,379,791.54	241,406.10	82	32,648.90	160,613.66	24,275.68
33	1,628,219.91	26,658,870.24	238,420.98	83	27,318.31	127,964.76	20,647.16
34	1,540,388.67	25,030,650.33	235,473.25	84	22,572.96	100,646.45	17,325.98
35	1,457,163.95	23,490,261.66	232,553.16	85	18,395.36	78,073.49	14,325.18
36	1,378,283.66	22,033,097.71	229,638.76	86	14,769.47	59,678.14	11,658.28
37	1,303,503.55	20,654,814.04	226,712.29	87	11,674.32	44,908.67	9,333.11
38	1,232,583.09	19,351,310.50	223,746.99	88	9,078.86	33,234.35	7,346.27
39	1,165,310.88	18,118,727.41	220,732.68	89	6,942.36	24,155.49	5,683.07
40	1,101,478.31	16,953,416.53	217,650.91	90	5,216.37	17,213.13	4,319.01
41	1,040,902.25	15,851,938.21	214,497.89	91	3,847.91	11,996.76	3,222.48
42	983,391.16	14,811,035.96	211,251.84	92	2,782.96	8,148.85	2,358.14
43	928,805.92	13,827,644.80	207,933.44	93	1,969.31	5,365.89	1,689.57
44	876,977.65	12,898,838.88	204,526.34	94	1,358.71	3,396.58	1,181.64
45	827,775.51	12,021,861.23	201,043.41	95	906.79	2,037.87	800.55
46	781,051.29	11,194,085.72	197,473.37	96	575.91	1,131.08	516.95
47	736,690.58	10,413,034.42	193,830.96	97	335.97	555.16	307.02
48	694,570.02	9,676,343.85	190,116.07	98	165.53	219.20	154.10
49	654,581.22	8,981,773.83	186,337.09	99	53.67	53.67	50.87

Table III	Commutation Columns at 5.75% based on 1980 CSO Mortality Table - Males						
Age	D_x	N_x	M_x	Age	D_x	N_x	M_x
0	10,000,000.00	176,461,857.52	405,147.23	50	547,784.66	7,227,074.59	154,823.16
1	9,416,737.59	166,461,857.52	365,620.04	51	514,523.90	6,679,289.93	151,347.38
2	8,895,188.59	157,045,119.93	356,092.24	52	482,995.63	6,164,766.03	147,795.59
3	8,403,198.51	148,149,931.34	347,764.89	53	453,097.85	5,681,770.40	144,159.98
4	7,938,499.60	139,746,732.83	339,977.48	54	424,729.42	5,228,672.55	140,428.08
5	7,499,723.56	131,808,233.24	332,845.63	55	397,795.75	4,803,943.12	136,588.44
6	7,085,554.20	124,308,509.67	326,462.66	56	372,227.73	4,406,147.37	132,649.97
7	6,694,525.58	117,222,955.47	320,700.58	57	347,954.61	4,033,919.65	128,616.19
8	6,325,456.14	110,528,429.89	315,636.08	58	324,925.45	3,685,965.04	124,506.55
9	5,976,972.73	104,202,973.74	311,090.00	59	303,082.46	3,361,039.59	120,330.90
10	5,647,801.42	98,226,001.01	306,907.75	60	282,369.66	3,057,957.13	116,097.76
11	5,336,811.94	92,578,199.59	303,009.12	61	262,722.61	2,775,587.47	111,804.14
12	5,042,744.64	87,241,387.65	299,123.09	62	244,079.86	2,512,864.86	107,446.54
13	4,764,499.58	82,198,643.01	295,069.81	63	226,379.17	2,268,785.00	103,017.33
14	4,500,976.37	77,434,143.44	290,609.24	64	209,561.83	2,042,405.84	98,509.03
15	4,251,347.83	72,933,167.06	285,714.63	65	193,581.62	1,832,844.01	93,923.44
16	4,014,840.34	68,681,819.23	280,367.90	66	178,402.63	1,639,262.39	89,270.16
17	3,790,806.42	64,666,978.89	274,634.99	67	164,003.89	1,460,859.76	84,571.80
18	3,578,700.59	60,876,172.47	268,648.66	68	150,365.59	1,296,855.86	79,850.97
19	3,378,090.23	57,297,471.88	262,624.86	69	137,470.41	1,146,490.28	75,131.69
20	3,188,469.81	53,919,381.64	256,683.10	70	125,293.71	1,009,019.87	70,429.74
21	3,009,372.76	50,730,911.83	250,954.38	71	113,799.86	883,726.16	65,748.56
22	2,840,307.20	47,721,539.08	245,519.03	72	102,952.56	769,926.29	61,088.96
23	2,680,793.45	44,881,231.87	240,442.78	73	92,715.72	666,973.73	56,450.01
24	2,530,313.99	42,200,438.42	235,727.50	74	83,059.26	574,258.01	51,834.83
25	2,388,377.07	39,670,124.43	231,372.67	75	73,972.62	491,198.75	47,264.41
26	2,254,514.92	37,281,747.36	227,374.99	76	65,460.34	417,226.13	42,774.29
27	2,128,240.85	35,027,232.44	223,686.84	77	57,535.15	351,765.79	38,408.41
28	2,009,079.48	32,898,991.59	220,245.42	78	50,210.92	294,230.63	34,212.56
29	1,896,608.96	30,889,912.11	217,015.63	79	43,497.14	244,019.71	30,228.93
30	1,790,416.77	28,993,303.14	213,948.75	80	37,386.98	200,522.58	26,483.86
31	1,690,136.57	27,202,886.37	211,019.82	81	31,859.71	163,135.60	22,989.45
32	1,595,393.01	25,512,749.80	208,174.94	82	26,889.30	131,275.89	19,751.36
33	1,505,885.04	23,917,356.79	205,414.10	83	22,445.89	104,386.59	16,770.02
34	1,421,284.95	22,411,471.75	202,694.29	84	18,503.06	81,940.71	14,047.65
35	1,341,316.74	20,990,186.80	200,006.35	85	15,043.03	63,437.65	11,593.70
36	1,265,708.27	19,648,870.06	197,329.99	86	12,049.36	48,394.62	9,417.97
37	1,194,206.17	18,383,161.79	194,648.91	87	9,501.73	36,345.26	7,525.51
38	1,126,562.73	17,188,955.62	191,938.67	88	7,371.82	26,843.52	5,912.24
39	1,062,559.02	16,062,392.89	189,190.14	89	5,623.70	19,471.70	4,564.96
40	1,001,980.56	14,999,833.87	186,386.75	90	4,215.57	13,848.00	3,462.60
41	944,637.92	13,997,853.31	183,525.33	91	3,102.30	9,632.43	2,578.55
42	890,335.75	13,053,215.39	180,586.45	92	2,238.41	6,530.13	1,883.34
43	838,927.77	12,162,879.65	177,589.16	93	1,580.22	4,291.72	1,346.86
44	790,242.18	11,323,951.88	174,519.03	94	1,087.68	2,711.50	940.25
45	744,142.89	10,533,709.71	171,387.99	95	724.20	1,623.82	635.90
46	700,479.45	9,789,566.82	168,186.22	96	458.86	899.62	409.94
47	659,132.98	9,089,087.36	164,927.29	97	267.05	440.77	243.08
48	619,977.67	8,429,954.38	161,611.35	98	131.26	173.72	121.82
49	582,902.12	7,809,976.71	158,246.18	99	42.46	42.46	40.15

Table III Commutation Columns at 6% based on 1980 CSO Mortality Table-Males

Age	D_x	N_x	M_x	Age	D_x	N_x	M_x
0	10,000,000.00	170,107,858.83	371,253.27	50	486,783.00	6,276,820.33	131,491.29
1	9,394,528.30	160,107,858.83	331,819.31	51	456,147.82	5,790,037.33	128,409.86
2	8,853,279.64	150,713,330.52	322,336.40	52	427,186.74	5,333,889.51	125,268.47
3	8,343,882.03	141,860,050.89	314,067.83	53	399,798.42	4,906,702.77	122,060.53
4	7,863,872.62	133,516,168.86	306,353.63	54	373,883.18	4,506,904.35	118,775.39
5	7,411,699.62	125,652,296.24	299,305.49	55	349,347.98	4,133,021.17	115,403.38
6	6,985,876.27	118,240,596.62	293,012.31	56	326,122.92	3,783,673.19	111,952.74
7	6,584,781.69	111,254,720.35	287,344.69	57	304,137.31	3,457,550.27	108,426.92
8	6,207,088.47	104,669,938.66	282,374.96	58	283,338.35	3,153,412.96	104,843.28
9	5,851,293.35	98,462,850.20	277,924.47	59	263,667.70	2,870,074.61	101,210.65
10	5,516,003.41	92,611,556.84	273,839.81	60	245,069.17	2,606,406.90	97,536.70
11	5,199,978.14	87,095,553.44	270,041.15	61	227,479.68	2,361,337.74	93,819.05
12	4,901,862.29	81,895,575.30	266,263.69	62	210,839.32	2,133,858.06	90,054.90
13	4,620,467.64	76,993,713.01	262,332.94	63	195,088.04	1,923,018.74	86,237.92
14	4,354,616.20	72,373,245.37	258,017.41	64	180,169.33	1,727,930.70	82,361.93
15	4,103,404.22	68,018,629.17	253,293.13	65	166,037.94	1,547,761.37	78,428.80
16	3,865,987.56	63,915,224.95	248,144.63	66	152,657.78	1,381,723.43	74,447.02
17	3,641,650.74	60,049,237.40	242,637.30	67	140,005.91	1,229,065.65	70,436.16
18	3,429,782.33	56,407,586.66	236,900.07	68	128,060.50	1,089,059.73	66,415.61
19	3,229,884.19	52,977,804.33	231,140.55	69	116,802.05	960,999.23	62,405.87
20	3,041,392.88	49,747,920.13	225,472.88	70	106,205.01	844,197.18	58,420.27
21	2,863,786.99	46,706,527.25	220,021.29	71	96,234.77	737,992.17	54,461.63
22	2,696,525.63	43,842,740.26	214,861.09	72	86,856.42	641,757.40	50,530.53
23	2,539,084.20	41,146,214.64	210,053.18	73	78,035.58	554,900.97	46,626.09
24	2,390,906.96	38,607,130.44	205,597.69	74	69,743.19	476,865.39	42,750.81
25	2,251,467.40	36,216,223.48	201,492.48	75	61,966.83	407,122.20	38,922.18
26	2,120,266.22	33,964,756.08	197,732.85	76	54,706.77	345,155.37	35,169.68
27	1,996,790.80	31,844,489.87	194,272.50	77	47,970.10	290,448.59	31,529.61
28	1,880,543.66	29,847,699.07	191,051.26	78	41,764.76	242,478.49	28,039.57
29	1,771,081.76	27,967,155.41	188,035.23	79	36,095.00	200,713.73	24,733.84
30	1,667,974.71	26,196,073.65	185,178.08	80	30,951.47	164,618.73	21,633.42
31	1,570,838.85	24,528,098.94	182,455.89	81	26,313.41	133,667.27	18,747.34
32	1,479,285.60	22,957,260.09	179,818.05	82	22,155.90	107,353.85	16,079.26
33	1,392,998.58	21,477,974.49	177,264.18	83	18,451.05	85,197.96	13,628.53
34	1,311,639.61	20,084,975.91	174,754.19	84	15,174.08	66,746.90	11,395.95
35	1,234,921.13	18,773,336.30	172,279.45	85	12,307.47	51,572.82	9,388.25
36	1,162,561.68	17,538,415.17	169,821.20	86	9,834.95	39,265.35	7,612.38
37	1,094,299.52	16,375,853.48	167,364.42	87	7,737.23	29,430.41	6,071.35
38	1,029,880.39	15,281,553.96	164,886.77	88	5,988.69	21,693.18	4,760.77
39	969,078.55	14,251,673.57	162,380.05	89	4,557.79	15,704.49	3,668.85
40	911,674.33	13,282,595.01	159,829.33	90	3,408.49	11,146.71	2,777.54
41	857,472.73	12,370,920.69	157,231.94	91	2,502.45	7,738.22	2,064.43
42	806,275.13	11,513,447.96	154,570.53	92	1,801.33	5,235.77	1,504.97
43	757,929.01	10,707,172.83	151,862.63	93	1,268.67	3,434.44	1,074.26
44	712,260.20	9,949,243.81	149,095.46	94	871.18	2,165.77	748.59
45	669,128.18	9,236,983.61	146,280.05	95	578.68	1,294.60	505.40
46	628,380.79	8,567,855.43	143,407.84	96	365.79	715.92	325.26
47	589,895.46	7,939,474.64	140,491.23	97	212.38	350.13	192.56
48	553,544.53	7,349,579.19	137,530.61	98	104.15	137.75	96.35
49	519,214.32	6,796,034.66	134,533.12	99	33.61	33.61	31.70

Table III	Commutation Columns at 6.25% based on 1980 CSO Mortality Table - Males						
Age	D_x	N_x	M_x	Age	D_x	N_x	M_x
0	10,000,000.00	164,190,331.53	341,745.20	50	432,694.86	5,455,378.85	111,790.22
1	9,372,423.53	154,190,331.53	302,404.03	51	404,509.62	5,022,683.99	109,057.62
2	8,811,666.16	144,817,908.00	292,965.69	52	377,935.72	4,618,174.37	106,278.41
3	8,285,122.51	136,006,241.84	284,755.35	53	352,872.80	4,240,238.65	103,447.00
4	7,790,120.51	127,721,119.32	277,113.50	54	329,222.84	3,887,365.85	100,554.27
5	7,324,912.52	119,930,998.81	270,147.89	55	306,894.56	3,558,143.00	97,592.03
6	6,887,830.46	112,606,086.29	263,943.03	56	285,817.76	3,251,248.44	94,567.85
7	6,477,089.03	105,718,255.83	258,368.10	57	265,922.15	2,965,430.68	91,485.05
8	6,091,206.85	99,241,166.80	253,491.15	58	247,153.69	2,699,508.53	88,359.07
9	5,728,543.44	93,149,959.95	249,134.03	59	229,453.98	2,452,354.84	85,197.82
10	5,387,580.74	87,421,416.51	245,144.47	60	212,767.00	2,222,900.86	82,008.13
11	5,066,962.73	82,033,835.78	241,442.98	61	197,031.25	2,010,133.86	78,788.09
12	4,765,233.92	76,966,873.05	237,770.80	62	182,188.54	1,813,102.60	75,535.45
13	4,481,113.85	72,201,639.13	233,958.61	63	168,181.03	1,630,914.06	72,244.91
14	4,213,343.39	67,720,525.28	229,783.08	64	154,954.49	1,462,733.03	68,911.37
15	3,960,939.41	63,507,181.89	225,222.83	65	142,464.79	1,307,778.54	65,536.64
16	3,722,984.91	59,546,242.48	220,264.77	66	130,676.08	1,165,313.75	62,128.22
17	3,498,694.66	55,823,257.57	214,973.63	67	119,564.01	1,034,637.66	58,702.97
18	3,287,390.05	52,324,562.91	209,474.59	68	109,105.39	915,073.66	55,277.53
19	3,088,506.75	49,037,172.85	203,967.17	69	99,279.23	805,968.27	51,869.33
20	2,901,423.05	45,948,666.10	198,560.33	70	90,059.57	706,689.04	48,489.63
21	2,725,562.65	43,047,243.06	193,371.88	71	81,413.01	616,629.47	45,140.68
22	2,560,335.84	40,321,680.41	188,472.29	72	73,306.19	535,216.46	41,822.87
23	2,405,173.51	37,761,344.57	183,917.95	73	65,706.50	461,910.27	38,535.30
24	2,259,482.14	35,356,171.05	179,707.37	74	58,586.07	396,203.77	35,279.97
25	2,122,701.01	33,096,688.91	175,836.95	75	51,931.25	337,617.70	32,071.39
26	1,994,299.97	30,973,987.90	172,300.69	76	45,739.09	285,686.45	28,934.00
27	1,873,741.10	28,979,687.93	169,053.58	77	40,012.34	239,947.36	25,897.79
28	1,760,505.41	27,105,946.83	166,037.95	78	34,754.44	199,935.02	22,993.56
29	1,654,129.40	25,345,441.42	163,221.08	79	29,965.69	165,180.58	20,249.18
30	1,554,165.47	23,691,312.02	160,558.88	80	25,635.12	135,214.89	17,681.30
31	1,460,213.49	22,137,146.55	158,028.40	81	21,742.44	109,579.77	15,296.57
32	1,371,872.28	20,676,933.06	155,582.10	82	18,264.06	87,837.33	13,097.16
33	1,288,811.05	19,305,060.78	153,219.24	83	15,174.21	69,573.28	11,081.66
34	1,210,681.84	18,016,249.73	150,902.44	84	12,449.85	54,399.06	9,249.91
35	1,137,186.38	16,805,567.90	148,623.57	85	10,074.13	41,949.21	7,606.53
36	1,068,034.69	15,668,381.51	146,365.19	86	8,031.33	31,875.08	6,156.33
37	1,002,957.41	14,600,346.82	144,113.48	87	6,303.44	23,843.75	4,900.87
38	941,694.42	13,597,389.41	141,847.99	88	4,867.45	17,540.31	3,835.66
39	884,013.95	12,655,694.99	139,561.30	89	3,695.73	12,672.86	2,950.27
40	829,691.78	11,771,681.04	137,239.96	90	2,757.31	8,977.13	2,229.24
41	778,528.12	10,941,989.26	134,881.70	91	2,019.60	6,219.82	1,653.73
42	730,321.65	10,163,461.13	132,471.00	92	1,450.34	4,200.22	1,203.27
43	684,914.52	9,433,139.48	130,023.96	93	1,019.06	2,749.88	857.31
44	642,130.72	8,748,224.96	127,529.25	94	698.13	1,730.81	596.32
45	601,826.10	8,106,094.24	124,997.02	95	462.64	1,032.68	401.89
46	563,847.32	7,504,268.14	122,419.78	96	291.75	570.04	258.22
47	528,068.91	6,940,420.82	119,808.86	97	169.00	278.29	152.63
48	494,361.95	6,412,351.92	117,164.78	98	82.68	109.29	76.25
49	462,611.12	5,917,989.97	114,494.06	99	26.61	26.61	25.05

Age	Table III D_x	Commutation Columns at 6.5% based on 1980 CSO MortalityTable-Males N_x	M_x	Age	D_x	N_x	M_x
0	10,000,000.00	158,669,005.07	315,976.22	50	384,723.11	4,744,747.60	95,137.58
1	9,350,422.54	148,669,005.07	276,727.39	51	358,818.42	4,360,024.49	92,713.64
2	8,770,345.39	139,318,582.54	267,333.31	52	334,459.21	4,001,206.06	90,254.14
3	8,226,913.44	130,548,237.15	259,180.65	53	311,546.39	3,666,746.86	87,754.33
4	7,717,231.00	122,321,323.71	251,610.30	54	289,983.86	3,355,200.47	85,206.37
5	7,239,342.05	114,604,092.71	244,726.07	55	269,682.27	3,065,216.60	82,603.32
6	6,791,386.30	107,364,750.66	238,608.09	56	250,571.54	2,795,534.33	79,952.07
7	6,371,404.58	100,573,364.36	233,124.13	57	232,582.15	2,544,962.79	77,255.78
8	5,977,753.40	94,201,959.78	228,338.01	58	215,659.35	2,312,380.64	74,528.14
9	5,608,648.06	88,224,206.38	224,072.08	59	199,745.10	2,096,721.29	71,776.19
10	5,262,439.32	82,615,558.32	220,175.19	60	184,783.90	1,896,976.19	69,006.01
11	4,937,650.54	77,353,119.00	216,568.16	61	170,716.03	1,712,192.29	66,216.03
12	4,632,721.53	72,415,468.46	212,998.10	62	157,485.14	1,541,476.26	63,404.43
13	4,346,275.78	67,782,746.93	209,300.62	63	145,035.68	1,383,991.12	60,566.74
14	4,076,969.75	63,436,471.14	205,260.24	64	133,315.72	1,238,955.44	57,698.72
15	3,823,738.32	59,359,501.39	200,857.95	65	122,282.43	1,105,639.73	54,802.07
16	3,585,589.53	55,535,763.08	196,082.87	66	111,900.48	983,357.30	51,883.37
17	3,361,666.83	51,950,173.54	190,998.96	67	102,144.65	871,456.82	48,957.15
18	3,151,223.41	48,588,506.71	185,727.69	68	92,990.95	769,312.16	46,037.63
19	2,953,628.32	45,437,283.31	180,460.79	69	84,417.45	676,321.21	43,139.63
20	2,768,201.34	42,483,654.99	175,302.21	70	76,398.19	591,903.76	40,272.60
21	2,594,311.49	39,715,453.65	170,363.61	71	68,901.12	515,505.57	37,438.34
22	2,431,320.53	37,121,142.16	165,710.91	72	61,894.56	446,604.45	34,637.01
23	2,278,615.38	34,689,821.63	161,396.22	73	55,347.69	384,709.89	31,867.74
24	2,135,565.30	32,411,206.25	157,416.56	74	49,233.97	329,362.21	29,132.06
25	2,001,576.07	30,275,640.95	153,767.00	75	43,539.02	280,128.23	26,441.99
26	1,876,087.49	28,274,064.88	150,440.34	76	38,257.51	236,589.21	23,817.79
27	1,758,537.03	26,397,977.39	147,392.87	77	33,388.93	198,331.71	21,284.17
28	1,648,384.91	24,639,440.35	144,569.30	78	28,933.31	164,942.78	18,866.38
29	1,545,147.98	22,991,055.44	141,938.02	79	24,888.08	136,009.47	16,587.03
30	1,448,362.21	21,445,907.46	139,457.06	80	21,241.34	111,121.39	14,459.28
31	1,357,611.84	19,997,545.25	137,104.38	81	17,973.56	89,880.06	12,487.92
32	1,272,483.82	18,639,933.42	134,835.30	82	15,062.69	71,906.50	10,674.03
33	1,192,633.96	17,367,449.59	132,648.77	83	12,485.06	56,843.82	9,015.72
34	1,117,705.22	16,174,815.64	130,509.90	84	10,219.46	44,358.76	7,512.12
35	1,047,389.54	15,057,110.41	128,410.97	85	8,249.94	34,139.30	6,166.32
36	981,389.20	14,009,720.87	126,335.81	86	6,561.61	25,889.36	4,981.50
37	919,428.02	13,028,331.67	124,271.63	87	5,137.83	19,327.75	3,958.20
38	861,240.76	12,108,903.65	122,199.69	88	3,958.06	14,189.92	3,092.01
39	806,590.35	11,247,662.89	120,113.28	89	2,998.20	10,231.86	2,373.72
40	755,248.77	10,441,072.53	118,000.21	90	2,231.65	7,233.66	1,790.15
41	707,012.15	9,685,823.76	115,858.58	91	1,630.74	5,002.02	1,325.45
42	661,677.06	8,978,811.62	113,674.47	92	1,168.34	3,371.28	962.58
43	619,081.18	8,317,134.56	111,462.64	93	818.99	2,202.94	684.54
44	579,047.26	7,698,053.38	109,213.01	94	559.75	1,383.95	475.28
45	541,428.24	7,119,006.12	106,934.91	95	370.07	824.20	319.76
46	506,070.17	6,577,577.87	104,621.76	96	232.83	454.13	205.11
47	472,845.38	6,071,507.70	102,283.88	97	134.55	221.31	121.04
48	441,624.26	5,598,662.32	99,921.86	98	65.67	86.76	60.37
49	412,290.46	5,157,038.06	97,541.65	99	21.09	21.09	19.80

Table III Commutation Columns at 6.75% based on 1980 CSO Mortality Table - Males

Age	D_x	N_x	M_x	Age	D_x	N_x	M_x
0	10,000,000.00	153,508,054.92	293,401.68	50	342,164.13	4,129,517.59	81,047.10
1	9,328,524.59	143,508,054.92	254,244.77	51	318,377.72	3,787,353.46	78,896.35
2	8,729,314.59	134,179,530.33	244,894.64	52	296,068.91	3,468,975.75	76,719.16
3	8,169,248.37	125,450,215.74	236,799.13	53	275,140.23	3,172,906.83	74,511.46
4	7,645,192.00	117,280,967.37	229,299.45	54	255,497.66	2,897,766.60	72,266.52
5	7,154,968.36	109,635,775.38	222,495.44	55	237,053.96	2,642,268.94	69,978.41
6	6,696,513.96	102,480,807.02	216,462.93	56	219,739.58	2,405,214.98	67,653.39
7	6,267,686.31	95,784,293.05	211,068.24	57	203,486.06	2,185,475.39	65,294.40
8	5,866,671.73	89,516,606.75	206,371.07	58	188,238.43	1,981,989.34	62,913.58
9	5,491,534.37	83,649,935.02	202,194.21	59	173,939.35	1,793,750.91	60,517.16
10	5,140,487.90	78,158,400.65	198,387.63	60	160,534.20	1,619,811.56	58,110.52
11	4,811,930.15	73,017,912.75	194,872.44	61	147,965.16	1,459,277.36	55,692.36
12	4,504,191.90	68,205,982.60	191,401.43	62	136,177.85	1,311,312.19	53,261.15
13	4,215,797.03	63,701,790.70	187,814.94	63	125,119.06	1,175,134.35	50,813.14
14	3,945,314.48	59,485,993.67	183,905.04	64	114,739.17	1,050,015.28	48,344.76
15	3,691,594.79	55,540,679.20	179,654.89	65	104,996.82	935,276.12	45,857.58
16	3,453,569.15	51,849,084.41	175,055.62	66	95,857.42	830,279.30	43,357.32
17	3,230,308.33	48,395,515.25	170,170.37	67	87,295.36	734,421.88	40,856.50
18	3,020,996.54	45,165,206.92	165,116.94	68	79,286.26	647,126.52	38,367.25
19	2,824,935.92	42,144,210.38	160,079.52	69	71,807.73	567,840.26	35,902.14
20	2,641,387.73	39,319,274.47	155,157.26	70	64,834.14	496,032.53	33,469.08
21	2,469,666.58	36,677,886.73	150,455.94	71	58,334.93	431,198.40	31,069.46
22	2,309,086.21	34,208,220.16	146,037.16	72	52,280.13	372,863.46	28,703.28
23	2,158,990.24	31,899,133.95	141,948.98	73	46,640.73	320,583.34	26,369.65
24	2,018,711.39	29,740,143.71	138,187.08	74	41,391.62	273,942.61	24,069.73
25	1,887,622.74	27,721,432.32	134,745.29	75	36,518.08	232,550.99	21,813.45
26	1,765,134.95	25,833,809.58	131,615.37	76	32,013.10	196,032.91	19,617.57
27	1,650,661.68	24,068,674.63	128,754.85	77	27,873.74	164,019.81	17,502.46
28	1,543,643.13	22,418,012.95	126,110.69	78	24,097.53	136,146.07	15,488.77
29	1,443,577.41	20,874,369.82	123,652.38	79	20,679.86	112,048.54	13,594.82
30	1,349,984.89	19,430,792.41	121,339.94	80	17,608.39	91,368.68	11,830.98
31	1,262,435.10	18,080,807.52	119,152.19	81	14,864.62	73,760.29	10,200.62
32	1,180,503.94	16,818,372.42	117,047.13	82	12,428.07	58,895.67	8,703.99
33	1,103,834.76	15,637,868.48	115,023.40	83	10,277.17	46,467.60	7,338.94
34	1,032,062.26	14,534,033.72	113,048.42	84	8,392.53	36,190.42	6,104.14
35	964,869.50	13,501,971.46	111,114.86	85	6,759.23	27,797.90	5,001.52
36	901,951.82	12,537,101.96	109,207.67	86	5,363.38	21,038.67	4,033.06
37	843,027.08	11,635,150.14	107,315.01	87	4,189.77	15,675.29	3,198.59
38	787,825.60	10,792,123.05	105,419.69	88	3,220.13	11,485.52	2,493.88
39	736,105.84	10,004,297.45	103,515.60	89	2,433.52	8,265.39	1,910.88
40	687,636.61	9,268,191.61	101,591.70	90	1,807.09	5,831.87	1,438.33
41	642,210.74	8,580,555.01	99,646.37	91	1,317.41	4,024.78	1,062.92
42	599,623.28	7,938,344.27	97,667.09	92	941.65	2,707.37	770.46
43	559,708.29	7,338,720.99	95,667.39	93	658.54	1,765.72	546.89
44	522,287.79	6,779,012.69	93,638.27	94	449.03	1,107.19	379.02
45	487,212.58	6,256,724.91	91,588.29	95	296.17	658.16	254.56
46	454,328.57	5,769,512.33	89,511.63	96	185.90	361.98	163.01
47	423,506.60	5,315,183.76	87,417.70	97	107.18	176.09	96.04
48	394,616.90	4,891,677.16	85,307.10	98	52.19	68.91	47.83
49	367,542.67	4,497,060.26	83,185.23	99	16.72	16.72	15.66

Table III Commutation Columns at 7% based on 1980 CSO MortalityTable-Males

Age	D_x	N_x	M_x	Age	D_x	N_x	M_x
0	10,000,000.00	148,675,548.57	273,562.24	50	304,396.59	3,596,488.36	69,112.31
1	9,306,728.97	138,675,548.57	234,496.82	51	282,573.92	3,292,091.77	67,203.43
2	8,688,571.05	129,368,819.60	225,190.33	52	262,159.93	3,009,517.85	65,275.59
3	8,112,120.96	120,680,248.55	217,151.43	53	243,059.01	2,747,357.92	63,325.31
4	7,573,991.62	112,568,127.59	209,721.59	54	225,179.40	2,504,298.91	61,346.76
5	7,071,771.96	104,994,135.96	202,996.71	55	208,436.15	2,279,119.51	59,334.88
6	6,603,184.23	97,922,364.00	197,048.27	56	192,760.58	2,070,683.36	57,295.32
7	6,165,893.15	91,319,179.77	191,741.20	57	178,085.56	1,877,922.78	55,230.80
8	5,757,906.86	85,153,286.62	187,131.10	58	164,356.33	1,699,837.22	53,152.03
9	5,377,131.53	79,395,379.76	183,041.27	59	151,516.56	1,535,480.89	51,064.54
10	5,021,638.00	74,018,248.23	179,322.69	60	139,512.76	1,383,964.33	48,973.04
11	4,689,693.73	68,996,610.23	175,896.80	61	128,289.15	1,244,451.57	46,876.44
12	4,379,516.40	64,306,916.50	172,521.86	62	117,793.42	1,116,162.41	44,773.45
13	4,089,526.92	59,927,400.10	169,042.80	63	107,974.74	998,368.99	42,660.88
14	3,818,203.84	55,837,873.18	165,258.87	64	98,785.79	890,394.25	40,535.70
15	3,564,311.19	52,019,669.34	161,155.25	65	90,186.81	791,608.47	38,399.34
16	3,326,701.65	48,455,358.15	156,724.94	66	82,144.17	701,421.66	36,256.77
17	3,104,372.16	45,128,656.50	152,030.15	67	74,632.20	619,277.49	34,118.72
18	2,896,437.33	42,024,284.34	147,185.08	68	67,626.54	544,645.29	31,995.54
19	2,702,132.34	39,127,847.02	142,366.64	69	61,104.69	477,018.75	29,897.85
20	2,520,660.04	36,425,714.68	137,669.36	70	55,041.61	415,914.06	27,832.28
21	2,351,281.09	33,905,054.64	133,193.40	71	49,408.34	360,872.45	25,799.86
22	2,193,261.82	31,553,773.55	128,996.26	72	44,176.60	311,464.12	23,800.44
23	2,045,903.37	29,360,511.73	125,122.22	73	39,319.24	267,287.52	21,833.14
24	1,908,502.69	27,314,608.36	121,565.70	74	34,812.59	227,968.28	19,898.78
25	1,780,401.09	25,406,105.67	118,319.41	75	30,641.91	193,155.70	18,005.56
26	1,660,981.02	23,625,704.58	115,374.18	76	26,799.07	162,513.78	16,167.33
27	1,549,633.27	21,964,723.55	112,688.74	77	23,279.38	135,714.71	14,400.84
28	1,445,778.87	20,415,090.29	110,212.21	78	20,078.57	112,435.33	12,722.99
29	1,348,898.13	18,969,311.42	107,915.14	79	17,190.64	92,356.76	11,148.60
30	1,258,496.72	17,620,413.29	105,759.40	80	14,603.21	75,166.12	9,685.80
31	1,174,130.44	16,361,916.57	103,724.68	81	12,298.90	60,562.92	8,336.84
32	1,095,364.93	15,187,786.13	101,771.45	82	10,258.89	48,264.02	7,101.43
33	1,021,832.15	14,092,421.20	99,898.06	83	8,463.59	38,005.12	5,977.27
34	953,159.33	13,070,589.05	98,074.06	84	6,895.37	29,541.54	4,962.75
35	889,021.55	12,117,429.72	96,292.50	85	5,540.46	22,646.17	4,058.94
36	829,108.10	11,228,408.17	94,539.34	86	4,386.03	17,105.70	3,266.96
37	773,131.64	10,399,300.07	92,803.60	87	3,418.27	12,719.67	2,586.15
38	720,818.82	9,626,168.43	91,069.48	88	2,621.05	9,301.40	2,012.55
39	671,924.38	8,905,349.60	89,331.41	89	1,976.15	6,680.35	1,539.12
40	626,214.66	8,233,425.23	87,579.36	90	1,464.03	4,704.20	1,156.28
41	583,479.91	7,607,210.57	85,811.93	91	1,064.82	3,240.17	852.84
42	543,514.26	7,023,730.66	84,017.86	92	759.32	2,175.35	617.01
43	506,148.90	6,480,216.40	82,209.51	93	529.79	1,416.03	437.15
44	471,205.70	5,974,067.50	80,378.85	94	360.40	886.24	302.42
45	438,533.99	5,502,861.80	78,533.69	95	237.16	525.84	202.75
46	407,980.05	5,064,327.81	76,668.88	96	148.51	288.69	129.62
47	379,413.84	4,656,347.76	74,792.96	97	85.42	140.18	76.25
48	352,705.93	4,276,933.92	72,906.52	98	41.50	54.76	37.91
49	327,739.63	3,924,227.99	71,014.43	99	13.26	13.26	12.40

Age	D_x	N_x	M_x	Age	D_x	N_x	M_x
Table III Commutation Columns at 7.25% based on 1980 CSO Mortality Table - Males							
0	10,000,000.00	144,142,967.01	256,069.83	50	270,871.69	3,134,341.80	58,993.11
1	9,285,034.97	134,142,967.01	217,095.47	51	250,866.34	2,863,470.11	57,298.43
2	8,648,112.11	124,857,932.05	207,832.32	52	232,200.48	2,612,603.77	55,590.90
3	8,055,524.97	116,209,819.94	199,849.50	53	214,780.57	2,380,403.29	53,867.53
4	7,503,618.19	108,154,294.97	192,488.70	54	198,517.32	2,165,622.72	52,123.24
5	6,989,733.73	100,650,676.77	185,841.83	55	183,328.20	1,967,105.40	50,353.71
6	6,511,368.52	93,660,943.04	179,976.10	56	169,145.69	1,783,777.20	48,564.01
7	6,065,985.00	87,149,574.52	174,755.02	57	155,904.22	1,614,631.51	46,756.64
8	5,651,405.25	81,083,589.52	170,230.20	58	143,549.64	1,458,727.29	44,941.03
9	5,265,370.69	75,432,184.28	166,225.36	59	132,026.84	1,315,177.66	43,122.06
10	4,905,803.74	70,166,813.59	162,592.57	60	121,283.73	1,183,150.81	41,303.84
11	4,570,836.91	65,261,009.85	159,253.50	61	111,266.65	1,061,867.08	39,485.43
12	4,258,570.87	60,690,172.93	155,971.77	62	101,925.44	950,600.43	37,665.74
13	3,967,320.35	56,431,602.07	152,596.67	63	93,211.65	848,674.99	35,842.01
14	3,695,470.89	52,464,281.71	148,934.37	64	85,080.30	755,463.34	34,011.68
15	3,441,698.06	48,768,810.82	144,971.92	65	77,493.27	670,383.04	32,176.01
16	3,204,774.54	45,327,112.76	140,703.98	66	70,418.08	592,889.77	30,339.29
17	2,983,622.59	42,122,338.22	136,191.80	67	63,829.32	522,471.68	28,510.72
18	2,777,286.73	39,138,715.63	131,546.04	68	57,702.89	458,642.37	26,699.09
19	2,584,935.28	36,361,428.91	126,936.59	69	52,016.53	400,939.47	24,913.40
20	2,405,712.99	33,776,493.63	122,453.52	70	46,746.01	348,922.94	23,159.14
21	2,238,827.14	31,370,780.63	118,191.62	71	41,863.94	302,176.93	21,437.06
22	2,083,497.43	29,131,953.49	114,204.54	72	37,343.81	260,312.99	19,746.89
23	1,938,983.37	27,048,456.06	110,532.96	73	33,160.26	222,969.19	18,087.75
24	1,804,547.10	25,109,472.69	107,170.16	74	29,291.09	189,808.93	16,460.19
25	1,679,499.09	23,304,925.59	104,107.85	75	25,721.82	160,517.84	14,870.96
26	1,563,194.68	21,625,426.50	101,336.01	76	22,443.57	134,796.02	13,331.49
27	1,455,002.72	20,062,231.82	98,814.56	77	19,450.47	112,352.45	11,855.55
28	1,354,326.02	18,607,229.10	96,494.69	78	16,737.02	92,901.98	10,456.93
29	1,260,628.09	17,252,903.08	94,347.93	79	14,296.30	76,164.96	9,147.62
30	1,173,400.83	15,992,274.99	92,337.96	80	12,116.20	61,868.66	7,933.93
31	1,092,187.32	14,818,874.16	90,445.24	81	10,180.54	49,752.47	6,817.32
32	1,016,543.79	13,726,686.84	88,632.56	82	8,472.11	39,571.92	5,797.08
33	946,091.85	12,710,143.05	86,898.03	83	6,973.20	31,099.81	4,870.88
34	880,452.06	11,764,051.20	85,213.17	84	5,667.89	24,126.62	4,036.96
35	819,292.49	10,883,599.14	83,571.34	85	4,543.56	18,458.72	3,295.77
36	762,297.20	10,064,306.65	81,959.45	86	3,588.46	13,915.16	2,647.81
37	709,174.47	9,302,009.45	80,367.30	87	2,790.17	10,326.69	2,092.09
38	659,647.98	8,592,834.98	78,780.35	88	2,134.45	7,536.52	1,624.98
39	613,469.53	7,933,187.00	77,193.48	89	1,605.52	5,402.08	1,240.35
40	570,403.66	7,319,717.47	75,597.58	90	1,186.68	3,796.56	930.03
41	530,238.74	6,749,313.81	73,991.42	91	861.08	2,609.88	684.66
42	492,768.53	6,219,075.07	72,364.86	92	612.61	1,748.80	494.39
43	457,822.15	5,726,306.53	70,729.17	93	426.43	1,136.19	349.62
44	425,221.79	5,268,484.38	69,077.16	94	289.41	709.76	241.43
45	394,815.98	4,843,262.59	67,415.94	95	190.00	420.36	161.58
46	366,451.80	4,448,446.61	65,740.96	96	118.70	230.36	103.13
47	339,998.95	4,081,994.81	64,059.91	97	68.12	111.66	60.57
48	315,328.81	3,741,995.86	62,373.38	98	33.01	43.54	30.07
49	292,325.24	3,426,667.05	60,685.75	99	10.53	10.53	9.82

Table III Commutation Columns at 7.5% based on 1980 CSO MortalityTable-Males

Age	D_x	N_x	M_x	Age	D_x	N_x	M_x
0	10,000,000.00	139,884,790.41	240,596.02	50	241,104.58	2,733,365.67	50,404.65
1	9,263,441.86	129,884,790.41	201,712.30	51	222,778.39	2,492,261.09	48,899.71
2	8,607,935.10	120,621,348.55	192,492.18	52	205,722.89	2,269,482.71	47,386.89
3	7,999,454.24	112,013,413.45	184,564.93	53	189,846.82	2,063,759.82	45,863.58
4	7,434,060.21	104,013,959.21	177,272.36	54	175,063.49	1,873,913.00	44,325.37
5	6,908,834.91	96,579,899.00	170,702.42	55	161,292.91	1,698,849.51	42,768.53
6	6,421,038.82	89,671,064.10	164,918.06	56	148,469.00	1,537,556.59	41,197.61
7	5,967,922.69	83,250,025.28	159,781.39	57	136,527.95	1,389,087.59	39,614.86
8	5,547,114.66	77,282,102.60	155,340.06	58	125,416.48	1,252,559.64	38,028.60
9	5,156,184.88	71,734,987.93	151,418.28	59	115,080.99	1,127,143.16	36,443.10
10	4,792,901.86	66,578,803.05	147,869.09	60	105,470.92	1,012,062.17	34,861.93
11	4,455,258.72	61,785,901.20	144,614.45	61	96,534.84	906,591.25	33,284.28
12	4,141,235.40	57,330,642.48	141,423.14	62	88,224.76	810,056.41	31,709.19
13	3,849,037.53	53,189,407.07	138,148.67	63	80,494.63	721,831.66	30,134.28
14	3,576,955.18	49,340,369.54	134,603.82	64	73,301.78	641,337.03	28,557.34
15	3,323,573.72	45,763,414.36	130,777.37	65	66,609.84	568,035.24	26,979.47
16	3,087,584.61	42,439,840.65	126,665.50	66	60,387.55	501,425.40	25,404.38
17	2,867,834.65	39,352,256.03	122,328.42	67	54,610.01	441,037.85	23,839.93
18	2,663,298.09	36,484,421.38	117,873.34	68	49,253.65	386,427.84	22,293.57
19	2,473,076.61	33,821,123.29	113,463.36	69	44,296.68	337,174.19	20,772.90
20	2,296,257.28	31,348,046.68	109,184.25	70	39,715.78	292,877.51	19,282.46
21	2,131,994.76	29,051,789.40	105,125.73	71	35,485.22	253,161.73	17,822.77
22	1,979,462.94	26,919,794.65	101,337.73	72	31,580.20	217,676.51	16,393.46
23	1,837,880.74	24,940,331.71	97,857.59	73	27,977.12	186,096.32	14,993.65
24	1,706,476.45	23,102,450.97	94,677.55	74	24,655.26	158,119.20	13,623.68
25	1,584,530.81	21,395,974.52	91,788.40	75	21,600.53	133,463.95	12,289.09
26	1,471,373.13	19,811,443.71	89,179.38	76	18,803.71	111,863.42	10,999.29
27	1,366,351.36	18,340,070.58	86,811.55	77	16,258.13	93,059.71	9,765.59
28	1,268,851.07	16,973,719.22	84,638.10	78	13,957.49	76,801.58	8,599.24
29	1,178,319.98	15,704,868.16	82,631.50	79	11,894.38	62,844.09	7,509.91
30	1,094,237.24	14,526,548.18	80,757.13	80	10,057.11	50,949.71	6,502.48
31	1,016,134.19	13,432,310.94	78,996.22	81	8,430.76	40,892.60	5,577.79
32	943,558.57	12,416,176.75	77,313.68	82	6,999.65	32,461.84	4,734.87
33	876,122.65	11,472,618.18	75,707.43	83	5,747.85	25,462.20	3,971.42
34	813,441.19	10,596,495.53	74,150.80	84	4,661.05	19,714.34	3,285.63
35	755,176.14	9,783,054.34	72,637.46	85	3,727.76	15,053.29	2,677.53
36	701,007.14	9,027,878.21	71,155.17	86	2,937.30	11,325.53	2,147.15
37	650,638.93	8,326,871.07	69,694.44	87	2,278.55	8,388.23	1,693.33
38	603,792.94	7,676,232.13	68,241.86	88	1,739.01	6,109.68	1,312.76
39	560,218.73	7,072,439.20	66,792.74	89	1,305.04	4,370.66	1,000.11
40	519,679.72	6,512,220.47	65,338.75	90	962.34	3,065.63	748.46
41	481,963.07	5,992,540.75	63,878.83	91	696.67	2,103.29	549.93
42	446,862.70	5,510,577.69	62,403.79	92	494.49	1,406.62	396.35
43	414,206.37	5,063,714.99	60,923.93	93	343.40	912.13	279.77
44	383,817.10	4,649,508.61	59,432.78	94	232.52	568.73	192.84
45	355,543.19	4,265,691.51	57,936.81	95	152.30	336.21	128.84
46	329,232.99	3,910,148.32	56,431.94	96	94.93	183.91	82.09
47	304,756.44	3,580,915.33	54,925.14	97	54.35	88.98	48.14
48	281,986.17	3,276,158.89	53,416.95	98	26.28	34.64	23.86
49	260,807.04	2,994,172.71	51,911.27	99	8.36	8.36	7.78

Age	Table III D_x	Commutation Columns at 7.75% based on 1980 CSO Mortality Table - Males N_x	M_x	Age	D_x	N_x	M_x
0	10,000,000.00	135,878,139.01	226,862.39	50	214,666.72	2,385,218.81	43,108.06
1	9,241,948.96	125,878,139.01	188,068.89	51	197,889.85	2,170,552.09	41,771.25
2	8,568,037.42	116,636,190.06	178,891.50	52	182,315.78	1,972,662.25	40,430.56
3	7,943,902.67	108,068,152.63	171,019.30	53	167,855.72	1,790,346.47	39,083.70
4	7,365,306.36	100,124,249.96	163,794.18	54	154,425.71	1,622,490.74	37,726.84
5	6,829,057.07	92,758,943.60	157,300.10	55	141,948.40	1,468,065.04	36,356.71
6	6,332,167.69	85,929,886.54	151,595.81	56	130,359.34	1,326,116.64	34,977.40
7	5,871,667.93	79,597,718.84	146,541.98	57	119,596.68	1,195,757.30	33,590.94
8	5,444,984.23	73,726,050.91	142,182.43	58	109,608.28	1,076,160.61	32,204.62
9	5,049,508.99	68,281,066.68	138,341.78	59	100,342.18	966,552.33	30,822.18
10	4,682,851.55	63,231,557.69	134,874.08	60	91,749.53	866,210.15	29,446.72
11	4,342,861.39	58,548,706.14	131,701.55	61	83,781.16	774,460.62	28,077.50
12	4,027,394.23	54,205,844.75	128,597.97	62	76,391.31	690,679.46	26,713.67
13	3,734,543.79	50,178,450.52	125,420.90	63	69,536.30	614,288.14	25,353.16
14	3,462,502.49	46,443,906.73	121,989.48	64	63,175.75	544,751.84	23,994.06
15	3,209,763.97	42,981,404.24	118,294.05	65	57,275.05	481,576.10	22,637.32
16	2,974,937.42	39,771,640.28	114,332.20	66	51,804.28	424,301.05	21,286.11
17	2,756,793.66	36,796,702.86	110,163.06	67	46,739.24	372,496.77	19,947.13
18	2,554,236.55	34,039,909.20	105,890.41	68	42,057.08	325,757.52	18,626.72
19	2,366,301.58	31,485,672.65	101,670.83	69	37,736.61	283,700.45	17,331.25
20	2,192,018.70	29,119,371.08	97,585.97	70	33,755.62	245,963.83	16,064.48
21	2,030,490.80	26,927,352.38	93,720.68	71	30,089.96	212,208.22	14,826.73
22	1,880,846.94	24,896,861.58	90,121.39	72	26,716.54	182,118.25	13,617.54
23	1,742,266.52	23,016,014.64	86,822.31	73	23,613.45	155,401.72	12,436.07
24	1,613,945.07	21,273,748.13	83,814.70	74	20,761.43	131,788.26	11,282.46
25	1,495,134.71	19,659,803.06	81,088.55	75	18,146.93	111,026.84	10,161.25
26	1,385,139.90	18,164,668.35	78,632.44	76	15,760.63	92,879.90	9,080.18
27	1,283,288.79	16,779,528.45	76,408.55	77	13,595.39	77,119.27	8,048.53
28	1,188,950.68	15,496,239.66	74,371.96	78	11,644.47	63,523.87	7,075.47
29	1,101,558.63	14,307,288.97	72,496.08	79	9,900.23	51,879.40	6,168.77
30	1,020,580.00	13,205,730.34	70,747.89	80	8,351.57	41,979.17	5,332.19
31	945,535.44	12,185,150.34	69,109.32	81	6,984.78	33,627.60	4,566.09
32	875,965.08	11,239,614.90	67,547.31	82	5,785.67	26,642.82	3,869.36
33	811,472.91	10,363,649.81	66,059.58	83	4,739.95	20,857.16	3,239.78
34	751,668.70	9,552,176.90	64,621.17	84	3,834.81	16,117.20	2,675.57
35	696,209.18	8,800,508.21	63,225.99	85	3,059.84	12,282.40	2,176.42
36	644,770.44	8,104,299.03	61,862.62	86	2,405.42	9,222.56	1,742.08
37	597,054.41	7,459,528.59	60,522.19	87	1,861.63	6,817.14	1,371.30
38	552,780.96	6,862,474.18	59,192.33	88	1,417.51	4,955.51	1,061.08
39	511,698.16	6,309,693.23	57,868.71	89	1,061.30	3,538.00	806.83
40	473,568.91	5,797,995.07	56,543.74	90	780.79	2,476.70	602.65
41	438,179.81	5,324,426.16	55,216.44	91	563.93	1,695.91	441.95
42	405,325.47	4,886,246.35	53,878.52	92	399.34	1,131.98	317.92
43	374,832.94	4,480,920.88	52,539.33	93	276.68	732.64	223.99
44	346,526.52	4,106,087.94	51,193.05	94	186.91	455.95	154.12
45	320,254.85	3,759,561.42	49,845.56	95	122.14	269.04	102.79
46	295,867.92	3,439,306.57	48,493.20	96	75.95	146.90	65.38
47	273,236.44	3,143,438.65	47,142.24	97	43.38	70.95	38.28
48	252,234.63	2,870,202.22	45,793.17	98	20.93	27.57	18.94
49	232,748.77	2,617,967.59	44,449.48	99	6.64	6.64	6.17

Table III Commutation Columns at 8% based on 1980 CSO Mortality Table-Males

Age	D_x	N_x	M_x	Age	D_x	N_x	M_x
0	10,000,000.00	132,102,461.23	214,632.50	50	191,179.31	2,082,731.81	36,902.88
1	9,220,555.56	122,102,461.23	175,928.80	51	175,830.09	1,891,552.50	35,715.09
2	8,528,416.50	112,881,905.67	166,793.85	52	161,617.16	1,715,722.41	34,526.61
3	7,888,864.28	104,353,489.18	158,976.19	53	148,454.34	1,554,105.25	33,335.43
4	7,297,345.51	96,464,624.90	151,817.74	54	136,260.46	1,405,650.91	32,138.17
5	6,750,382.13	89,167,279.39	145,398.48	55	124,960.94	1,269,390.45	30,932.01
6	6,244,728.30	82,416,897.25	139,772.95	56	114,493.14	1,144,429.51	29,720.59
7	5,777,183.33	76,172,168.95	134,800.45	57	104,797.27	1,029,936.37	28,505.69
8	5,344,964.36	70,394,985.62	130,520.98	58	95,822.55	925,139.10	27,293.73
9	4,945,279.70	65,050,021.26	126,759.61	59	87,518.81	829,316.55	26,087.96
10	4,575,574.42	60,104,741.56	123,371.34	60	79,839.03	741,797.74	24,891.05
11	4,233,550.31	55,529,167.14	120,278.67	61	72,736.32	661,958.71	23,702.34
12	3,916,935.53	51,295,616.83	117,260.21	62	66,167.15	589,222.39	22,521.04
13	3,623,709.38	47,378,681.30	114,177.43	63	60,090.19	523,055.25	21,345.35
14	3,351,964.58	43,754,971.92	110,855.55	64	54,467.31	462,965.06	20,173.60
15	3,100,101.74	40,403,007.34	107,286.38	65	49,265.68	408,497.75	19,006.59
16	2,866,646.94	37,302,905.60	103,468.75	66	44,456.80	359,232.08	17,847.02
17	2,650,294.63	34,436,258.66	99,460.66	67	40,017.30	314,775.28	16,700.61
18	2,449,878.44	31,785,964.02	95,362.58	68	35,925.16	274,757.98	15,572.71
19	2,264,368.14	29,336,085.58	91,324.77	69	32,160.00	238,832.82	14,468.68
20	2,092,737.33	27,071,717.44	87,424.93	70	28,700.71	206,672.82	13,391.62
21	1,934,038.06	24,978,980.11	83,743.24	71	25,524.77	177,972.11	12,341.65
22	1,787,355.61	23,044,942.05	80,322.86	72	22,610.69	152,447.34	11,318.30
23	1,651,831.06	21,257,586.44	77,195.02	73	19,938.23	129,836.65	10,320.70
24	1,526,628.30	19,605,755.38	74,350.13	74	17,489.52	109,898.42	9,348.90
25	1,410,972.04	18,079,127.08	71,777.44	75	15,251.67	92,408.89	8,406.57
26	1,304,143.10	16,668,155.04	69,464.95	76	13,215.43	77,157.22	7,500.08
27	1,205,450.91	15,364,011.94	67,375.96	77	11,373.47	63,941.78	6,637.04
28	1,114,249.62	14,158,561.03	65,467.32	78	9,718.84	52,568.31	5,824.89
29	1,029,958.66	13,044,311.41	63,713.37	79	8,243.92	42,849.47	5,069.88
30	952,034.65	12,014,352.75	62,082.59	80	6,938.25	34,605.55	4,374.88
31	879,988.58	11,062,318.10	60,557.61	81	5,789.33	27,667.30	3,739.90
32	813,353.88	10,182,329.52	59,107.25	82	4,784.34	21,877.97	3,163.75
33	751,727.26	9,368,975.64	57,729.06	83	3,910.54	17,093.63	2,644.34
34	694,714.33	8,617,248.38	56,399.64	84	3,156.45	13,183.09	2,179.93
35	641,967.53	7,922,534.05	55,113.16	85	2,512.74	10,026.64	1,770.03
36	593,160.14	7,280,566.52	53,858.91	86	1,970.76	7,513.89	1,414.17
37	547,992.07	6,687,406.38	52,628.63	87	1,521.70	5,543.14	1,111.10
38	506,182.31	6,139,414.32	51,410.88	88	1,156.00	4,021.44	858.11
39	467,478.09	5,633,232.01	50,201.65	89	863.50	2,865.44	651.24
40	431,642.42	5,165,753.92	48,993.98	90	633.80	2,001.94	485.51
41	398,461.92	4,734,111.50	47,786.99	91	456.71	1,368.14	355.36
42	367,732.38	4,335,649.58	46,573.15	92	322.66	911.43	255.15
43	339,280.78	3,967,917.20	45,360.98	93	223.04	588.77	179.43
44	312,933.11	3,628,636.43	44,145.22	94	150.32	365.73	123.23
45	288,538.83	3,315,703.32	42,931.17	95	98.00	215.41	82.05
46	265,949.97	3,027,164.49	41,715.56	96	60.80	117.41	52.10
47	245,038.44	2,761,214.52	40,504.03	97	34.65	56.61	30.46
48	225,680.39	2,516,176.09	39,296.98	98	16.68	21.96	15.05
49	207,763.88	2,290,495.69	38,097.53	99	5.28	5.28	4.89

Table IV	Commutation Columns at 2% based on 1980 CSO Mortality Table Females						
Age	D_x	N_x	M_x	Age	D_x	N_x	M_x
0	10,000,000.00	389,421,242.25	2,364,289.37	50	3,425,163.84	75,952,744.49	1,935,894.34
1	9,775,588.24	379,421,242.25	2,335,956.03	51	3,341,348.03	72,527,580.64	1,919,238.60
2	9,575,571.90	369,645,654.02	2,327,617.90	52	3,258,436.65	69,186,232.62	1,901,843.86
3	9,380,211.04	360,070,082.12	2,320,013.35	53	3,176,336.74	65,927,795.96	1,883,634.86
4	9,189,020.22	350,689,871.08	2,312,748.23	54	3,094,904.18	62,751,459.23	1,864,483.41
5	9,001,906.36	341,500,850.86	2,305,811.24	55	3,014,163.46	59,656,555.05	1,844,427.08
6	8,818,690.65	332,498,944.51	2,299,103.51	56	2,934,110.86	56,642,391.59	1,823,475.74
7	8,639,463.59	323,680,253.85	2,292,791.94	57	2,854,803.53	53,708,280.72	1,801,699.99
8	8,463,964.15	315,040,790.27	2,286,693.76	58	2,776,352.54	50,853,477.19	1,779,225.53
9	8,292,195.32	306,576,826.11	2,280,885.00	59	2,698,859.54	48,077,124.65	1,756,170.83
10	8,123,993.71	298,284,630.80	2,275,275.46	60	2,622,285.97	45,378,265.11	1,732,516.06
11	7,959,283.81	290,160,637.09	2,269,859.55	61	2,546,522.47	42,755,979.14	1,708,169.94
12	7,797,835.59	282,201,353.28	2,264,475.72	62	2,471,300.23	40,209,456.67	1,682,879.51
13	7,639,432.86	274,403,517.69	2,258,971.73	63	2,396,289.14	37,738,156.44	1,656,325.29
14	7,484,022.69	266,764,084.83	2,253,354.36	64	2,321,064.53	35,341,867.30	1,628,086.74
15	7,331,407.33	259,280,062.14	2,247,484.54	65	2,245,402.38	33,020,802.77	1,597,935.66
16	7,181,544.77	251,948,654.81	2,241,375.07	66	2,169,256.92	30,775,400.39	1,565,817.70
17	7,034,393.40	244,767,110.04	2,235,038.30	67	2,092,694.80	28,606,143.47	1,531,790.02
18	6,889,912.73	237,732,716.64	2,228,486.91	68	2,015,901.15	26,513,448.67	1,496,029.61
19	6,748,196.46	230,842,803.91	2,221,866.97	69	1,939,138.80	24,497,547.52	1,458,794.73
20	6,609,131.00	224,094,607.45	2,215,119.09	70	1,862,409.69	22,558,408.73	1,420,087.95
21	6,472,736.59	217,485,476.45	2,208,315.48	71	1,785,521.47	20,695,999.04	1,379,717.56
22	6,339,030.32	211,012,739.86	2,201,525.61	72	1,708,096.43	18,910,477.57	1,337,302.75
23	6,207,961.55	204,673,709.55	2,194,751.56	73	1,629,607.72	17,202,381.14	1,292,306.13
24	6,079,481.19	198,465,748.00	2,187,995.93	74	1,549,549.30	15,572,773.43	1,244,200.80
25	5,953,481.23	192,386,266.81	2,181,201.49	75	1,467,620.58	14,023,224.13	1,192,655.40
26	5,829,975.73	186,432,785.58	2,174,430.92	76	1,383,822.25	12,555,603.54	1,137,633.95
27	5,708,860.63	180,602,809.84	2,167,629.06	77	1,298,391.51	11,171,781.29	1,079,336.98
28	5,590,094.02	174,893,949.22	2,160,800.90	78	1,211,781.20	9,873,389.78	1,018,185.32
29	5,473,578.89	169,303,855.20	2,153,895.45	79	1,124,520.99	8,661,608.58	954,685.53
30	5,359,277.84	163,830,276.31	2,146,919.48	80	1,037,039.91	7,537,087.59	889,253.88
31	5,247,100.93	158,470,998.47	2,139,826.45	81	949,613.40	6,500,047.68	822,161.48
32	5,137,014.85	153,223,897.54	2,132,624.70	82	862,472.41	5,550,434.28	753,640.36
33	5,028,986.61	148,086,882.69	2,125,322.25	83	775,887.03	4,687,961.87	683,966.21
34	4,922,983.63	143,057,896.08	2,117,926.84	84	690,288.37	3,912,074.84	613,581.02
35	4,818,828.61	138,134,912.45	2,110,300.92	85	606,499.63	3,221,786.47	543,327.35
36	4,716,546.74	133,316,083.84	2,102,505.88	86	525,573.61	2,615,286.84	474,293.47
37	4,615,927.25	128,599,537.10	2,094,367.70	87	448,649.16	2,089,713.23	407,674.39
38	4,516,865.88	123,983,609.85	2,085,814.71	88	376,812.43	1,641,064.07	344,634.70
39	4,419,266.02	119,466,743.97	2,076,780.85	89	310,988.51	1,264,251.64	286,199.26
40	4,322,995.26	115,047,477.95	2,067,162.36	90	251,858.06	953,263.13	233,166.63
41	4,227,974.01	110,724,482.69	2,056,905.72	91	199,819.67	701,405.07	186,066.63
42	4,134,129.45	106,496,508.68	2,045,962.61	92	154,983.74	501,585.40	145,148.73
43	4,041,435.65	102,362,379.23	2,034,330.18	93	117,178.32	346,601.66	110,382.21
44	3,949,948.57	98,320,943.58	2,022,086.93	94	85,987.05	229,423.34	81,488.55
45	3,859,641.80	94,370,995.01	2,009,230.14	95	60,754.84	143,436.29	57,942.37
46	3,770,491.61	90,511,353.20	1,995,759.19	96	40,662.89	82,681.45	39,041.68
47	3,682,513.41	86,740,861.60	1,981,712.20	97	24,886.51	42,018.56	24,062.61
48	3,595,685.70	83,058,348.19	1,967,090.64	98	12,809.97	17,132.05	12,474.04
49	3,509,918.00	79,462,662.48	1,951,826.57	99	4,322.08	4,322.08	4,237.34

Table IV Commutation Columns at 2.25% based on 1980 CSO Mortality Table Females

Age	D_x	N_x	M_x	Age	D_x	N_x	M_x
0	10,000,000.00	363,453,466.30	2,002,246.46	50	3,030,568.50	64,930,057.90	1,601,789.72
1	9,751,687.04	353,453,466.30	1,973,982.40	51	2,949,180.29	61,899,489.40	1,587,088.83
2	9,528,804.83	343,701,779.26	1,965,684.99	52	2,868,968.28	58,950,309.11	1,571,773.21
3	9,311,575.63	334,172,974.44	1,958,136.09	53	2,789,843.60	56,081,340.83	1,555,779.86
4	9,099,481.10	324,861,398.81	1,950,941.76	54	2,711,673.41	53,291,497.23	1,538,999.87
5	8,892,395.42	315,761,917.71	1,944,089.16	55	2,634,473.47	50,579,823.82	1,521,470.01
6	8,690,109.30	306,869,522.29	1,937,479.23	56	2,558,234.81	47,945,350.35	1,503,202.65
7	8,492,680.07	298,179,412.99	1,931,274.90	57	2,483,001.41	45,387,115.54	1,484,262.92
8	8,299,819.66	289,686,732.92	1,925,294.97	58	2,408,863.58	42,904,114.13	1,464,763.27
9	8,111,500.86	281,386,913.26	1,919,612.80	59	2,335,902.62	40,495,250.55	1,444,809.08
10	7,927,534.29	273,275,412.39	1,914,138.91	60	2,264,077.85	38,159,347.94	1,424,385.60
11	7,747,817.76	265,347,878.10	1,908,866.89	61	2,193,288.05	35,895,270.08	1,403,416.58
12	7,572,099.90	257,600,060.35	1,903,638.92	62	2,123,295.92	33,701,982.03	1,381,687.51
13	7,400,145.08	250,027,960.45	1,898,307.32	63	2,053,813.90	31,578,686.11	1,358,928.39
14	7,231,877.59	242,627,815.36	1,892,879.21	64	1,984,476.41	29,524,872.21	1,334,784.84
15	7,067,082.73	235,395,937.77	1,887,221.02	65	1,915,092.51	27,540,395.80	1,309,069.13
16	6,905,697.55	228,328,855.03	1,881,346.22	66	1,845,624.84	25,625,303.29	1,281,742.87
17	6,747,659.97	221,423,157.49	1,875,267.75	67	1,776,131.78	23,779,678.44	1,252,862.57
18	6,592,909.47	214,675,497.52	1,868,998.77	68	1,706,771.48	22,003,546.67	1,222,585.86
19	6,441,514.14	208,082,588.05	1,862,679.68	69	1,637,766.17	20,296,775.18	1,191,137.87
20	6,293,343.86	201,641,073.91	1,856,254.22	70	1,569,116.10	18,659,009.01	1,158,526.66
21	6,148,396.83	195,347,730.05	1,849,791.53	71	1,500,658.20	17,089,892.91	1,124,596.99
22	6,006,668.17	189,199,333.21	1,843,357.66	72	1,432,075.61	15,589,234.71	1,089,036.22
23	5,868,088.91	183,192,665.04	1,836,954.47	73	1,362,929.82	14,157,159.09	1,051,403.09
24	5,732,592.10	177,324,576.14	1,830,584.31	74	1,292,803.96	12,794,229.27	1,011,268.35
25	5,600,055.95	171,591,984.03	1,824,193.22	75	1,221,456.27	11,501,425.31	968,368.68
26	5,470,474.26	165,991,928.08	1,817,840.15	76	1,148,897.53	10,279,969.04	922,687.94
27	5,343,730.26	160,521,453.82	1,811,473.33	77	1,075,334.33	9,131,071.50	874,406.10
28	5,219,766.24	155,177,723.56	1,805,097.51	78	1,001,149.43	8,055,737.18	823,883.82
29	5,098,473.65	149,957,957.32	1,798,665.30	79	926,785.26	7,054,587.74	771,549.83
30	4,979,800.28	144,859,483.67	1,792,183.28	80	852,597.15	6,127,802.49	717,755.53
31	4,863,645.65	139,879,683.39	1,785,608.61	81	778,811.03	5,275,205.33	662,730.72
32	4,749,962.53	135,016,037.74	1,778,949.47	82	705,614.23	4,496,394.30	606,671.56
33	4,638,704.39	130,266,075.21	1,772,213.74	83	633,224.14	3,790,780.07	549,808.44
34	4,529,825.41	125,627,370.82	1,765,408.94	84	561,987.15	3,157,555.93	492,505.48
35	4,423,147.36	121,097,545.41	1,758,409.20	85	492,564.63	2,595,568.79	435,449.43
36	4,318,679.01	116,674,398.05	1,751,271.71	86	425,797.48	2,103,004.16	379,521.10
37	4,216,213.49	112,355,719.04	1,743,838.25	87	362,587.85	1,677,206.68	325,681.10
38	4,115,642.93	108,139,505.55	1,736,045.00	88	303,786.50	1,314,618.83	274,858.46
39	4,016,867.38	104,023,862.63	1,727,833.73	89	250,106.19	1,010,832.33	227,862.94
40	3,919,755.36	100,006,995.25	1,719,112.43	90	202,056.50	760,726.13	185,316.80
41	3,824,224.39	96,087,239.89	1,709,835.25	91	159,916.05	558,669.64	147,622.58
42	3,730,198.86	92,263,015.50	1,699,961.35	92	123,730.51	398,753.59	114,955.98
43	3,637,646.03	88,532,816.64	1,689,491.14	93	93,320.01	275,023.08	87,268.16
44	3,546,606.98	84,895,170.61	1,678,498.10	94	68,312.07	181,703.06	64,313.71
45	3,457,048.56	81,348,563.62	1,666,982.37	95	48,148.43	113,391.00	45,653.27
46	3,368,940.28	77,891,515.06	1,654,946.06	96	32,146.69	65,242.57	30,711.04
47	3,282,286.80	74,522,574.78	1,642,425.74	97	19,626.32	33,095.87	18,898.05
48	3,197,059.86	71,240,287.99	1,629,425.16	98	10,077.66	13,469.55	9,781.27
49	3,113,170.23	68,043,228.13	1,615,886.48	99	3,391.89	3,391.89	3,317.25

	Table IV	Commutation Columns at 2.5% based on 1980 CSO Mortality Table Females					
Age	D_x	N_x	M_x	*Age*	D_x	N_x	M_x
0	10,000,000.00	340,217,895.92	1,702,002.54	50	2,682,234.04	55,563,838.59	1,327,018.46
1	9,727,902.44	330,217,895.92	1,673,807.42	51	2,603,834.27	52,881,604.55	1,314,039.03
2	9,482,379.54	320,489,993.48	1,665,550.43	52	2,526,836.92	50,277,770.28	1,300,549.84
3	9,243,608.19	311,007,613.94	1,658,056.63	53	2,451,155.00	47,750,933.36	1,286,498.09
4	9,011,029.93	301,764,005.75	1,650,932.23	54	2,376,663.80	45,299,778.36	1,271,791.16
5	8,784,479.28	292,752,975.82	1,644,162.79	55	2,303,369.71	42,923,114.56	1,256,464.47
6	8,563,709.89	283,968,496.54	1,637,649.00	56	2,231,257.41	40,619,744.86	1,240,531.93
7	8,348,739.75	275,404,786.64	1,631,549.83	57	2,160,357.82	38,388,487.44	1,224,053.25
8	8,139,247.72	267,056,046.90	1,625,685.60	58	2,090,741.70	36,228,129.62	1,207,128.78
9	7,935,170.82	258,916,799.18	1,620,126.94	59	2,022,471.26	34,137,387.93	1,189,852.04
10	7,736,288.24	250,981,628.36	1,614,785.11	60	1,955,502.76	32,114,916.67	1,172,212.11
11	7,542,466.00	243,245,340.12	1,609,652.82	61	1,889,740.63	30,159,413.91	1,154,145.17
12	7,353,426.41	235,702,874.13	1,604,575.82	62	1,824,973.26	28,269,673.27	1,135,469.03
13	7,168,909.55	228,349,447.72	1,599,410.83	63	1,760,947.95	26,444,700.02	1,115,955.27
14	6,988,812.43	221,180,538.16	1,594,165.16	64	1,697,347.73	24,683,752.06	1,095,304.99
15	6,812,898.91	214,191,725.73	1,588,710.48	65	1,634,007.68	22,986,404.34	1,073,363.67
16	6,641,080.95	207,378,826.82	1,583,060.79	66	1,570,895.20	21,352,396.66	1,050,105.04
17	6,473,272.08	200,737,745.87	1,577,229.49	67	1,508,059.31	19,781,501.46	1,025,583.66
18	6,309,388.00	194,264,473.80	1,571,230.10	68	1,445,633.04	18,273,442.15	999,939.33
19	6,149,467.90	187,955,085.80	1,565,197.51	69	1,383,802.26	16,827,809.11	973,367.89
20	5,993,361.69	181,805,617.90	1,559,078.33	70	1,322,563.91	15,444,006.86	945,880.82
21	5,841,042.51	175,812,256.20	1,552,938.70	71	1,261,777.64	14,121,442.94	917,352.20
22	5,692,480.73	169,971,213.69	1,546,841.37	72	1,201,175.44	12,859,665.30	887,525.06
23	5,547,586.27	164,278,732.96	1,540,787.91	73	1,140,390.10	11,658,489.86	856,036.69
24	5,406,271.75	158,731,146.69	1,534,780.36	74	1,079,076.09	10,518,099.76	822,537.07
25	5,268,398.87	153,324,874.94	1,528,767.78	75	1,017,037.04	9,439,023.67	786,816.95
26	5,133,939.10	148,056,476.07	1,522,805.54	76	954,288.29	8,421,986.63	748,873.98
27	5,002,760.50	142,922,536.97	1,516,844.97	77	891,007.29	7,467,698.34	708,868.31
28	4,874,787.52	137,919,776.46	1,510,890.53	78	827,515.44	6,576,691.05	667,108.34
29	4,749,897.81	133,044,988.94	1,504,898.08	79	764,180.18	5,749,175.61	623,956.38
30	4,628,022.52	128,295,091.13	1,498,873.96	80	701,293.77	4,984,995.43	579,708.52
31	4,509,048.60	123,667,068.61	1,492,778.63	81	639,039.42	4,283,701.66	534,558.89
32	4,392,913.23	119,158,020.02	1,486,620.06	82	577,566.95	3,644,662.24	488,672.75
33	4,279,554.78	114,765,106.79	1,480,405.83	83	517,049.26	3,067,095.29	442,242.06
34	4,168,912.77	110,485,552.01	1,474,143.21	84	457,762.58	2,550,046.03	395,566.34
35	4,060,805.64	106,316,639.24	1,467,716.88	85	400,236.40	2,092,283.44	349,205.10
36	3,955,224.80	102,255,833.60	1,461,180.07	86	345,140.48	1,692,047.04	303,871.04
37	3,851,964.63	98,300,608.81	1,454,388.81	87	293,187.54	1,346,906.57	260,336.16
38	3,750,911.67	94,448,644.18	1,447,286.20	88	245,041.80	1,053,719.03	219,341.34
39	3,651,960.69	90,697,732.51	1,439,820.87	89	201,249.87	808,677.22	181,526.04
40	3,554,978.79	87,045,771.82	1,431,911.19	90	162,189.76	607,427.35	147,374.46
41	3,459,878.68	83,490,793.03	1,423,517.87	91	128,050.75	445,237.59	117,191.29
42	3,366,580.00	80,030,914.35	1,414,606.48	92	98,833.99	317,186.84	91,097.72
43	3,275,041.76	76,664,334.35	1,405,179.95	93	74,360.75	218,352.85	69,035.07
44	3,185,289.59	73,389,292.59	1,395,306.84	94	54,300.75	143,992.10	50,788.75
45	3,097,282.28	70,204,003.01	1,384,989.53	95	38,179.48	89,691.35	35,991.89
46	3,010,981.39	67,106,720.72	1,374,232.11	96	25,428.67	51,511.87	24,172.28
47	2,926,380.11	64,095,739.33	1,363,069.39	97	15,486.94	26,083.20	14,850.77
48	2,843,442.36	61,169,359.22	1,351,506.77	98	7,932.79	10,596.26	7,674.35
49	2,762,078.28	58,325,916.86	1,339,494.94	99	2,663.47	2,663.47	2,598.50

Age	D_x	N_x	M_x	Age	D_x	N_x	M_x
Table IV	Commutation Columns at 2.75% based on 1980 CSO Mortality Table Females						
0	10,000,000.00	319,365,495.29	1,452,504.99	50	2,374,643.46	47,596,555.50	1,100,769.71
1	9,704,233.58	309,365,495.29	1,424,378.47	51	2,299,625.50	45,221,912.04	1,089,306.68
2	9,436,292.70	299,661,261.71	1,416,161.61	52	2,226,194.09	42,922,286.55	1,077,422.43
3	9,176,300.63	290,224,969.00	1,408,722.38	53	2,154,262.50	40,696,092.46	1,065,072.68
4	8,923,650.90	281,048,668.38	1,401,667.07	54	2,083,711.71	38,541,829.95	1,052,178.55
5	8,678,130.90	272,125,017.48	1,394,979.58	55	2,014,538.48	36,458,118.24	1,038,773.76
6	8,439,450.22	263,446,886.58	1,388,560.31	56	1,946,720.61	34,443,579.76	1,024,872.98
7	8,207,580.80	255,007,436.36	1,382,564.26	57	1,880,276.31	32,496,859.15	1,010,530.69
8	7,982,162.14	246,799,855.56	1,376,813.20	58	1,815,258.18	30,616,582.84	995,836.25
9	7,763,089.51	238,817,693.42	1,371,375.09	59	1,751,710.83	28,801,324.66	980,872.46
10	7,550,104.99	231,054,603.91	1,366,161.82	60	1,689,586.86	27,049,613.83	965,631.26
11	7,343,037.47	223,504,498.92	1,361,165.23	61	1,628,794.62	25,360,026.97	950,059.11
12	7,141,577.75	216,161,461.45	1,356,234.50	62	1,569,143.51	23,731,232.34	934,001.04
13	6,945,436.65	209,019,883.70	1,351,230.51	63	1,510,409.52	22,162,088.83	917,263.59
14	6,754,479.26	202,074,447.06	1,346,160.72	64	1,452,315.77	20,651,679.31	899,594.42
15	6,568,443.48	195,319,967.80	1,340,901.77	65	1,394,717.84	19,199,363.54	880,866.26
16	6,387,212.01	188,751,524.31	1,335,468.05	66	1,337,585.38	17,804,645.70	861,062.01
17	6,210,669.99	182,364,312.31	1,329,873.31	67	1,280,957.60	16,467,060.32	840,233.36
18	6,038,705.68	176,153,642.31	1,324,131.31	68	1,224,944.56	15,186,102.72	818,503.86
19	5,871,326.08	170,114,936.64	1,318,371.57	69	1,169,699.86	13,961,158.15	796,043.56
20	5,708,357.77	164,243,610.56	1,312,543.37	70	1,115,216.30	12,791,458.29	772,865.83
21	5,549,745.90	158,535,252.79	1,306,709.94	71	1,061,371.21	11,676,241.99	748,868.38
22	5,395,433.39	152,985,506.89	1,300,930.77	72	1,007,935.99	10,614,870.78	723,839.69
23	5,245,306.45	147,590,073.50	1,295,207.16	73	954,601.21	9,606,934.80	697,481.30
24	5,099,254.74	142,344,767.05	1,289,540.78	74	901,078.54	8,652,333.59	669,507.57
25	4,957,120.99	137,245,512.31	1,283,883.44	75	847,206.71	7,751,255.04	639,752.19
26	4,818,852.34	132,288,391.32	1,278,287.12	76	793,001.93	6,904,048.34	608,222.04
27	4,684,299.48	127,469,538.98	1,272,705.98	77	738,614.69	6,111,046.41	575,058.70
28	4,553,367.11	122,785,239.49	1,267,144.15	78	684,313.05	5,372,431.72	540,525.34
29	4,425,917.12	118,231,872.38	1,261,560.43	79	630,400.46	4,688,118.67	504,927.70
30	4,301,862.34	113,805,955.27	1,255,960.86	80	577,115.54	4,057,718.21	468,514.81
31	4,181,075.37	109,504,092.92	1,250,308.89	81	524,605.06	3,480,602.66	431,450.24
32	4,063,476.38	105,323,017.56	1,244,612.16	82	472,986.98	2,955,997.61	393,872.69
33	3,948,987.32	101,259,541.18	1,238,877.95	83	422,396.99	2,483,010.62	355,941.71
34	3,837,531.85	97,310,553.85	1,233,113.13	84	373,053.59	2,060,613.64	317,903.34
35	3,728,923.06	93,473,022.01	1,227,212.01	85	325,379.02	1,687,560.05	280,213.18
36	3,623,134.25	89,744,098.94	1,221,224.06	86	279,905.15	1,362,181.03	243,447.75
37	3,519,958.80	86,120,964.70	1,215,018.14	87	237,193.37	1,082,275.88	208,227.35
38	3,419,276.03	82,601,005.90	1,208,543.51	88	197,760.36	845,082.51	175,142.58
39	3,320,973.82	79,181,729.87	1,201,754.78	89	162,023.02	647,322.15	144,698.09
40	3,224,916.01	75,860,756.05	1,194,579.47	90	130,258.65	485,299.13	117,270.11
41	3,131,008.88	72,635,840.04	1,186,983.96	91	102,590.53	355,040.49	93,088.23
42	3,039,165.86	69,504,831.16	1,178,939.24	92	78,990.26	252,449.95	72,233.69
43	2,949,336.60	66,465,665.30	1,170,450.18	93	59,286.12	173,459.69	54,643.64
44	2,861,531.00	63,516,328.70	1,161,580.59	94	43,187.41	114,173.58	40,131.67
45	2,775,698.92	60,654,797.70	1,152,334.50	95	30,291.68	70,986.17	28,391.80
46	2,691,793.09	57,879,098.79	1,142,717.46	96	20,126.07	40,694.49	19,036.92
47	2,609,794.87	55,187,305.69	1,132,762.36	97	12,227.65	20,568.42	11,677.16
48	2,529,659.70	52,577,510.82	1,122,475.71	98	6,248.06	8,340.77	6,024.83
49	2,451,295.62	50,047,851.12	1,111,815.42	99	2,092.71	2,092.71	2,036.70

Table IV Commutation Columns at 3% based on 1980 CSO Mortality Table Females							
Age	D_x	N_x	M_x	*Age*	D_x	N_x	M_x
0	10,000,000.00	300,597,181.53	1,244,742.29	50	2,102,948.82	40,812,046.11	914,248.45
1	9,680,679.61	290,597,181.53	1,216,684.03	51	2,031,571.04	38,709,097.29	904,121.60
2	9,390,541.05	280,916,501.92	1,208,507.01	52	1,961,925.58	36,677,526.25	893,648.11
3	9,109,644.95	271,525,960.87	1,201,121.82	53	1,893,924.80	34,715,600.67	882,790.80
4	8,837,328.43	262,416,315.91	1,194,134.76	54	1,827,453.56	32,821,675.87	871,482.41
5	8,573,323.79	253,578,987.49	1,187,528.04	55	1,762,499.04	30,994,222.31	859,754.70
6	8,317,288.97	245,005,663.70	1,181,201.68	56	1,699,031.98	29,231,723.28	847,622.57
7	8,069,142.93	236,688,374.72	1,175,306.77	57	1,637,058.53	27,532,691.29	835,135.49
8	7,828,479.02	228,619,231.80	1,169,666.44	58	1,576,614.59	25,895,632.76	822,372.86
9	7,595,144.60	220,790,752.78	1,164,345.97	59	1,517,728.75	24,319,018.17	809,407.83
10	7,368,838.68	213,195,608.18	1,159,257.86	60	1,460,349.73	22,801,289.42	796,234.50
11	7,149,347.53	205,826,769.50	1,154,393.08	61	1,404,388.56	21,340,939.70	782,807.79
12	6,936,325.08	198,677,421.98	1,149,604.05	62	1,349,671.97	19,936,551.14	768,995.72
13	6,729,447.82	191,741,096.90	1,144,755.68	63	1,295,999.65	18,586,879.17	754,634.24
14	6,528,544.28	185,011,649.07	1,139,855.47	64	1,243,127.94	17,290,879.52	739,510.09
15	6,333,321.80	178,483,104.79	1,134,784.77	65	1,190,928.63	16,047,751.58	723,518.39
16	6,143,629.61	172,149,782.99	1,129,558.26	66	1,139,371.88	14,856,822.95	706,648.88
17	5,959,320.63	166,006,153.38	1,124,189.95	67	1,088,487.25	13,717,451.07	688,949.84
18	5,780,251.94	160,046,832.74	1,118,693.70	68	1,038,364.02	12,628,963.81	670,530.12
19	5,606,395.25	154,266,580.81	1,113,193.86	69	989,127.42	11,590,599.79	651,537.14
20	5,437,550.47	148,660,185.56	1,107,642.16	70	940,765.79	10,601,472.37	631,985.04
21	5,273,632.01	143,222,635.09	1,102,098.95	71	893,170.39	9,660,706.58	611,790.59
22	5,114,552.77	137,949,003.08	1,096,620.64	72	846,144.57	8,767,536.19	590,779.44
23	4,960,172.73	132,834,450.31	1,091,208.16	73	799,425.89	7,921,391.62	568,705.74
24	4,810,356.33	127,874,277.57	1,085,862.81	74	752,772.04	7,121,965.73	545,336.15
25	4,664,925.00	123,063,921.24	1,080,538.94	75	706,048.97	6,369,193.69	520,538.47
26	4,523,799.73	118,398,996.25	1,075,285.28	76	659,271.48	5,663,144.72	494,325.51
27	4,386,811.88	113,875,196.51	1,070,058.58	77	612,565.58	5,003,873.24	466,821.70
28	4,253,844.69	109,488,384.63	1,064,862.62	78	566,153.35	4,391,307.66	438,251.18
29	4,124,742.54	105,234,539.94	1,059,658.86	79	520,283.90	3,825,154.31	408,871.63
30	3,999,398.55	101,109,797.40	1,054,452.99	80	475,150.55	3,304,870.42	378,892.19
31	3,877,669.38	97,110,398.85	1,049,211.16	81	430,869.29	2,829,719.87	348,450.27
32	3,759,457.04	93,232,729.47	1,043,940.65	82	387,531.38	2,398,850.57	317,661.94
33	3,644,665.98	89,473,272.43	1,038,648.33	83	345,241.58	2,011,319.20	286,659.47
34	3,533,203.01	85,828,606.45	1,033,340.68	84	304,171.21	1,666,077.62	255,644.68
35	3,424,874.23	82,295,403.45	1,027,920.73	85	264,655.57	1,361,906.41	224,988.39
36	3,319,634.25	78,870,529.22	1,022,434.37	86	227,115.62	1,097,250.83	195,156.85
37	3,217,273.61	75,550,894.97	1,016,762.10	87	191,992.05	870,135.22	166,648.31
38	3,117,663.10	72,333,621.36	1,010,858.60	88	159,685.16	678,143.17	139,933.42
39	3,020,682.51	69,215,958.26	1,004,683.72	89	130,510.85	518,458.01	115,410.14
40	2,926,190.81	66,195,275.76	998,173.07	90	104,669.73	387,947.15	93,370.30
41	2,834,086.74	63,269,084.95	991,297.86	91	82,236.84	283,277.42	73,986.04
42	2,744,276.38	60,434,998.21	984,033.71	92	63,165.11	201,040.58	57,309.56
43	2,656,699.24	57,690,721.83	976,386.96	93	47,293.49	137,875.47	43,277.70
44	2,571,349.51	55,034,022.59	968,416.81	94	34,367.67	90,581.98	31,729.36
45	2,488,167.53	52,462,673.08	960,128.51	95	24,047.00	56,214.31	22,409.69
46	2,407,096.73	49,974,505.54	951,528.61	96	15,938.27	32,167.32	15,001.36
47	2,328,106.53	47,567,408.81	942,648.02	97	9,659.84	16,229.05	9,187.15
48	2,251,143.51	45,239,302.28	933,493.93	98	4,923.99	6,569.21	4,732.65
49	2,176,112.66	42,988,158.77	924,030.37	99	1,645.22	1,645.22	1,597.30

Table IV	Commutation Columns at 3.25% based on 1980 CSO Mortality Table Females						
Age	D_x	N_x	M_x	Age	D_x	N_x	M_x
0	10,000,000.00	283,656,108.39	1,071,357.36	50	1,862,888.76	35,028,661.21	760,291.68
1	9,657,239.71	273,656,108.39	1,043,367.05	51	1,795,301.51	33,165,772.45	751,342.57
2	9,345,121.33	263,998,868.68	1,035,229.58	52	1,729,557.79	31,370,470.94	742,109.55
3	9,043,633.29	254,653,747.35	1,027,897.90	53	1,665,568.29	29,640,913.15	732,561.34
4	8,752,047.23	245,610,114.06	1,020,978.27	54	1,603,220.38	27,975,344.86	722,640.52
5	8,470,031.94	236,858,066.83	1,014,451.14	55	1,542,492.03	26,372,124.48	712,376.73
6	8,197,185.76	228,388,034.89	1,008,216.14	56	1,483,347.01	24,829,632.45	701,784.73
7	7,933,367.24	220,190,849.13	1,002,420.42	57	1,425,780.19	23,346,285.44	690,909.22
8	7,678,116.67	212,257,481.88	996,888.42	58	1,369,812.34	21,920,505.25	679,820.65
9	7,431,226.97	204,579,365.22	991,682.78	59	1,315,457.61	20,550,692.91	668,583.50
10	7,192,348.01	197,148,138.25	986,716.54	60	1,262,660.91	19,235,235.30	657,193.46
11	6,961,217.71	189,955,790.24	981,979.76	61	1,211,335.12	17,972,574.38	645,612.44
12	6,737,447.77	182,994,572.53	977,328.05	62	1,161,321.37	16,761,239.26	633,727.89
13	6,520,675.18	176,257,124.76	972,630.09	63	1,112,439.09	15,599,917.89	621,400.51
14	6,310,687.22	169,736,449.58	967,893.41	64	1,064,472.26	14,487,478.80	608,449.92
15	6,107,156.10	163,425,762.36	963,003.78	65	1,017,305.57	13,423,006.55	594,789.62
16	5,909,893.51	157,318,606.26	957,976.12	66	970,908.60	12,405,700.98	580,414.38
17	5,718,716.23	151,408,712.75	952,824.55	67	925,301.70	11,434,792.38	565,368.77
18	5,533,446.64	145,689,996.52	947,562.98	68	880,555.65	10,509,490.68	549,748.44
19	5,354,018.10	140,156,549.87	942,310.72	69	836,770.93	9,628,935.03	533,680.97
20	5,180,200.73	134,802,531.77	937,021.76	70	793,931.48	8,792,164.10	517,180.55
21	5,011,875.49	129,622,331.04	931,753.69	71	751,939.65	7,998,232.61	500,179.30
22	4,848,922.92	124,610,455.55	926,559.91	72	710,624.88	7,246,292.96	482,533.33
23	4,691,174.43	119,761,532.63	921,440.96	73	669,763.09	6,535,668.08	464,039.88
24	4,538,467.12	115,070,358.20	916,397.73	74	629,149.19	5,865,905.00	444,508.12
25	4,390,599.01	110,531,891.08	911,386.94	75	588,670.34	5,236,755.80	423,832.99
26	4,247,463.39	106,141,292.08	906,454.20	76	548,338.55	4,648,085.47	402,030.78
27	4,108,870.46	101,893,828.68	901,558.66	77	508,258.03	4,099,746.92	379,210.31
28	3,974,680.57	97,784,958.23	896,703.68	78	468,611.46	3,591,488.89	355,562.17
29	3,844,719.08	93,810,277.65	891,853.20	79	429,602.07	3,122,877.43	331,303.27
30	3,718,858.17	89,965,558.57	887,012.50	80	391,385.19	2,693,275.36	306,608.97
31	3,596,937.34	86,246,700.40	882,150.16	81	354,051.03	2,301,890.17	281,594.44
32	3,478,839.45	82,649,763.06	877,273.06	82	317,668.64	1,947,839.14	256,356.52
33	3,364,450.61	79,170,923.60	872,387.64	83	282,317.46	1,630,170.50	231,004.59
34	3,253,660.09	75,806,472.99	867,499.92	84	248,130.37	1,347,853.04	205,704.00
35	3,146,265.59	72,552,812.91	862,520.87	85	215,372.38	1,099,722.67	180,756.42
36	3,042,202.75	69,406,547.32	857,493.03	86	184,375.47	884,350.29	156,538.77
37	2,941,257.71	66,364,344.57	852,307.39	87	155,484.30	699,974.82	133,451.20
38	2,843,291.78	63,423,086.86	846,923.43	88	129,007.52	544,490.52	111,868.59
39	2,748,175.67	60,579,795.08	841,305.61	89	105,182.69	415,483.00	92,104.53
40	2,655,762.39	57,831,619.41	835,396.65	90	84,152.28	310,300.31	74,384.96
41	2,565,942.25	55,175,857.02	829,171.93	91	65,956.62	226,148.03	58,838.16
42	2,478,613.16	52,609,914.77	822,611.00	92	50,537.81	160,191.41	45,495.46
43	2,393,704.08	50,131,301.60	815,721.22	93	37,747.45	109,653.61	34,295.88
44	2,311,193.70	47,737,597.52	808,557.46	94	27,364.25	71,906.16	25,100.86
45	2,231,012.56	45,426,403.83	801,125.76	95	19,100.36	44,541.91	17,698.31
46	2,153,094.56	43,195,391.26	793,433.34	96	12,629.00	25,441.55	11,828.18
47	2,077,397.35	41,042,296.70	785,509.08	97	7,635.63	12,812.55	7,232.33
48	2,003,858.60	38,964,899.35	777,360.55	98	3,882.75	5,176.92	3,719.79
49	1,932,379.54	36,961,040.75	768,956.95	99	1,294.18	1,294.18	1,253.44

Table IV	Commutation Columns at 3.5% based on 1980 CSO Mortality Table Females						
Age	D_x	N_x	M_x	Age	D_x	N_x	M_x
0	10,000,000.00	268,321,210.59	926,335.87	50	1,650,716.34	30,093,579.47	633,059.06
1	9,633,913.04	258,321,210.59	898,413.17	51	1,586,984.31	28,442,863.14	625,148.36
2	9,300,030.34	248,687,297.55	890,314.96	52	1,525,176.22	26,855,878.82	617,006.40
3	8,978,257.88	239,387,267.21	883,036.29	53	1,465,200.65	25,330,702.61	608,606.84
4	8,667,792.27	230,409,009.33	876,183.26	54	1,406,946.53	23,865,501.96	599,900.57
5	8,368,229.85	221,741,217.05	869,734.59	55	1,350,383.15	22,458,555.42	590,915.09
6	8,079,101.05	213,372,987.20	863,589.41	56	1,295,467.58	21,108,172.28	581,664.66
7	7,800,196.30	205,293,886.15	857,890.97	57	1,242,184.42	19,812,704.70	572,189.58
8	7,530,995.55	197,493,689.85	852,464.98	58	1,190,540.81	18,570,520.28	562,552.20
9	7,271,230.64	189,962,694.29	847,371.42	59	1,140,538.05	17,379,979.47	552,809.28
10	7,020,496.03	182,691,463.65	842,523.83	60	1,092,117.51	16,239,441.42	542,957.65
11	6,778,475.53	175,670,967.62	837,911.41	61	1,045,193.39	15,147,323.91	532,965.04
12	6,544,733.06	168,892,492.09	833,392.75	62	999,618.93	14,102,130.53	522,735.29
13	6,318,861.03	162,347,759.03	828,840.19	63	955,230.11	13,102,511.59	512,150.01
14	6,100,600.74	156,028,898.00	824,261.19	64	911,834.08	12,147,281.49	501,056.45
15	5,889,584.80	149,928,297.26	819,545.76	65	869,325.87	11,235,447.40	489,383.21
16	5,685,583.26	144,038,712.47	814,708.92	66	827,673.87	10,366,121.53	477,128.69
17	5,488,373.09	138,353,129.21	809,764.86	67	786,889.89	9,538,447.66	464,333.69
18	5,297,738.49	132,864,756.12	804,727.41	68	747,028.42	8,751,557.77	451,082.02
19	5,113,571.53	127,567,017.63	799,711.03	69	708,168.51	8,004,529.35	437,483.94
20	4,935,609.59	122,453,446.10	794,671.80	70	670,290.03	7,296,360.84	423,553.19
21	4,763,697.71	117,517,836.51	789,664.59	71	633,304.30	6,626,070.81	409,234.27
22	4,597,681.81	112,754,138.80	784,739.92	72	597,062.18	5,992,766.51	394,408.24
23	4,437,362.65	108,156,456.99	779,897.92	73	561,371.13	5,395,704.32	378,907.70
24	4,282,548.07	103,719,094.34	775,139.08	74	526,056.30	4,834,333.19	362,576.43
25	4,133,010.78	99,436,546.27	770,422.26	75	491,021.42	4,308,276.89	345,330.90
26	3,988,615.00	95,303,535.49	765,790.13	76	456,275.11	3,817,255.47	327,189.18
27	3,849,148.21	91,314,920.49	761,204.04	77	421,902.36	3,360,980.36	308,246.02
28	3,714,446.67	87,465,772.28	756,666.93	78	388,052.35	2,939,077.99	288,663.24
29	3,584,315.40	83,751,325.61	752,144.97	79	354,889.79	2,551,025.64	268,623.22
30	3,458,604.74	80,167,010.20	747,643.04	80	322,538.25	2,196,135.85	248,272.79
31	3,337,135.95	76,708,405.47	743,131.90	81	291,066.63	1,873,597.60	227,708.25
32	3,219,772.03	73,371,269.52	738,617.99	82	260,525.72	1,582,530.97	207,010.18
33	3,106,380.17	70,151,497.49	734,107.30	83	230,974.33	1,322,005.25	186,268.84
34	2,996,831.59	67,045,117.32	729,605.40	84	202,514.26	1,091,030.92	165,619.49
35	2,890,914.51	64,048,285.73	725,030.45	85	175,353.89	888,516.66	145,307.43
36	2,788,545.49	61,157,371.21	720,421.82	86	149,753.93	713,162.77	125,637.32
37	2,689,505.07	58,368,825.73	715,680.05	87	125,982.83	563,408.84	106,930.36
38	2,593,644.39	55,679,320.65	710,768.81	88	104,277.25	437,426.00	89,485.06
39	2,500,824.42	53,085,676.27	705,656.63	89	84,814.19	333,148.76	73,548.29
40	2,410,891.35	50,584,851.84	700,292.49	90	67,692.39	248,334.57	59,294.60
41	2,323,726.50	48,173,960.50	694,655.37	91	52,927.58	180,642.18	46,818.91
42	2,239,219.12	45,850,234.00	688,728.12	92	40,456.64	127,714.60	36,137.78
43	2,157,287.43	43,611,014.88	682,518.81	93	30,144.68	87,257.97	27,193.93
44	2,077,895.05	41,453,727.45	676,078.18	94	21,799.99	57,113.29	19,868.62
45	2,000,962.69	39,375,832.40	669,412.80	95	15,179.73	35,313.29	13,985.56
46	1,926,414.72	37,374,869.72	662,530.24	96	10,012.47	20,133.57	9,331.62
47	1,854,197.41	35,448,454.99	655,457.38	97	6,039.02	10,121.10	5,696.76
48	1,784,239.62	33,594,257.59	648,201.92	98	3,063.45	4,082.08	2,925.41
49	1,716,438.50	31,810,017.97	640,737.41	99	1,018.63	1,018.63	984.18

Table IV Commutation Columns at 3.75% based on 1980 CSO Mortality Table Females

Age	D_x	N_x	M_x	Age	D_x	N_x	M_x
0	10,000,000.00	254,401,793.72	804,754.44	50	1,463,135.91	25,878,086.71	527,783.38
1	9,610,698.80	244,401,793.72	776,899.02	51	1,403,256.62	24,414,950.80	520,788.51
2	9,255,264.92	234,791,094.92	768,839.80	52	1,345,354.49	23,011,694.18	513,606.51
3	8,913,511.08	225,535,830.01	761,613.61	53	1,289,335.84	21,666,339.69	506,215.13
4	8,584,548.77	216,622,318.93	754,826.40	54	1,235,090.53	20,377,003.84	498,572.32
5	8,267,892.53	208,037,770.15	748,455.05	55	1,182,579.79	19,141,913.31	490,703.40
6	7,962,996.17	199,769,877.63	742,398.18	56	1,131,754.52	17,959,333.53	482,621.98
7	7,669,574.02	191,806,881.46	736,795.17	57	1,082,590.00	16,827,579.01	474,364.25
8	7,387,038.23	184,137,307.44	731,472.90	58	1,035,081.30	15,744,989.02	465,985.31
9	7,115,052.69	176,750,269.21	726,488.74	59	989,218.43	14,709,907.72	457,535.02
10	6,853,150.07	169,635,216.53	721,756.70	60	944,939.57	13,720,689.29	449,011.04
11	6,600,954.22	162,782,066.45	717,265.07	61	902,159.99	12,775,749.72	440,385.90
12	6,357,975.80	156,181,112.23	712,875.36	62	860,743.25	11,873,589.73	431,577.36
13	6,123,757.46	149,823,136.43	708,463.37	63	820,539.33	11,012,846.48	422,484.63
14	5,897,989.90	143,699,378.97	704,036.45	64	781,374.91	10,192,307.16	412,978.27
15	5,680,261.70	137,801,389.07	699,488.60	65	743,153.44	9,410,932.24	402,999.26
16	5,470,297.35	132,121,127.37	694,834.92	66	705,841.80	8,667,778.80	392,548.59
17	5,267,830.36	126,650,830.01	690,089.52	67	669,444.14	7,961,937.01	381,663.28
18	5,072,603.48	121,382,999.65	685,266.15	68	634,000.71	7,292,492.87	370,416.63
19	4,884,464.74	116,310,396.17	680,474.52	69	599,572.18	6,658,492.16	358,903.79
20	4,703,115.97	111,425,931.42	675,672.67	70	566,134.82	6,058,919.97	347,137.72
21	4,528,363.98	106,722,815.46	670,912.82	71	533,607.33	5,492,785.15	335,072.93
22	4,360,018.07	102,194,451.47	666,242.71	72	501,858.36	4,959,177.82	322,610.97
23	4,197,846.41	97,834,433.40	661,662.07	73	470,721.37	4,457,319.46	309,613.44
24	4,041,625.90	93,636,586.99	657,170.95	74	440,046.24	3,986,598.09	295,952.33
25	3,891,102.30	89,594,961.09	652,730.21	75	409,749.82	3,546,551.85	281,561.20
26	3,746,109.54	85,703,858.79	648,379.71	76	379,837.08	3,136,802.03	266,458.69
27	3,606,411.11	81,957,749.25	644,082.83	77	350,376.35	2,756,964.95	250,727.01
28	3,471,818.16	78,351,338.14	639,842.09	78	321,488.46	2,406,588.61	234,503.33
29	3,342,114.36	74,879,519.97	635,625.69	79	293,305.91	2,085,100.15	217,940.84
30	3,217,127.43	71,537,405.61	631,438.07	80	265,925.99	1,791,794.24	201,162.34
31	3,096,659.66	68,320,278.18	627,252.01	81	239,400.04	1,525,868.25	184,248.17
32	2,980,553.66	65,223,618.52	623,073.47	82	213,764.04	1,286,468.22	167,265.19
33	2,868,657.31	62,243,064.86	618,907.98	83	189,060.15	1,072,704.17	150,287.71
34	2,760,823.53	59,374,407.56	614,760.61	84	165,365.20	883,644.02	133,426.26
35	2,656,830.23	56,613,584.02	610,556.11	85	142,842.09	718,278.82	116,880.20
36	2,556,574.97	53,956,753.80	606,330.85	86	121,694.59	575,436.73	100,895.67
37	2,459,831.80	51,400,178.83	601,994.01	87	102,130.78	453,742.15	85,730.46
38	2,366,441.20	48,940,347.03	597,512.99	88	84,330.96	351,611.37	71,622.12
39	2,276,254.06	46,573,905.84	592,859.88	89	68,425.54	267,280.41	58,764.81
40	2,189,109.14	44,297,651.77	587,989.20	90	54,480.59	198,854.86	47,293.07
41	2,104,878.49	42,108,542.63	582,882.98	91	42,494.85	144,374.27	37,276.50
42	2,023,442.46	40,003,664.14	577,526.89	92	32,403.82	101,879.42	28,721.43
43	1,944,708.55	37,980,221.68	571,929.45	93	24,086.26	69,475.60	21,575.10
44	1,868,625.89	36,035,513.13	566,137.46	94	17,376.70	45,389.34	15,736.12
45	1,795,105.55	34,166,887.24	560,157.81	95	12,070.55	28,012.64	11,058.05
46	1,724,062.60	32,371,781.70	553,998.21	96	7,942.49	15,942.08	7,366.27
47	1,655,432.42	30,647,719.09	547,683.54	97	4,778.97	7,999.59	4,489.83
48	1,589,135.43	28,992,286.67	541,221.45	98	2,418.41	3,220.62	2,302.01
49	1,525,064.54	27,403,151.25	534,589.19	99	802.21	802.21	773.21

Table IV	Commutation Columns at 4% based on 1980 CSO Mortality Table Females						
Age	D_x	N_x	M_x	Age	D_x	N_x	M_x
0	10,000,000.00	241,732,992.38	702,577.22	50	1,297,247.89	22,273,651.42	440,568.99
1	9,587,596.15	231,732,992.38	674,788.75	51	1,241,166.85	20,976,403.53	434,382.10
2	9,210,821.93	222,145,396.22	666,768.23	52	1,187,092.53	19,735,236.68	428,044.96
3	8,849,385.35	212,934,574.29	659,594.03	53	1,134,928.91	18,548,144.15	421,538.75
4	8,502,302.19	204,085,188.95	652,871.85	54	1,084,566.44	17,413,215.24	414,827.39
5	8,168,995.43	195,582,886.75	646,576.71	55	1,035,959.04	16,328,648.81	407,934.09
6	7,848,833.26	187,413,891.32	640,606.67	56	989,052.03	15,292,689.76	400,871.65
7	7,541,445.66	179,565,058.06	635,097.27	57	943,812.39	14,303,637.74	393,672.48
8	7,246,169.27	172,023,612.40	629,876.49	58	900,224.64	13,359,825.34	386,385.20
9	6,962,593.10	164,777,443.13	624,999.13	59	858,268.95	12,459,600.71	379,053.53
10	6,690,181.55	157,814,850.04	620,379.62	60	817,880.77	11,601,331.76	371,675.71
11	6,428,492.59	151,124,668.49	616,005.34	61	778,976.39	10,783,450.99	364,228.27
12	6,176,978.13	144,696,175.89	611,740.60	62	741,428.24	10,004,474.60	356,640.75
13	5,935,125.96	138,519,197.76	607,464.51	63	705,098.29	9,263,046.36	348,827.28
14	5,702,571.64	132,584,071.80	603,184.26	64	669,829.84	8,557,948.07	340,677.99
15	5,478,855.37	126,881,500.16	598,797.67	65	635,533.27	7,888,118.23	332,144.10
16	5,263,652.28	121,402,644.79	594,319.79	66	602,173.91	7,252,584.96	323,228.33
17	5,056,648.95	116,138,992.51	589,764.62	67	569,749.13	6,650,411.05	313,964.09
18	4,857,543.58	111,082,343.56	585,145.75	68	538,286.94	6,080,661.92	304,415.32
19	4,666,137.53	106,224,799.98	580,568.29	69	507,832.32	5,542,374.99	294,664.05
20	4,482,094.51	101,558,662.46	575,992.10	70	478,358.50	5,034,542.67	284,722.24
21	4,305,181.00	97,076,567.95	571,466.85	71	449,790.40	4,556,184.17	274,552.55
22	4,135,167.85	92,771,386.95	567,037.59	72	422,011.53	4,106,393.77	264,073.31
23	3,971,788.97	88,636,219.10	562,703.62	73	394,877.00	3,684,382.24	253,169.99
24	3,814,788.79	84,664,430.13	558,464.56	74	368,256.99	3,289,505.23	241,737.56
25	3,663,884.72	80,849,641.34	554,283.13	75	342,078.85	2,921,248.24	229,723.15
26	3,518,879.47	77,185,756.62	550,196.52	76	316,343.98	2,579,169.39	217,145.16
27	3,379,511.41	73,666,877.15	546,169.98	77	291,106.41	2,262,825.40	204,074.66
28	3,245,565.82	70,287,365.74	542,205.60	78	266,463.14	1,971,718.99	190,627.79
29	3,116,804.22	67,041,799.92	538,273.45	79	242,519.87	1,705,255.85	176,933.11
30	2,993,031.22	63,924,995.70	534,377.54	80	219,352.24	1,462,735.98	163,093.16
31	2,874,029.52	60,931,964.49	530,492.43	81	196,997.30	1,243,383.74	149,174.84
32	2,759,621.12	58,057,934.97	526,623.62	82	175,479.13	1,046,386.45	135,233.50
33	2,649,634.38	55,298,313.85	522,776.16	83	154,826.60	870,907.32	121,330.17
34	2,543,903.86	52,648,679.46	518,954.65	84	135,096.62	716,080.72	107,555.05
35	2,442,196.57	50,104,775.60	515,089.81	85	116,415.63	580,984.10	94,070.09
36	2,344,391.35	47,662,579.03	511,215.24	86	98,942.10	464,568.46	81,074.09
37	2,250,255.11	45,318,187.68	507,247.89	87	82,836.41	365,626.36	68,773.86
38	2,159,617.45	43,067,932.57	503,158.51	88	68,234.89	282,789.95	57,358.35
39	2,072,319.01	40,908,315.11	498,922.28	89	55,232.21	214,555.06	46,980.09
40	1,988,190.79	38,835,996.10	494,498.63	90	43,870.32	159,322.85	37,742.52
41	1,907,095.49	36,847,805.31	489,872.21	91	34,136.58	115,452.53	29,696.10
42	1,828,904.53	34,940,709.82	485,031.07	92	25,967.77	81,315.95	22,840.24
43	1,753,514.91	33,111,805.29	479,983.94	93	19,255.85	55,348.18	17,127.07
44	1,680,862.04	31,358,290.38	474,773.95	94	13,858.47	36,092.33	12,470.30
45	1,610,847.63	29,677,428.34	469,408.08	95	9,603.51	22,233.86	8,748.36
46	1,543,377.88	28,066,580.71	463,894.01	96	6,303.97	12,630.35	5,818.19
47	1,478,377.90	26,523,202.83	458,254.72	97	3,783.96	6,326.38	3,540.64
48	1,415,760.14	25,044,824.92	452,497.65	98	1,910.28	2,542.42	1,812.50
49	1,355,413.36	23,629,064.78	446,603.17	99	632.13	632.13	607.82

Table IV	Commutation Columns at 4.25% based on 1980 CSO Mortality Table Females						
Age	D_x	N_x	M_x	Age	D_x	N_x	M_x
0	10,000,000.00	230,171,950.85	616,491.21	50	1,150,500.35	19,188,658.65	368,228.90
1	9,564,604.32	220,171,950.85	588,769.39	51	1,098,123.61	18,038,158.29	362,755.04
2	9,166,698.30	210,607,346.53	580,787.29	52	1,047,762.63	16,940,034.69	357,161.69
3	8,785,873.26	201,440,648.23	573,664.58	53	999,319.28	15,892,272.06	351,432.89
4	8,421,038.24	192,654,774.98	567,006.64	54	952,684.38	14,892,952.77	345,537.63
5	8,071,514.50	184,233,736.74	560,786.63	55	907,805.37	13,940,268.39	339,497.06
6	7,736,575.30	176,162,222.24	554,901.98	56	864,622.58	13,032,463.02	333,323.13
7	7,415,757.77	168,425,646.93	549,484.39	57	823,095.81	12,167,840.44	327,044.76
8	7,108,315.25	161,009,889.16	544,362.93	58	783,200.37	11,344,744.64	320,704.78
9	6,813,754.72	153,901,573.92	539,589.84	59	744,908.04	10,561,544.27	314,341.49
10	6,531,465.84	147,087,819.19	535,079.92	60	708,152.08	9,816,636.23	307,953.49
11	6,260,934.78	140,556,353.36	530,819.66	61	672,849.77	9,108,484.14	301,520.68
12	6,001,549.21	134,295,418.57	526,676.03	62	638,881.35	8,435,634.38	294,982.59
13	5,752,737.06	128,293,869.36	522,531.35	63	606,119.18	7,796,753.02	288,265.94
14	5,514,074.24	122,541,132.30	518,392.59	64	574,420.77	7,190,633.84	281,277.42
15	5,285,048.43	117,027,058.06	514,161.17	65	543,702.34	6,616,213.07	273,976.63
16	5,065,281.69	111,742,009.64	509,852.04	66	513,927.82	6,072,510.72	266,367.43
17	4,854,410.41	106,676,727.95	505,479.06	67	485,088.68	5,558,582.90	258,479.78
18	4,652,085.28	101,822,317.54	501,055.54	68	457,202.49	5,073,494.23	250,369.39
19	4,458,058.59	97,170,232.26	496,682.22	69	430,301.00	4,616,291.74	242,106.85
20	4,271,953.56	92,712,173.67	492,320.58	70	404,354.98	4,185,990.74	233,703.07
21	4,093,494.43	88,440,220.11	488,017.83	71	379,294.68	3,781,635.76	225,127.28
22	3,922,411.99	84,346,725.68	483,816.45	72	355,016.20	3,402,341.08	216,311.65
23	3,758,404.38	80,424,313.70	479,715.33	73	331,392.72	3,047,324.88	207,161.25
24	3,601,182.36	76,665,909.32	475,713.63	74	308,311.27	2,715,932.16	197,589.81
25	3,450,433.74	73,064,726.96	471,775.81	75	285,707.67	2,407,620.89	187,555.26
26	3,305,929.27	69,614,293.22	467,936.50	76	263,580.04	2,121,913.22	177,075.19
27	3,167,381.38	66,308,363.95	464,162.71	77	241,970.26	1,858,333.18	166,210.87
28	3,034,548.89	63,140,982.57	460,456.08	78	220,955.41	1,616,362.92	155,060.52
29	2,907,170.59	60,106,433.67	456,788.40	79	200,619.02	1,395,407.51	143,731.90
30	2,785,027.68	57,199,263.08	453,163.24	80	181,018.98	1,194,788.48	132,310.58
31	2,667,882.94	54,414,235.40	449,556.80	81	162,180.86	1,013,769.50	120,852.12
32	2,555,537.63	51,746,352.46	445,974.10	82	144,119.28	851,588.64	109,402.23
33	2,447,800.66	49,190,814.83	442,419.72	83	126,852.63	707,469.37	98,010.95
34	2,344,488.29	46,743,014.17	438,897.78	84	110,422.01	580,616.74	86,751.78
35	2,245,356.29	44,398,525.88	435,344.45	85	94,924.81	470,194.73	75,756.20
36	2,150,265.24	42,153,169.59	431,790.70	86	80,483.50	375,269.92	65,184.72
37	2,058,974.44	40,002,904.35	428,160.59	87	67,220.89	294,786.42	55,203.22
38	1,971,302.64	37,943,929.91	424,427.80	88	55,239.12	227,565.53	45,961.87
39	1,887,080.22	35,972,627.27	420,570.23	89	44,605.66	172,326.42	37,580.36
40	1,806,130.32	34,085,547.05	416,551.67	90	35,344.81	127,720.76	30,137.96
41	1,728,306.41	32,279,416.73	412,358.97	91	27,436.71	92,375.95	23,670.79
42	1,653,471.11	30,551,110.31	407,982.20	92	20,821.12	64,939.24	18,173.72
43	1,581,511.37	28,897,639.20	403,430.15	93	15,402.43	44,118.12	13,603.85
44	1,512,349.61	27,316,127.83	398,742.48	94	11,058.58	28,715.69	9,887.91
45	1,445,878.72	25,803,778.22	393,926.13	95	7,644.89	17,657.11	6,925.05
46	1,381,996.53	24,357,899.50	388,988.63	96	5,006.25	10,012.22	4,598.08
47	1,320,618.63	22,975,902.97	383,951.12	97	2,997.80	5,005.97	2,793.72
48	1,261,650.06	21,655,284.35	378,820.72	98	1,509.77	2,008.17	1,427.90
49	1,204,975.64	20,393,634.29	373,580.48	99	498.40	498.40	478.08

Table IV	Commutation Columns at 4.5% based on 1980 CSO Mortality Table Females						
Age	D_x	N_x	M_x	Age	D_x	N_x	M_x
0	10,000,000.00	219,594,605.55	543,772.97	50	1,020,646.66	16,545,688.40	308,152.90
1	9,541,722.49	209,594,605.55	516,117.46	51	971,850.94	15,525,041.74	303,308.48
2	9,122,890.96	200,052,883.06	508,173.51	52	925,062.57	14,553,190.80	298,370.15
3	8,722,967.48	190,929,992.10	501,101.79	53	880,181.52	13,628,128.23	293,324.33
4	8,340,742.86	182,207,024.62	494,507.35	54	837,098.95	12,747,946.70	288,144.31
5	7,975,426.14	173,866,281.76	488,361.37	55	795,756.64	11,910,847.75	282,849.32
6	7,626,186.05	165,890,855.62	482,560.69	56	756,090.66	11,115,091.11	277,450.37
7	7,292,458.19	158,264,669.57	477,233.18	57	718,054.59	10,359,000.45	271,973.23
8	6,973,404.63	150,972,211.38	472,208.93	58	681,615.92	9,640,945.86	266,455.58
9	6,668,443.18	143,998,806.75	467,537.62	59	646,739.34	8,959,329.94	260,930.88
10	6,376,882.16	137,330,363.57	463,134.45	60	613,356.44	8,312,590.60	255,397.99
11	6,098,130.10	130,953,481.41	458,984.96	61	581,385.60	7,699,234.16	249,839.63
12	5,831,504.97	124,855,351.32	454,958.74	62	550,714.04	7,117,848.55	244,203.82
13	5,576,369.91	119,023,846.35	450,941.12	63	521,223.20	6,567,134.51	238,427.94
14	5,332,236.87	113,447,476.44	446,938.85	64	492,782.89	6,045,911.30	232,432.64
15	5,098,536.93	108,115,239.56	442,856.75	65	465,314.37	5,553,128.42	226,184.44
16	4,874,835.59	103,016,702.63	438,709.64	66	438,780.34	5,087,814.05	219,687.87
17	4,660,715.95	98,141,867.04	434,511.15	67	413,167.30	4,649,033.71	212,969.68
18	4,455,778.41	93,481,151.10	430,274.29	68	388,484.02	4,235,866.41	206,078.29
19	4,259,724.02	89,025,372.69	426,095.53	69	364,751.18	3,847,382.39	199,074.43
20	4,072,133.30	84,765,648.67	421,937.91	70	341,937.64	3,482,631.20	191,967.88
21	3,892,686.61	80,693,515.37	417,846.23	71	319,978.39	3,140,693.56	184,733.21
22	3,721,073.24	76,800,828.76	413,860.51	72	298,780.22	2,820,715.17	177,314.01
23	3,556,954.33	73,079,755.52	409,979.21	73	278,231.57	2,521,934.95	169,631.50
24	3,400,005.90	69,522,801.19	406,201.06	74	258,233.52	2,243,703.38	161,614.71
25	3,249,885.21	66,122,795.30	402,492.11	75	238,728.84	1,985,469.86	153,230.14
26	3,106,330.50	62,872,910.08	398,884.61	76	219,712.77	1,746,741.02	144,494.25
27	2,969,027.61	59,766,579.58	395,347.15	77	201,216.93	1,527,028.25	135,459.74
28	2,837,708.55	56,797,551.97	391,880.96	78	183,301.89	1,325,811.32	126,209.54
29	2,712,089.01	53,959,843.42	388,459.39	79	166,032.91	1,142,509.43	116,833.94
30	2,591,926.68	51,247,754.40	385,085.58	80	149,453.46	976,476.51	107,404.23
31	2,476,964.25	48,655,827.72	381,737.22	81	133,579.93	827,023.05	97,966.50
32	2,366,982.37	46,178,863.47	378,418.87	82	118,419.57	693,443.12	88,558.38
33	2,261,770.64	43,811,881.10	375,134.62	83	103,982.59	575,023.55	79,220.81
34	2,161,127.33	41,550,110.46	371,888.13	84	90,297.68	471,040.96	70,013.62
35	2,064,796.84	39,388,983.12	368,620.53	85	77,439.13	380,743.28	61,043.48
36	1,972,621.99	37,324,186.29	365,360.38	86	65,500.91	303,304.15	52,439.97
37	1,884,354.30	35,351,564.30	362,038.14	87	54,576.36	237,803.24	44,336.02
38	1,799,801.81	33,467,210.00	358,630.09	88	44,741.11	183,226.88	36,850.96
39	1,718,784.85	31,667,408.18	355,116.55	89	36,042.07	138,485.77	30,078.57
40	1,641,118.77	29,948,623.33	351,465.13	90	28,490.84	102,443.70	24,079.38
41	1,566,648.05	28,307,504.57	347,664.60	91	22,063.35	73,952.86	18,878.78
42	1,495,226.85	26,740,856.52	343,706.71	92	16,703.34	51,889.51	14,468.86
43	1,426,732.53	25,245,629.67	339,600.16	93	12,326.74	35,186.17	10,811.55
44	1,361,075.51	23,818,897.13	335,381.37	94	8,829.13	22,859.44	7,844.75
45	1,298,140.38	22,457,821.62	331,057.16	95	6,089.05	14,030.31	5,484.88
46	1,237,817.22	21,159,681.24	326,634.77	96	3,977.87	7,941.25	3,635.90
47	1,180,012.91	19,921,864.02	322,133.60	97	2,376.30	3,963.38	2,205.62
48	1,124,625.76	18,741,851.11	317,560.40	98	1,193.90	1,587.09	1,125.56
49	1,071,536.96	17,617,225.36	312,900.46	99	393.19	393.19	376.25

Table IV	Commutation Columns at 4.75% based on 1980 CSO Mortality Table Females						
Age	D_x	N_x	M_x	Age	D_x	N_x	M_x
0	10,000,000.00	209,892,969.83	482,180.37	50	905,708.32	14,279,245.42	258,200.77
1	9,518,949.88	199,892,969.83	454,590.87	51	860,349.40	13,373,537.10	253,912.16
2	9,079,396.90	190,374,019.95	446,684.78	52	816,974.63	12,513,187.70	249,550.84
3	8,660,660.81	181,294,623.04	439,663.58	53	775,482.43	11,696,213.08	245,105.23
4	8,261,402.23	172,633,962.24	433,131.86	54	735,764.41	10,920,730.64	240,552.28
5	7,880,707.19	164,372,560.01	427,058.88	55	697,757.49	10,184,966.24	235,909.38
6	7,517,630.04	156,491,852.82	421,340.77	56	661,394.18	9,487,208.75	231,186.62
7	7,171,495.98	148,974,222.78	416,101.63	57	626,622.83	8,825,814.57	226,406.89
8	6,841,367.75	141,802,726.80	411,172.51	58	593,404.37	8,199,191.74	221,603.31
9	6,526,566.75	134,961,359.04	406,600.59	59	561,697.58	7,605,787.37	216,805.07
10	6,226,313.44	128,434,792.29	402,301.38	60	531,432.93	7,044,089.80	212,011.19
11	5,939,932.81	122,208,478.85	398,259.54	61	502,530.08	6,512,656.87	207,206.73
12	5,666,667.83	116,268,546.05	394,347.12	62	474,882.54	6,010,126.79	202,346.95
13	5,405,812.02	110,601,878.22	390,452.39	63	448,379.81	5,535,244.26	197,378.28
14	5,156,809.13	105,196,066.20	386,581.79	64	422,902.43	5,086,864.45	192,233.16
15	4,919,029.77	100,039,257.06	382,643.41	65	398,376.11	4,663,962.02	186,883.80
16	4,691,979.59	95,120,227.29	378,651.86	66	374,762.60	4,265,585.91	181,335.07
17	4,475,185.43	90,428,247.70	374,620.50	67	352,044.27	3,890,823.31	175,610.76
18	4,268,194.91	85,953,062.27	370,562.01	68	330,222.58	3,538,779.04	169,752.88
19	4,070,655.79	81,684,867.37	366,568.72	69	309,309.01	3,208,556.46	163,813.60
20	3,882,103.97	77,614,211.58	362,605.12	70	289,271.09	2,899,247.46	157,801.63
21	3,702,174.42	73,732,107.61	358,713.70	71	270,048.04	2,609,976.37	151,695.89
22	3,530,513.78	70,029,933.19	354,932.09	72	251,555.88	2,339,928.33	145,449.34
23	3,366,745.13	66,499,419.40	351,258.34	73	233,696.01	2,088,372.45	138,996.55
24	3,210,508.93	63,132,674.27	347,690.76	74	216,381.32	1,854,676.44	132,279.05
25	3,061,431.12	59,922,165.34	344,196.89	75	199,560.37	1,638,295.12	125,270.14
26	2,919,217.07	56,860,734.22	340,806.69	76	183,225.94	1,438,734.75	117,984.99
27	2,783,525.63	53,941,517.15	337,490.24	77	167,401.16	1,255,508.81	110,468.78
28	2,654,061.83	51,157,991.53	334,248.37	78	152,132.90	1,088,107.65	102,791.50
29	2,530,518.08	48,503,929.70	331,055.88	79	137,471.49	935,974.75	95,028.72
30	2,412,628.63	45,973,411.62	327,915.45	80	123,448.75	798,503.26	87,239.77
31	2,300,116.12	43,560,782.99	324,806.15	81	110,073.86	675,054.52	79,462.79
32	2,192,740.84	41,260,666.87	321,732.08	82	97,348.37	564,980.66	71,728.73
33	2,090,273.45	39,067,926.03	318,696.85	83	85,276.26	467,632.29	64,070.97
34	1,992,494.61	36,977,652.58	315,703.68	84	73,876.50	382,356.03	56,538.16
35	1,899,137.39	34,985,157.97	312,698.25	85	63,205.15	308,479.53	49,216.82
36	1,810,027.56	33,086,020.58	309,706.82	86	53,333.69	245,274.38	42,211.46
37	1,724,908.81	31,275,993.02	306,665.69	87	44,332.38	191,940.69	35,628.63
38	1,643,578.76	29,551,084.21	303,553.46	88	36,256.47	147,608.31	29,563.02
39	1,565,848.03	27,907,505.46	300,352.55	89	29,137.40	111,351.84	24,088.03
40	1,491,524.41	26,341,657.43	297,033.97	90	22,977.80	82,214.44	19,249.70
41	1,420,443.80	24,850,133.02	293,588.12	91	17,751.58	59,236.64	15,065.43
42	1,352,452.30	23,429,689.23	290,008.16	92	13,406.98	41,485.06	11,525.80
43	1,287,418.35	22,077,236.93	286,302.59	93	9,870.48	28,078.08	8,597.25
44	1,225,241.24	20,789,818.58	282,504.84	94	7,052.94	18,207.60	6,227.30
45	1,165,798.00	19,564,577.34	278,621.47	95	4,852.49	11,154.66	4,346.67
46	1,108,971.60	18,398,779.34	274,659.41	96	3,162.48	6,302.17	2,876.70
47	1,054,661.09	17,289,807.74	270,636.39	97	1,884.69	3,139.69	1,742.32
48	1,002,758.72	16,235,146.65	266,558.76	98	944.65	1,255.00	887.74
49	953,142.50	15,232,387.92	262,413.69	99	310.36	310.36	296.28

Table IV	Commutation Columns at 5% based on 1980 CSO Mortality Table Females						
Age	D_x	N_x	M_x	Age	D_x	N_x	M_x
0	10,000,000.00	200,972,838.51	429,864.83	50	803,942.50	12,333,863.16	216,615.68
1	9,496,285.71	190,972,838.51	402,341.02	51	761,861.84	11,529,920.66	212,818.00
2	9,036,213.15	181,476,552.80	394,472.54	52	721,729.84	10,768,058.83	208,965.14
3	8,598,946.12	172,440,339.65	387,501.37	53	683,443.77	10,046,328.98	205,047.15
4	8,183,002.76	163,841,393.53	381,031.64	54	646,895.80	9,362,885.21	201,044.13
5	7,787,334.94	155,658,390.77	375,030.61	55	612,018.85	8,715,989.41	196,971.74
6	7,410,872.55	147,871,055.83	369,393.70	56	578,742.52	8,103,970.56	192,839.16
7	7,052,821.42	140,460,183.28	364,241.26	57	547,010.89	7,525,228.04	188,666.69
8	6,712,136.76	133,407,361.86	359,405.25	58	516,779.44	6,978,217.16	184,483.39
9	6,388,036.33	126,695,225.10	354,930.37	59	488,002.19	6,461,437.71	180,314.68
10	6,079,646.18	120,307,188.77	350,732.43	60	460,608.99	5,973,435.52	176,159.68
11	5,786,201.99	114,227,542.59	346,795.20	61	434,520.98	5,512,826.53	172,005.43
12	5,506,866.48	108,441,340.60	342,993.12	62	409,637.42	5,078,305.55	167,813.34
13	5,240,858.85	102,934,474.12	339,217.23	63	385,855.06	4,668,668.14	163,537.53
14	4,987,550.58	97,693,615.27	335,473.66	64	363,063.90	4,282,813.08	159,120.42
15	4,746,248.14	92,706,064.68	331,673.63	65	341,193.62	3,919,749.18	154,538.90
16	4,516,394.14	87,959,816.55	327,831.45	66	320,205.35	3,578,555.56	149,797.95
17	4,297,456.49	83,443,422.41	323,960.19	67	300,078.14	3,258,350.21	144,918.61
18	4,088,927.68	79,145,965.91	320,072.16	68	280,807.42	2,958,272.06	139,937.32
19	3,890,400.38	75,057,038.23	316,255.70	69	262,397.15	2,677,464.64	134,898.83
20	3,701,364.15	71,166,637.85	312,476.63	70	244,814.03	2,415,067.50	129,810.82
21	3,521,407.30	67,465,273.71	308,775.22	71	228,001.15	2,170,253.46	124,655.75
22	3,350,132.84	63,943,866.41	305,186.82	72	211,882.56	1,942,252.31	119,394.36
23	3,187,124.95	60,593,733.57	301,709.07	73	196,370.74	1,730,369.75	113,972.18
24	3,031,987.90	57,406,608.62	298,339.87	74	181,388.60	1,533,999.01	108,341.02
25	2,884,315.78	54,374,620.72	295,048.13	75	166,889.59	1,352,610.41	102,479.57
26	2,743,780.95	51,490,304.94	291,861.67	76	152,864.50	1,185,720.82	96,401.60
27	2,610,015.00	48,746,523.98	288,751.95	77	139,329.43	1,032,856.32	90,145.80
28	2,482,696.02	46,136,508.98	285,719.40	78	126,320.05	893,526.89	83,771.15
29	2,361,493.15	43,653,812.97	282,740.15	79	113,874.51	767,206.85	77,340.85
30	2,246,117.41	41,292,319.82	279,816.46	80	102,015.29	653,332.34	70,904.23
31	2,136,271.62	39,046,202.41	276,928.65	81	90,746.01	551,317.04	64,492.81
32	2,031,696.10	36,909,930.79	274,080.35	82	80,063.91	460,571.04	58,131.95
33	1,932,143.06	34,878,234.69	271,274.74	83	69,968.24	380,507.13	51,848.85
34	1,837,376.10	32,946,091.63	268,514.59	84	60,470.54	310,538.90	45,682.98
35	1,747,117.14	31,108,715.53	265,749.73	85	51,612.48	250,068.35	39,704.46
36	1,661,175.66	29,361,598.39	263,004.31	86	43,447.88	198,455.88	33,997.60
37	1,579,287.67	27,700,422.73	260,219.92	87	36,029.05	155,007.99	28,647.71
38	1,501,240.79	26,121,135.06	257,377.22	88	29,395.58	118,978.95	23,729.91
39	1,426,836.40	24,619,894.27	254,460.48	89	23,567.42	89,583.37	19,301.54
40	1,355,875.04	23,193,057.87	251,443.72	90	18,541.05	66,015.95	15,397.44
41	1,288,184.56	21,837,182.82	248,318.71	91	14,289.85	47,474.90	12,029.14
42	1,223,603.54	20,548,998.26	245,079.81	92	10,766.79	33,185.05	9,186.55
43	1,161,992.15	19,325,394.72	241,735.26	93	7,907.85	22,418.25	6,840.31
44	1,103,239.59	18,163,402.57	238,315.66	94	5,637.09	14,510.40	4,946.12
45	1,047,216.01	17,060,162.98	234,827.30	95	3,869.13	8,873.31	3,446.60
46	993,798.01	16,012,946.97	231,276.72	96	2,515.60	5,004.18	2,277.31
47	942,877.68	15,019,148.96	227,680.11	97	1,495.61	2,488.57	1,377.11
48	894,341.97	14,076,271.29	224,043.34	98	747.85	992.96	700.56
49	848,066.16	13,181,929.31	220,355.24	99	245.11	245.11	233.44

Table IV	Commutation Columns at 5.25% based on 1980 CSO Mortality Table Females						
Age	D_x	N_x	M_x	Age	D_x	N_x	M_x
0	10,000,000.00	192,751,843.94	385,299.95	50	713,813.43	10,662,518.50	181,953.84
1	9,473,729.22	182,751,843.94	357,841.51	51	674,843.61	9,948,705.08	178,589.92
2	8,993,336.76	173,278,114.72	350,010.37	52	637,776.89	9,273,861.46	175,185.23
3	8,537,816.40	164,284,777.97	343,088.76	53	602,509.78	8,636,084.57	171,731.22
4	8,105,531.09	155,746,961.57	336,680.28	54	568,935.25	8,033,574.79	168,210.61
5	7,695,287.09	147,641,430.48	330,750.18	55	536,982.95	7,464,639.54	164,637.51
6	7,305,879.62	139,946,143.38	325,193.13	56	506,580.29	6,927,656.59	161,020.22
7	6,936,385.94	132,640,263.76	320,125.75	57	477,667.90	6,421,076.30	157,376.69
8	6,585,645.56	125,703,877.82	315,380.87	58	450,196.91	5,943,408.40	153,732.36
9	6,252,765.31	119,118,232.26	311,000.76	59	424,117.56	5,493,211.49	150,109.38
10	5,936,770.37	112,865,466.95	306,901.47	60	399,359.56	5,069,093.93	146,506.89
11	5,636,801.36	106,928,696.58	303,065.90	61	375,845.73	4,669,734.37	142,913.61
12	5,351,935.63	101,291,895.22	299,370.78	62	353,480.68	4,293,888.64	139,296.21
13	5,081,313.53	95,939,959.59	295,709.84	63	332,167.74	3,940,407.96	135,615.33
14	4,824,230.35	90,858,646.06	292,088.86	64	311,805.32	3,608,240.22	131,821.84
15	4,579,924.92	86,034,415.71	288,421.99	65	292,326.74	3,296,434.90	127,896.50
16	4,347,773.87	81,454,490.79	284,723.26	66	273,692.84	3,004,108.16	123,844.21
17	4,127,183.67	77,106,716.92	281,005.39	67	255,880.03	2,730,415.32	119,683.55
18	3,917,589.54	72,979,533.26	277,280.28	68	238,878.91	2,474,535.29	115,446.03
19	3,718,527.48	69,061,943.71	273,632.43	69	222,687.34	2,235,656.38	111,170.04
20	3,529,439.20	65,343,416.23	270,028.89	70	207,271.66	2,012,969.03	106,862.28
21	3,349,865.31	61,813,977.03	266,507.79	71	192,578.52	1,805,697.37	102,508.11
22	3,179,364.41	58,464,111.72	263,102.30	72	178,539.05	1,613,118.85	98,074.69
23	3,017,481.15	55,284,747.31	259,809.67	73	165,075.26	1,434,579.80	93,516.64
24	2,863,783.17	52,267,266.16	256,627.38	74	152,118.62	1,269,504.54	88,794.16
25	2,717,832.39	49,403,482.99	253,525.64	75	139,626.82	1,117,385.92	83,890.23
26	2,579,268.15	46,685,650.59	250,530.24	76	127,589.06	977,759.10	78,817.22
27	2,447,694.75	44,106,382.44	247,613.92	77	116,015.72	850,170.05	73,608.19
28	2,322,763.51	41,658,687.69	244,776.72	78	104,933.33	734,154.33	68,312.81
29	2,204,120.49	39,335,924.18	241,996.01	79	94,370.20	629,221.00	62,983.88
30	2,091,453.87	37,131,803.69	239,273.64	80	84,341.41	534,850.79	57,662.39
31	1,984,446.99	35,040,349.82	236,591.07	81	74,846.29	450,509.39	52,374.33
32	1,882,820.74	33,055,902.82	233,951.48	82	65,878.96	375,663.09	47,140.42
33	1,786,309.46	31,173,082.09	231,357.63	83	57,435.19	309,784.14	41,982.78
34	1,694,660.39	29,386,772.62	228,811.87	84	49,520.86	252,348.94	36,933.39
35	1,607,584.59	27,692,112.23	226,267.83	85	42,166.37	202,828.08	32,049.06
36	1,524,876.12	26,084,527.65	223,747.66	86	35,411.75	160,661.71	27,397.74
37	1,446,263.56	24,559,651.52	221,197.81	87	29,295.35	125,249.96	23,047.73
38	1,371,525.07	23,113,387.97	218,600.73	88	23,844.88	95,954.61	19,058.55
39	1,300,453.32	21,741,862.89	215,942.35	89	19,071.83	72,109.73	15,474.91
40	1,232,842.08	20,441,409.57	213,199.32	90	14,968.63	53,037.90	12,323.03
41	1,168,511.70	19,208,567.49	210,364.63	91	11,509.13	38,069.28	9,610.18
42	1,107,293.87	18,040,055.79	207,433.61	92	8,651.04	26,560.15	7,326.19
43	1,049,041.24	16,932,761.92	204,414.16	93	6,338.81	17,909.11	5,445.48
44	993,633.90	15,883,720.68	201,334.29	94	4,507.87	11,570.30	3,930.73
45	940,935.88	14,890,086.78	198,199.96	95	3,086.72	7,062.43	2,734.44
46	890,818.18	13,949,150.90	195,017.30	96	2,002.13	3,975.71	1,803.82
47	843,166.80	13,058,332.72	191,801.03	97	1,187.51	1,973.58	1,089.06
48	797,864.15	12,215,165.92	188,556.59	98	592.38	786.07	553.17
49	754,783.27	11,417,301.77	185,274.16	99	193.70	193.70	184.03

Table IV	Commutation Columns at 5.5% based on 1980 CSO Mortality Table Females						
Age	D_x	N_x	M_x	Age	D_x	N_x	M_x
0	10,000,000.00	185,157,806.66	347,223.35	50	633,967.45	9,225,304.89	153,027.38
1	9,451,279.62	175,157,806.66	319,829.98	51	597,936.45	8,591,337.44	150,046.82
2	8,950,764.81	165,706,527.03	312,035.91	52	563,754.88	7,993,400.99	147,037.29
3	8,477,264.74	156,755,762.23	305,163.39	53	531,318.92	7,429,646.11	143,991.40
4	8,028,974.07	148,278,497.49	298,815.44	54	500,522.57	6,898,327.19	140,894.13
5	7,604,541.80	140,249,523.42	292,955.27	55	471,292.98	6,397,804.62	137,758.14
6	7,202,617.99	132,644,981.62	287,476.76	56	443,555.94	5,926,511.65	134,590.88
7	6,822,142.14	125,442,363.63	282,492.85	57	417,249.49	5,482,955.71	131,408.20
8	6,461,829.76	118,620,221.50	277,837.17	58	392,321.32	5,065,706.22	128,232.37
9	6,120,669.54	112,158,391.73	273,549.59	59	368,718.81	4,673,384.89	125,082.64
10	5,797,579.33	106,037,722.19	269,546.42	60	346,372.00	4,304,666.08	121,958.13
11	5,491,599.09	100,240,142.87	265,809.65	61	325,205.55	3,958,294.08	118,848.99
12	5,201,715.79	94,748,543.78	262,218.25	62	305,129.12	3,633,088.52	115,726.40
13	4,926,986.54	89,546,827.99	258,668.50	63	286,052.06	3,327,959.40	112,556.55
14	4,666,626.74	84,619,841.44	255,165.81	64	267,880.31	3,041,907.34	109,297.46
15	4,419,804.21	79,953,214.70	251,627.14	65	250,550.61	2,774,027.03	105,933.09
16	4,185,826.92	75,533,410.49	248,066.18	66	234,023.78	2,523,476.42	102,468.14
17	3,964,037.54	71,347,583.58	244,495.27	67	218,274.30	2,289,452.63	98,918.95
18	3,753,812.18	67,383,546.04	240,925.89	68	203,288.90	2,071,178.33	95,312.77
19	3,554,628.74	63,629,733.86	237,438.83	69	189,060.60	1,867,889.44	91,682.48
20	3,365,879.80	60,075,105.11	234,002.28	70	175,555.76	1,678,828.84	88,033.87
21	3,187,057.42	56,709,225.31	230,652.31	71	162,724.38	1,503,273.08	84,354.70
22	3,017,675.21	53,522,167.89	227,420.01	72	150,503.87	1,340,548.70	80,617.44
23	2,857,237.87	50,504,492.68	224,302.23	73	138,824.48	1,190,044.83	76,784.23
24	2,705,276.19	47,647,254.81	221,296.08	74	127,625.10	1,051,220.35	72,822.14
25	2,561,319.71	44,941,978.62	218,372.96	75	116,867.08	923,595.25	68,717.57
26	2,424,974.98	42,380,658.91	215,556.75	76	106,538.46	806,728.17	64,481.55
27	2,295,819.12	39,955,683.93	212,821.38	77	96,645.02	700,189.70	60,142.24
28	2,173,477.02	37,659,864.81	210,166.53	78	87,205.88	603,544.68	55,741.46
29	2,057,571.96	35,486,387.79	207,570.70	79	78,241.44	516,338.80	51,323.30
30	1,947,769.84	33,428,815.83	205,035.36	80	69,760.96	438,097.36	46,921.76
31	1,843,734.98	31,481,045.99	202,543.00	81	61,760.60	368,336.40	42,558.23
32	1,745,169.48	29,637,311.02	200,096.39	82	54,232.25	306,575.80	38,249.62
33	1,651,790.56	27,892,141.54	197,697.87	83	47,169.21	252,343.55	34,013.86
34	1,563,329.79	26,240,350.98	195,349.41	84	40,573.12	205,174.35	29,876.82
35	1,479,487.86	24,677,021.19	193,008.08	85	34,465.62	164,601.23	25,884.51
36	1,400,044.30	23,197,533.33	190,694.23	86	28,875.98	130,135.61	22,091.66
37	1,324,720.64	21,797,489.03	188,358.66	87	23,831.85	101,259.63	18,552.92
38	1,253,286.20	20,472,768.38	185,985.47	88	19,351.91	77,427.77	15,315.39
39	1,185,525.55	19,219,482.19	183,562.02	89	15,441.54	58,075.86	12,413.89
40	1,121,226.22	18,033,956.64	181,067.34	90	12,090.66	42,634.32	9,868.02
41	1,060,201.72	16,912,730.42	178,495.40	91	9,274.28	30,543.66	7,681.95
42	1,002,277.50	15,852,528.70	175,842.35	92	6,954.66	21,269.38	5,845.83
43	947,299.46	14,850,251.20	173,115.75	93	5,083.75	14,314.73	4,337.49
44	895,139.61	13,902,951.74	170,341.18	94	3,606.77	9,230.97	3,125.53
45	845,656.61	13,007,812.13	167,524.22	95	2,463.85	5,624.21	2,170.64
46	798,716.65	12,162,155.52	164,670.63	96	1,594.33	3,160.36	1,429.57
47	754,200.48	11,363,438.87	161,793.72	97	943.39	1,566.03	861.75
48	711,986.74	10,609,238.39	158,898.48	98	469.49	622.64	437.03
49	671,946.76	9,897,251.65	155,976.29	99	153.15	153.15	145.17

Table IV	Commutation Columns at 5.75% based on 1980 CSO Mortality Table Females						
Age	D_x	N_x	M_x	Age	D_x	N_x	M_x
0	10,000,000.00	178,127,333.61	314,589.43	50	563,211.01	7,988,320.38	128,857.89
1	9,428,936.17	168,127,333.61	287,260.82	51	529,945.60	7,425,109.37	126,216.25
2	8,908,494.43	158,698,397.44	279,503.55	52	498,469.58	6,895,163.76	123,555.24
3	8,417,284.31	149,789,903.01	272,679.66	53	468,679.23	6,396,694.18	120,868.44
4	7,953,318.78	141,372,618.70	266,391.52	54	440,469.84	5,928,014.95	118,142.79
5	7,515,077.60	133,419,299.92	260,600.30	55	413,766.73	5,487,545.11	115,389.57
6	7,101,055.10	125,904,222.32	255,199.04	56	388,494.68	5,073,778.38	112,615.48
7	6,710,043.70	118,803,167.22	250,297.02	57	364,589.86	4,685,283.70	109,834.48
8	6,340,626.64	112,093,123.51	245,728.67	58	341,997.37	4,320,693.85	107,066.02
9	5,991,667.22	105,752,496.87	241,531.46	59	320,662.54	3,978,696.48	104,326.80
10	5,661,969.63	99,760,829.65	237,621.92	60	300,516.13	3,658,033.94	101,615.94
11	5,350,467.66	94,098,860.02	233,981.18	61	281,484.87	3,357,517.80	98,924.80
12	5,056,053.06	88,748,392.36	230,490.35	62	263,483.15	3,076,032.93	96,228.40
13	4,777,695.45	83,692,339.30	227,048.16	63	246,425.90	2,812,549.79	93,497.66
14	4,514,526.80	78,914,643.85	223,659.64	64	230,225.88	2,566,123.89	90,696.68
15	4,265,640.83	74,400,117.05	220,244.40	65	214,823.06	2,335,898.01	87,812.06
16	4,030,274.29	70,134,476.22	216,815.77	66	200,178.54	2,121,074.95	84,848.22
17	3,807,704.00	66,104,201.93	213,385.69	67	186,265.41	1,920,896.41	81,819.50
18	3,597,245.22	62,296,497.94	209,965.19	68	173,067.43	1,734,631.00	78,749.43
19	3,398,316.60	58,699,252.72	206,631.46	69	160,573.84	1,561,563.58	75,666.13
20	3,210,260.50	55,300,936.12	203,353.80	70	148,751.35	1,400,989.73	72,574.61
21	3,032,519.79	52,090,675.63	200,166.27	71	137,553.16	1,252,238.38	69,464.55
22	2,864,562.71	49,058,155.84	197,097.97	72	126,922.22	1,114,685.22	66,312.86
23	2,705,853.75	46,193,593.13	194,145.38	73	116,796.05	987,763.00	63,087.90
24	2,555,886.81	43,487,739.37	191,305.23	74	107,119.93	870,966.94	59,762.39
25	2,414,159.07	40,931,852.57	188,550.06	75	97,858.48	763,847.01	56,325.43
26	2,280,244.60	38,517,693.50	185,901.93	76	88,998.93	665,988.53	52,786.79
27	2,153,693.64	36,237,448.90	183,335.89	77	80,543.40	576,989.60	49,170.44
28	2,034,105.12	34,083,755.27	180,851.28	78	72,505.05	496,446.21	45,511.52
29	1,921,080.03	32,049,650.15	178,427.66	79	64,898.02	423,941.15	41,846.84
30	1,814,262.58	30,128,570.12	176,066.10	80	57,727.02	359,043.14	38,204.58
31	1,713,298.70	28,314,307.54	173,750.06	81	50,985.92	301,316.12	34,602.30
32	1,617,872.47	26,601,008.84	171,481.91	82	44,665.12	250,330.19	31,053.78
33	1,527,684.73	24,983,136.37	169,263.61	83	38,756.23	205,665.08	27,573.50
34	1,442,452.25	23,455,451.64	167,096.72	84	33,257.79	166,908.85	24,182.37
35	1,361,865.85	22,012,999.39	164,941.54	85	28,184.68	133,651.06	20,917.61
36	1,285,691.55	20,651,133.54	162,816.67	86	23,557.87	105,466.37	17,823.29
37	1,213,644.23	19,365,441.99	160,676.94	87	19,396.76	81,908.51	14,943.10
38	1,145,485.07	18,151,797.76	158,507.88	88	15,713.29	62,511.75	12,314.31
39	1,080,991.25	17,006,312.69	156,298.12	89	12,508.53	46,798.46	9,963.93
40	1,019,944.61	15,925,321.44	154,028.79	90	9,770.97	34,289.93	7,906.50
41	962,152.55	14,905,376.83	151,694.70	91	7,477.21	24,518.96	6,144.03
42	907,434.93	13,943,224.28	149,292.71	92	5,593.81	17,041.75	4,667.19
43	855,631.74	13,035,789.36	146,829.95	93	4,079.32	11,447.95	3,456.86
44	806,607.87	12,180,157.62	144,329.79	94	2,887.31	7,368.62	2,486.65
45	760,217.41	11,373,549.76	141,797.44	95	1,967.71	4,481.31	1,724.05
46	716,322.48	10,613,332.35	139,238.22	96	1,270.28	2,513.60	1,133.60
47	674,799.47	9,897,009.87	136,664.19	97	749.87	1,243.32	682.26
48	635,523.94	9,222,210.40	134,079.88	98	372.30	493.45	345.46
49	598,366.07	8,586,686.45	131,477.68	99	121.16	121.16	114.57

Table IV	Commutation Columns at 6% based on 1980 CSO Mortality Table Females						
Age	D_x	N_x	M_x	Age	D_x	N_x	M_x
0	10,000,000.00	171,604,624.55	286,530.69	50	500,491.47	6,922,734.92	108,638.54
1	9,406,698.11	161,604,624.55	259,266.53	51	469,819.83	6,422,243.46	106,296.62
2	8,866,522.78	152,197,926.44	251,545.82	52	440,872.71	5,952,423.63	103,943.07
3	8,357,868.41	143,331,403.65	244,770.09	53	413,546.91	5,511,550.91	101,572.33
4	7,878,552.49	134,973,535.24	238,541.06	54	387,739.24	5,098,004.00	99,172.98
5	7,426,873.44	127,094,982.75	232,817.81	55	363,373.84	4,710,264.76	96,755.08
6	7,001,159.11	119,668,109.31	227,492.54	56	340,375.02	4,346,890.92	94,324.59
7	6,600,045.42	112,666,950.20	222,670.88	57	318,677.72	4,006,515.90	91,893.80
8	6,221,975.08	106,066,904.79	218,188.02	58	298,225.23	3,687,838.18	89,479.68
9	5,865,678.85	99,844,929.71	214,079.06	59	278,961.56	3,389,612.95	87,096.68
10	5,529,840.99	93,979,250.86	210,260.75	60	260,818.53	3,110,651.38	84,743.92
11	5,213,283.74	88,449,409.87	206,713.37	61	243,725.07	2,849,832.86	82,413.78
12	4,914,798.90	83,236,126.13	203,320.06	62	227,600.13	2,606,107.78	80,084.60
13	4,633,264.60	78,321,327.23	199,981.93	63	212,363.82	2,378,507.65	77,731.31
14	4,367,726.00	73,688,062.63	196,703.59	64	197,935.11	2,166,143.83	75,323.19
15	4,117,199.84	69,320,336.63	193,407.20	65	184,257.04	1,968,208.72	72,849.00
16	3,880,849.28	65,203,136.79	190,105.69	66	171,291.27	1,783,951.68	70,312.87
17	3,657,883.45	61,322,287.52	186,810.57	67	159,010.00	1,612,660.41	67,727.33
18	3,447,555.27	57,664,404.07	183,532.40	68	147,394.77	1,453,650.41	65,112.67
19	3,249,223.17	54,216,848.79	180,344.93	69	136,431.94	1,306,255.64	62,492.94
20	3,062,178.41	50,967,625.63	177,218.47	70	126,088.85	1,169,823.71	59,872.41
21	2,885,814.23	47,905,447.22	174,185.14	71	116,321.73	1,043,734.86	57,242.39
22	2,719,553.28	45,019,632.99	171,272.17	72	107,078.54	927,413.13	54,583.46
23	2,562,819.78	42,300,079.71	168,475.65	73	98,303.15	820,334.59	51,869.12
24	2,415,070.84	39,737,259.93	165,791.98	74	89,946.46	722,031.44	49,076.76
25	2,275,771.48	37,322,189.08	163,194.74	75	81,976.01	632,084.98	46,197.62
26	2,144,463.78	35,046,417.60	160,704.30	76	74,378.53	550,108.97	43,240.29
27	2,020,671.50	32,901,953.82	158,296.75	77	67,153.29	475,730.43	40,225.15
28	1,903,968.22	30,881,282.32	155,971.10	78	60,308.72	408,577.15	37,181.71
29	1,793,933.21	28,977,314.10	153,707.88	79	53,853.98	348,268.43	34,140.67
30	1,690,189.77	27,183,380.89	151,507.83	80	47,790.33	294,414.45	31,125.36
31	1,592,366.09	25,493,191.13	149,355.27	81	42,110.04	246,624.12	28,150.18
32	1,500,129.08	23,900,825.04	147,252.19	82	36,802.59	204,514.08	25,226.32
33	1,413,164.09	22,400,695.96	145,200.17	83	31,858.54	167,711.50	22,365.44
34	1,331,173.96	20,987,531.87	143,200.45	84	27,274.22	135,852.95	19,584.43
35	1,253,840.25	19,656,357.91	141,216.22	85	23,059.33	108,578.73	16,913.36
36	1,180,916.46	18,402,517.66	139,264.52	86	19,228.43	85,519.41	14,387.71
37	1,112,111.41	17,221,601.20	137,303.79	87	15,794.71	66,290.98	12,042.39
38	1,047,178.80	16,109,489.79	135,320.89	88	12,765.10	50,496.27	9,906.82
39	985,889.18	15,062,310.99	133,305.53	89	10,137.66	37,731.17	8,001.93
40	928,019.32	14,076,421.82	131,240.73	90	7,900.30	27,593.51	6,338.40
41	873,371.22	13,148,402.49	129,122.02	91	6,031.43	19,693.22	4,916.72
42	821,759.90	12,275,031.27	126,946.81	92	4,501.55	13,661.79	3,728.24
43	773,020.21	11,453,271.38	124,721.83	93	3,275.05	9,160.24	2,756.54
44	727,010.91	10,680,251.17	122,468.39	94	2,312.59	5,885.19	1,979.46
45	683,582.28	9,953,240.26	120,191.32	95	1,572.32	3,572.60	1,370.09
46	642,593.13	9,269,657.98	117,895.51	96	1,012.63	2,000.28	899.41
47	603,916.29	8,627,064.85	115,591.86	97	596.36	987.65	540.46
48	567,424.96	8,023,148.57	113,284.47	98	295.39	391.29	273.24
49	532,988.68	7,455,723.61	110,966.59	99	95.90	95.90	90.47

Table IV Commutation Columns at 6.25% based on 1980 CSO Mortality Table Females

Age	D_x	N_x	M_x	Age	D_x	N_x	M_x
0	10,000,000.00	165,540,453.84	262,326.24	50	444,880.12	6,004,006.98	91,703.24
1	9,384,564.71	155,540,453.84	235,126.24	51	416,633.89	5,559,126.86	89,626.43
2	8,824,847.06	146,155,889.13	227,441.82	52	390,043.82	5,142,492.97	87,544.23
3	8,299,010.40	137,331,042.07	220,713.80	53	365,007.58	4,752,449.15	85,451.75
4	7,804,662.71	129,032,031.68	214,543.20	54	341,423.80	4,387,441.56	83,339.00
5	7,339,908.67	121,227,368.97	208,886.97	55	319,215.98	4,046,017.76	81,214.93
6	6,902,898.80	113,887,460.30	203,636.43	56	298,308.46	3,726,801.79	79,084.83
7	6,492,103.13	106,984,561.49	198,893.63	57	278,635.54	3,428,493.33	76,959.46
8	6,105,815.54	100,492,458.37	194,494.46	58	260,139.40	3,149,857.79	74,853.64
9	5,742,627.16	94,386,642.83	190,471.70	59	242,763.30	2,889,718.39	72,779.87
10	5,401,096.15	88,644,015.68	186,742.29	60	226,440.46	2,646,955.09	70,737.22
11	5,079,927.97	83,242,919.52	183,285.64	61	211,102.18	2,420,514.63	68,718.97
12	4,777,809.95	78,162,991.55	179,986.92	62	196,671.74	2,209,412.44	66,706.30
13	4,493,524.85	73,385,181.60	176,749.46	63	183,074.10	2,012,740.70	64,677.58
14	4,226,027.88	68,891,656.75	173,577.48	64	170,233.93	1,829,666.61	62,606.48
15	3,974,256.06	64,665,628.87	170,395.54	65	158,097.25	1,659,432.68	60,483.57
16	3,737,296.90	60,691,372.81	167,216.15	66	146,626.47	1,501,335.42	58,312.62
17	3,514,290.14	56,954,075.91	164,050.38	67	135,793.35	1,354,708.95	56,104.59
18	3,304,425.12	53,439,785.77	160,908.32	68	125,577.86	1,218,915.60	53,876.94
19	3,106,999.23	50,135,360.64	157,860.37	69	115,964.21	1,093,337.74	51,650.23
20	2,921,251.99	47,028,361.41	154,877.79	70	106,920.64	977,373.53	49,428.08
21	2,746,526.71	44,107,109.42	151,990.86	71	98,406.24	870,452.90	47,203.12
22	2,582,200.47	41,360,582.71	149,225.01	72	90,373.51	772,046.66	44,959.00
23	2,427,657.29	38,778,382.24	146,575.98	73	82,771.93	681,673.15	42,673.51
24	2,282,317.77	36,350,724.95	144,039.83	74	75,557.34	598,901.22	40,327.85
25	2,145,615.09	34,068,407.18	141,591.14	75	68,699.93	523,343.88	37,914.99
26	2,017,059.95	31,922,792.09	139,248.65	76	62,186.20	454,643.95	35,442.44
27	1,896,150.18	29,905,732.14	136,989.47	77	56,013.23	392,457.75	32,927.48
28	1,782,434.74	28,009,581.95	134,812.27	78	50,185.75	336,444.52	30,394.89
29	1,675,471.87	26,227,147.22	132,698.50	79	44,709.00	286,258.78	27,870.25
30	1,574,864.76	24,551,675.35	130,648.56	80	39,581.67	241,549.77	25,372.86
31	1,480,224.69	22,976,810.60	128,647.59	81	34,794.99	201,968.10	22,914.51
32	1,391,202.27	21,496,585.91	126,697.22	82	30,337.96	167,173.11	20,504.24
33	1,307,468.31	20,105,383.64	124,798.68	83	26,200.58	136,835.16	18,151.45
34	1,228,712.61	18,797,915.33	122,952.88	84	22,377.63	110,634.58	15,869.72
35	1,154,608.20	17,569,202.73	121,125.68	85	18,874.93	88,256.95	13,683.34
36	1,084,897.06	16,414,594.53	119,332.68	86	15,702.17	69,382.02	11,620.87
37	1,019,282.52	15,329,697.47	117,535.61	87	12,867.79	53,679.85	9,710.16
38	957,511.61	14,310,414.95	115,722.50	88	10,375.13	40,812.06	7,974.42
39	899,348.95	13,352,903.33	113,884.05	89	8,220.23	30,436.92	6,429.82
40	844,566.95	12,453,554.38	112,004.92	90	6,390.97	22,216.69	5,084.11
41	792,962.89	11,608,987.43	110,081.28	91	4,867.66	15,825.72	3,936.74
42	744,347.71	10,816,024.54	108,110.98	92	3,624.43	10,958.06	2,979.84
43	698,551.92	10,071,676.83	106,100.34	93	2,630.70	7,333.63	2,199.31
44	655,429.06	9,373,124.91	104,068.77	94	1,853.23	4,702.93	1,576.59
45	614,826.37	8,717,695.85	102,020.74	95	1,257.04	2,849.70	1,089.41
46	576,600.08	8,102,869.47	99,960.70	96	807.67	1,592.66	713.99
47	540,620.23	7,526,269.39	97,898.50	97	474.54	784.99	428.37
48	506,758.35	6,985,649.17	95,837.81	98	234.49	310.45	216.23
49	474,883.84	6,478,890.82	93,772.61	99	75.95	75.95	71.49

Table IV	Commutation Columns at 6.5% based on 1980 CSO Mortality Table Females						
Age	D_x	N_x	M_x	Age	D_x	N_x	M_x
0	10,000,000.00	159,891,300.33	241,376.04	50	395,557.43	5,211,224.97	77,501.45
1	9,362,535.21	149,891,300.33	214,239.89	51	369,573.21	4,815,667.54	75,659.22
2	8,783,464.48	140,528,765.12	206,591.49	52	345,174.42	4,446,094.33	73,816.55
3	8,240,703.75	131,745,300.64	199,910.75	53	322,260.02	4,100,919.91	71,969.13
4	7,731,637.13	123,504,596.90	193,797.88	54	300,730.63	3,778,659.89	70,108.19
5	7,254,163.01	115,772,959.77	188,207.72	55	280,509.66	3,477,929.26	68,241.68
6	6,806,243.66	108,518,796.75	183,030.71	56	261,521.92	3,197,419.60	66,374.25
7	6,386,173.69	101,712,553.10	178,365.29	57	243,701.60	2,935,897.68	64,515.35
8	5,992,089.99	95,326,379.40	174,048.06	58	226,990.31	2,692,196.08	62,677.88
9	5,622,437.01	89,334,289.41	170,109.49	59	211,331.17	2,465,205.77	60,872.60
10	5,275,640.80	83,711,852.40	166,466.71	60	196,659.03	2,253,874.60	59,098.61
11	4,950,284.90	78,436,211.60	163,098.28	61	182,907.67	2,057,215.57	57,349.91
12	4,644,947.85	73,485,926.69	159,891.29	62	170,004.52	1,874,307.90	55,610.14
13	4,358,313.34	68,840,978.85	156,751.25	63	157,879.14	1,704,303.38	53,860.62
14	4,089,243.68	64,482,665.51	153,681.94	64	146,461.44	1,546,424.24	52,078.74
15	3,836,593.70	60,393,421.83	150,610.21	65	135,700.31	1,399,962.80	50,256.57
16	3,599,373.34	56,556,828.13	147,548.15	66	125,559.11	1,264,262.49	48,397.55
17	3,376,651.50	52,957,454.79	144,506.38	67	116,009.53	1,138,703.38	46,511.21
18	3,167,552.87	49,580,803.28	141,494.46	68	107,030.51	1,022,693.85	44,612.57
19	2,971,313.21	46,413,250.41	138,579.62	69	98,604.74	915,663.34	42,719.19
20	2,787,119.82	43,441,937.20	135,733.98	70	90,701.55	817,058.60	40,834.12
21	2,614,266.02	40,654,817.38	132,986.08	71	83,282.76	726,357.05	38,951.11
22	2,452,083.40	38,040,551.36	130,359.60	72	76,305.00	643,074.28	37,056.33
23	2,299,916.08	35,588,467.97	127,849.96	73	69,722.71	566,769.29	35,131.15
24	2,157,148.56	33,288,551.89	125,452.90	74	63,496.11	497,046.58	33,159.94
25	2,023,182.63	31,131,403.33	123,143.93	75	57,597.83	433,550.47	31,137.00
26	1,897,498.37	29,108,220.70	120,940.30	76	52,014.35	375,952.64	29,068.88
27	1,779,568.33	27,210,722.33	118,820.02	77	46,741.12	323,938.29	26,970.24
28	1,668,917.64	25,431,154.00	116,781.48	78	41,779.98	277,197.17	24,861.84
29	1,565,084.31	23,762,236.36	114,806.97	79	37,133.18	235,417.19	22,764.99
30	1,467,652.34	22,197,152.06	112,896.58	80	32,797.49	198,284.02	20,695.65
31	1,376,216.95	20,729,499.71	111,036.21	81	28,763.55	165,486.53	18,663.44
32	1,290,413.41	19,353,282.76	109,227.14	82	25,020.24	136,722.98	16,675.64
33	1,209,898.93	18,062,869.35	107,470.28	83	21,557.35	111,702.74	14,739.81
34	1,134,351.28	16,852,970.43	105,766.23	84	18,368.68	90,145.39	12,866.85
35	1,063,435.66	15,718,619.14	104,083.32	85	15,457.12	71,776.71	11,076.38
36	996,883.59	14,655,183.48	102,435.77	86	12,828.68	56,319.59	9,391.34
37	934,393.53	13,658,299.89	100,788.37	87	10,488.32	43,490.91	7,833.94
38	875,706.61	12,723,906.36	99,130.16	88	8,436.74	33,002.59	6,422.50
39	820,582.29	11,848,199.76	97,452.73	89	6,668.75	24,565.85	5,169.43
40	768,789.28	11,027,617.46	95,742.20	90	5,172.57	17,897.10	4,080.26
41	720,120.93	10,258,828.18	93,995.26	91	3,930.43	12,724.52	3,153.81
42	674,384.78	9,538,707.26	92,210.16	92	2,919.70	8,794.10	2,382.97
43	631,407.77	8,864,322.48	90,392.78	93	2,114.22	5,874.40	1,755.69
44	591,039.16	8,232,914.71	88,560.80	94	1,485.89	3,760.18	1,256.39
45	553,123.84	7,641,875.55	86,718.29	95	1,005.51	2,274.29	866.70
46	517,516.17	7,088,751.71	84,869.35	96	644.54	1,268.79	567.11
47	484,084.13	6,571,235.54	83,022.80	97	377.81	624.25	339.71
48	452,698.23	6,087,151.41	81,181.95	98	186.25	246.44	171.21
49	423,228.21	5,634,453.18	79,341.39	99	60.19	60.19	56.51

Table IV	Commutation Columns at 6.75% based on 1980 CSO Mortality Table Females						
Age	D_x	N_x	M_x	Age	D_x	N_x	M_x
0	10,000,000.00	154,618,602.42	223,179.71	50	351,799.93	4,526,552.99	65,577.61
1	9,340,608.90	144,618,602.42	196,107.11	51	327,920.38	4,174,753.05	63,943.02
2	8,742,372.31	135,277,993.52	188,494.49	52	305,554.20	3,846,832.67	62,311.86
3	8,182,942.02	126,535,621.22	181,860.58	53	284,601.91	3,541,278.47	60,680.32
4	7,659,463.65	118,352,679.20	175,804.77	54	264,966.37	3,256,676.56	59,040.69
5	7,169,616.58	110,693,215.56	170,279.77	55	246,571.37	2,991,710.19	57,400.00
6	6,711,163.77	103,523,598.97	165,175.08	56	229,342.56	2,745,138.82	55,762.35
7	6,282,214.99	96,812,435.20	160,585.60	57	213,214.46	2,515,796.27	54,136.00
8	5,880,741.91	90,530,220.21	156,348.60	58	198,128.67	2,302,581.81	52,532.16
9	5,505,035.39	84,649,478.30	152,492.27	59	184,028.58	2,104,453.14	50,960.12
10	5,153,383.46	79,144,442.90	148,933.91	60	170,850.93	1,920,424.56	49,418.93
11	4,824,242.82	73,991,059.45	145,651.24	61	158,532.05	1,749,573.64	47,903.28
12	4,516,079.00	69,166,816.62	142,533.22	62	147,003.40	1,591,041.58	46,398.89
13	4,227,473.20	64,650,737.62	139,487.45	63	136,198.83	1,444,038.19	44,889.62
14	3,957,192.05	60,423,264.42	136,517.25	64	126,053.13	1,307,839.36	43,356.03
15	3,704,005.90	56,466,072.36	133,551.68	65	116,517.97	1,181,786.22	41,791.44
16	3,466,845.45	52,762,066.46	130,602.37	66	107,557.82	1,065,268.26	40,198.94
17	3,244,707.48	49,295,221.01	127,679.45	67	99,144.63	957,710.43	38,586.83
18	3,036,651.17	46,050,513.53	124,792.00	68	91,256.71	858,565.80	36,968.01
19	2,841,850.26	43,013,862.36	122,004.16	69	83,875.82	767,309.09	35,357.45
20	2,659,439.54	40,172,012.10	119,288.89	70	76,972.46	683,433.27	33,757.71
21	2,488,662.38	37,512,572.56	116,673.02	71	70,511.11	606,460.81	32,163.47
22	2,328,805.22	35,023,910.18	114,178.58	72	64,452.11	535,949.70	30,563.02
23	2,179,172.67	32,695,104.96	111,800.70	73	58,754.36	471,497.59	28,940.70
24	2,039,113.65	30,515,932.28	109,534.80	74	53,381.98	412,743.23	27,283.47
25	1,907,999.21	28,476,818.63	107,357.28	75	48,309.82	359,361.25	25,586.74
26	1,785,279.58	26,568,819.42	105,283.97	76	43,524.54	311,051.43	23,856.18
27	1,670,402.83	24,783,539.84	103,293.76	77	39,020.41	267,526.89	22,104.19
28	1,562,871.16	23,113,137.01	101,384.75	78	34,797.07	228,506.48	20,348.18
29	1,462,203.22	21,550,265.85	99,540.04	79	30,854.49	193,709.41	18,605.88
30	1,367,964.78	20,088,062.63	97,759.41	80	27,188.08	162,854.92	16,890.46
31	1,279,735.89	18,720,097.85	96,029.46	81	23,788.23	135,666.84	15,209.77
32	1,197,137.51	17,440,361.97	94,351.16	82	20,643.95	111,878.61	13,569.66
33	1,119,814.24	16,243,224.46	92,725.11	83	17,745.10	91,234.66	11,976.16
34	1,047,432.84	15,123,410.22	91,151.63	84	15,084.91	73,489.56	10,438.03
35	979,651.40	14,075,977.38	89,601.31	85	12,664.12	58,404.65	8,971.09
36	916,192.04	13,096,325.98	88,087.12	86	10,486.01	45,740.53	7,593.75
37	856,749.01	12,180,133.94	86,576.61	87	8,552.95	35,254.52	6,323.74
38	801,058.33	11,323,384.93	85,059.75	88	6,863.83	26,701.57	5,175.44
39	748,875.08	10,522,326.60	83,528.90	89	5,412.75	19,837.74	4,158.37
40	699,964.93	9,773,451.52	81,971.51	90	4,188.53	14,424.99	3,276.41
41	654,118.03	9,073,486.58	80,384.69	91	3,175.24	10,236.46	2,527.97
42	611,139.24	8,419,368.56	78,766.99	92	2,353.19	7,061.22	1,906.69
43	570,852.69	7,808,229.32	77,123.91	93	1,700.00	4,708.03	1,402.31
44	533,104.21	7,237,376.62	75,471.50	94	1,191.98	3,008.02	1,001.77
45	497,737.04	6,704,272.41	73,813.50	95	804.73	1,816.05	689.89
46	464,604.30	6,206,535.36	72,153.59	96	514.63	1,011.32	450.69
47	433,572.65	5,741,931.06	70,499.72	97	300.95	496.69	269.54
48	404,512.13	5,308,358.41	68,854.81	98	148.02	195.74	135.64
49	377,293.29	4,903,846.28	67,214.02	99	47.72	47.72	44.70

Table IV Commutation Columns at 7% based on 1980 CSO Mortality Table Females

Age	D_x	N_x	M_x	Age	D_x	N_x	M_x
0	10,000,000.00	149,688,119.06	207,319.31	50	312,968.81	3,934,763.81	55,554.35
1	9,318,785.05	139,688,119.06	180,309.97	51	291,043.44	3,621,795.00	54,103.58
2	8,701,567.82	130,369,334.01	172,732.89	52	270,558.87	3,330,751.56	52,659.23
3	8,125,718.85	121,667,766.19	166,145.36	53	251,417.45	3,060,192.69	51,217.93
4	7,588,130.36	113,542,047.34	160,145.96	54	233,524.52	2,808,775.24	49,772.87
5	7,086,249.86	105,953,916.98	154,685.19	55	216,804.59	2,575,250.72	48,330.24
6	6,617,629.87	98,867,667.12	149,651.65	56	201,184.53	2,358,446.13	46,893.66
7	6,180,185.88	92,250,037.25	145,136.71	57	186,599.59	2,157,261.59	45,470.33
8	5,771,716.19	86,069,851.38	140,978.25	58	172,991.78	1,970,662.00	44,069.97
9	5,390,351.30	80,298,135.18	137,202.27	59	160,305.17	1,797,670.22	42,700.58
10	5,034,235.40	74,907,783.88	133,726.17	60	148,478.54	1,637,365.05	41,361.20
11	4,701,693.62	69,873,548.48	130,526.90	61	137,450.89	1,488,886.51	40,047.10
12	4,391,074.48	65,171,854.86	127,495.19	62	127,157.49	1,351,435.62	38,745.81
13	4,100,853.37	60,780,780.39	124,540.64	63	117,536.31	1,224,278.13	37,443.35
14	3,829,698.75	56,679,927.01	121,666.14	64	108,526.66	1,106,741.82	36,122.99
15	3,576,294.38	52,850,228.27	118,802.81	65	100,082.88	998,215.16	34,779.08
16	3,339,490.23	49,273,933.88	115,961.85	66	92,170.72	898,132.28	33,414.41
17	3,118,209.94	45,934,443.65	113,152.88	67	84,762.61	805,961.56	32,036.15
18	2,911,446.50	42,816,233.71	110,384.48	68	77,836.63	721,198.95	30,655.39
19	2,718,311.39	39,904,787.21	107,717.84	69	71,374.01	643,362.32	29,284.88
20	2,537,886.77	37,186,475.82	105,126.67	70	65,346.57	571,988.32	27,926.78
21	2,369,366.31	34,648,589.05	102,636.19	71	59,721.27	506,641.74	26,576.49
22	2,211,991.73	32,279,222.74	100,266.88	72	54,461.90	446,920.47	25,224.11
23	2,065,028.65	30,067,231.01	98,013.54	73	49,531.31	392,458.57	23,856.45
24	1,927,791.13	28,002,202.35	95,871.35	74	44,897.13	342,927.26	22,462.63
25	1,799,620.12	26,074,411.23	93,817.52	75	40,536.23	298,030.13	21,038.93
26	1,679,936.99	24,274,791.10	91,866.55	76	36,435.63	257,493.90	19,590.23
27	1,568,166.17	22,594,854.11	89,998.14	77	32,588.77	221,058.28	18,127.02
28	1,463,787.88	21,026,687.94	88,210.16	78	28,993.65	188,469.50	16,663.87
29	1,366,302.33	19,562,900.07	86,486.44	79	25,648.54	159,475.85	15,215.54
30	1,275,258.11	18,196,597.73	84,826.49	80	22,547.95	133,827.31	13,792.89
31	1,190,221.07	16,921,339.62	83,217.55	81	19,682.25	111,279.36	12,402.29
32	1,110,798.88	15,731,118.55	81,660.28	82	17,040.78	91,597.11	11,048.45
33	1,036,624.54	14,620,319.67	80,155.03	83	14,613.67	74,556.33	9,736.15
34	967,354.80	13,583,695.14	78,701.84	84	12,393.89	59,942.66	8,472.41
35	902,641.45	12,616,340.34	77,273.39	85	10,380.64	47,548.77	7,269.97
36	842,198.24	11,713,698.89	75,881.49	86	8,575.18	37,168.13	6,143.62
37	785,715.89	10,871,500.65	74,496.22	87	6,978.03	28,592.95	5,107.47
38	732,926.07	10,085,784.76	73,108.38	88	5,586.86	21,614.91	4,172.80
39	683,580.26	9,352,858.69	71,711.00	89	4,395.45	16,028.05	3,346.89
40	637,441.78	8,669,278.43	70,292.72	90	3,393.37	11,632.60	2,632.36
41	594,298.27	8,031,836.65	68,851.01	91	2,566.43	8,239.23	2,027.42
42	553,952.62	7,437,538.38	67,384.69	92	1,897.56	5,672.80	1,526.44
43	516,226.88	6,883,585.76	65,898.84	93	1,367.64	3,775.24	1,120.66
44	480,964.23	6,367,358.88	64,408.04	94	956.70	2,407.60	799.19
45	448,006.93	5,886,394.65	62,915.69	95	644.37	1,450.90	549.46
46	417,207.50	5,438,387.71	61,425.13	96	411.12	806.53	358.36
47	388,431.87	5,021,180.22	59,943.45	97	239.86	395.41	213.99
48	361,550.23	4,632,748.34	58,473.23	98	117.69	155.55	107.52
49	336,434.31	4,271,198.12	57,010.13	99	37.85	37.85	35.38

Table IV Commutation Columns at 7.25% based on 1980 CSO Mortality Table Females

Age	D_x	N_x	M_x	Age	D_x	N_x	M_x
0	10,000,000.00	145,069,380.69	193,445.13	50	278,499.81	3,422,845.05	47,118.67
1	9,297,062.94	135,069,380.69	166,498.74	51	258,385.49	3,144,345.25	45,830.69
2	8,661,048.35	125,772,317.75	158,956.94	52	239,639.59	2,885,959.75	44,551.40
3	8,069,028.00	117,111,269.40	152,415.38	53	222,166.56	2,646,320.16	43,277.79
4	7,517,625.56	109,042,241.40	146,471.71	54	205,874.35	2,424,153.60	42,003.82
5	7,004,043.67	101,524,615.84	141,074.30	55	190,688.59	2,218,279.25	40,734.98
6	6,525,613.29	94,520,572.17	136,110.74	56	176,537.63	2,027,590.67	39,474.39
7	6,080,046.14	87,994,958.88	131,668.97	57	163,357.80	1,851,053.04	38,228.34
8	5,664,959.15	81,914,912.74	127,587.43	58	151,091.88	1,687,695.24	37,005.25
9	5,278,315.69	76,249,953.59	123,889.92	59	139,684.97	1,536,603.36	35,812.01
10	4,918,110.56	70,971,637.90	120,494.01	60	129,078.02	1,396,918.39	34,647.64
11	4,582,532.68	66,053,527.34	117,375.82	61	119,212.73	1,267,840.37	33,507.90
12	4,269,809.75	61,470,994.66	114,427.83	62	110,028.07	1,148,627.64	32,381.92
13	3,978,308.34	57,201,184.91	111,561.57	63	101,465.89	1,038,599.57	31,257.53
14	3,706,596.30	53,222,876.57	108,779.47	64	93,469.72	937,133.67	30,120.36
15	3,453,269.02	49,516,280.26	106,014.65	65	85,996.50	843,663.95	28,965.61
16	3,217,094.41	46,063,011.24	103,277.80	66	79,013.35	757,667.45	27,795.74
17	2,996,922.13	42,845,916.83	100,578.10	67	72,493.36	678,654.10	26,616.98
18	2,791,678.47	39,848,994.70	97,923.58	68	66,414.74	606,160.74	25,438.84
19	2,600,412.62	37,057,316.23	95,372.60	69	60,758.49	539,746.00	24,272.17
20	2,422,154.15	34,456,903.61	92,899.59	70	55,497.85	478,987.51	23,118.74
21	2,256,047.41	32,034,749.46	90,528.22	71	50,602.14	423,489.65	21,974.64
22	2,101,289.98	29,778,702.05	88,277.48	72	46,038.28	372,887.51	20,831.43
23	1,957,109.16	27,677,412.07	86,141.91	73	41,772.71	326,849.23	19,678.01
24	1,822,784.89	25,720,302.92	84,116.40	74	37,776.16	285,076.52	18,505.26
25	1,697,628.91	23,897,518.02	82,178.97	75	34,027.43	247,300.36	17,310.15
26	1,581,034.66	22,199,889.12	80,342.85	76	30,513.95	213,272.93	16,096.90
27	1,472,403.89	20,618,854.46	78,588.55	77	27,228.69	182,758.98	14,874.35
28	1,371,195.87	19,146,450.57	76,913.67	78	24,168.41	155,530.29	13,654.71
29	1,276,893.39	17,775,254.70	75,302.74	79	21,330.17	131,361.87	12,450.23
30	1,189,028.87	16,498,361.31	73,755.03	80	18,707.90	110,031.70	11,269.86
31	1,107,154.98	15,309,332.44	72,258.38	81	16,292.19	91,323.80	10,118.78
32	1,030,867.12	14,202,177.46	70,813.17	82	14,072.80	75,031.61	9,000.74
33	959,787.79	13,171,310.34	69,419.49	83	12,040.29	60,958.81	7,919.53
34	893,564.70	12,211,522.55	68,077.16	84	10,187.60	48,918.52	6,880.75
35	831,844.14	11,317,957.85	66,760.74	85	8,512.84	38,730.93	5,894.67
36	774,332.51	10,486,113.71	65,481.00	86	7,015.85	30,218.08	4,973.14
37	720,717.68	9,711,781.20	64,210.33	87	5,695.82	23,202.23	4,127.38
38	670,727.77	8,991,063.52	62,940.26	88	4,549.65	17,506.41	3,366.23
39	624,111.39	8,320,335.74	61,664.45	89	3,571.08	12,956.76	2,695.22
40	580,630.17	7,696,224.35	60,372.58	90	2,750.52	9,385.68	2,116.05
41	540,069.95	7,115,594.18	59,062.42	91	2,075.39	6,635.16	1,626.86
42	502,232.31	6,575,524.23	57,733.00	92	1,530.91	4,559.77	1,222.68
43	466,937.89	6,073,291.92	56,389.02	93	1,100.81	3,028.86	896.07
44	434,028.01	5,606,354.03	55,043.71	94	768.25	1,928.04	637.92
45	403,344.55	5,172,326.02	53,700.13	95	516.24	1,159.79	437.84
46	374,739.99	4,768,981.47	52,361.29	96	328.60	643.55	285.10
47	348,080.16	4,394,241.48	51,033.53	97	191.27	314.95	169.98
48	323,235.85	4,046,161.32	49,719.12	98	93.63	123.68	85.27
49	300,080.41	3,722,925.46	48,414.12	99	30.05	30.05	28.01

Table IV	Commutation Columns at 7.5% based on 1980 CSO Mortality Table Females						
Age	D_x	N_x	M_x	Age	D_x	N_x	M_x
0	10,000,000.00	140,735,216.56	181,263.96	50	247,894.41	2,979,666.67	40,010.68
1	9,275,441.86	130,735,216.56	154,380.24	51	229,455.67	2,731,772.26	38,866.91
2	8,620,811.25	121,459,774.70	146,873.48	52	212,313.73	2,502,316.59	37,733.50
3	8,012,863.27	112,838,963.45	140,377.45	53	196,375.38	2,290,002.87	36,607.74
4	7,447,937.73	104,826,100.18	134,488.88	54	181,551.32	2,093,627.49	35,484.29
5	6,922,979.23	97,378,162.45	129,153.94	55	167,768.61	1,912,076.17	34,367.95
6	6,435,085.97	90,455,183.23	124,259.24	56	154,957.34	1,744,307.55	33,261.46
7	5,981,756.51	84,020,097.26	119,889.26	57	143,055.17	1,589,350.22	32,170.27
8	5,560,418.45	78,038,340.74	115,883.05	58	132,005.99	1,446,295.05	31,101.69
9	5,168,861.44	72,477,922.30	112,262.21	59	121,756.18	1,314,289.06	30,061.60
10	4,804,925.45	67,309,060.85	108,944.46	60	112,249.01	1,192,532.87	29,049.04
11	4,466,658.75	62,504,135.41	105,905.11	61	103,428.84	1,080,283.87	28,060.20
12	4,152,164.63	58,037,476.66	103,038.35	62	95,238.24	976,855.02	27,085.57
13	3,859,697.92	53,885,312.03	100,257.55	63	87,622.73	881,616.78	26,114.58
14	3,587,723.79	50,025,614.11	97,564.67	64	80,529.78	793,994.05	25,134.84
15	3,334,747.55	46,437,890.32	94,894.74	65	73,918.84	713,464.27	24,142.27
16	3,099,453.98	43,103,142.77	92,257.97	66	67,758.49	639,545.43	23,139.04
17	2,880,618.07	40,003,688.79	89,663.04	67	62,022.65	571,786.94	22,130.53
18	2,677,099.15	37,123,070.72	87,117.47	68	56,689.85	509,764.29	21,124.90
19	2,487,884.19	34,445,971.57	84,676.87	69	51,741.23	453,074.44	20,131.38
20	2,311,950.39	31,958,087.38	82,316.38	70	47,151.42	401,333.21	19,151.43
21	2,148,393.31	29,646,136.99	80,058.17	71	42,892.00	354,181.80	18,181.64
22	1,996,367.05	27,497,743.68	77,919.82	72	38,932.77	311,289.80	17,214.88
23	1,855,061.41	25,501,376.63	75,895.60	73	35,243.39	272,357.02	16,241.74
24	1,723,723.09	23,646,315.22	73,980.17	74	31,797.41	237,113.63	15,254.60
25	1,601,635.46	21,922,592.13	72,152.29	75	28,575.37	205,316.22	14,250.98
26	1,488,165.19	20,320,956.67	70,424.03	76	25,565.25	176,740.85	13,234.50
27	1,382,692.30	18,832,791.47	68,776.61	77	22,759.73	151,175.60	12,212.60
28	1,284,656.22	17,450,099.18	67,207.44	78	20,154.75	128,415.87	11,195.50
29	1,193,523.29	16,165,442.96	65,701.69	79	17,746.49	108,261.12	10,193.39
30	1,108,810.93	14,971,919.67	64,258.39	80	15,528.59	90,514.63	9,213.62
31	1,030,059.59	13,863,108.74	62,865.96	81	13,491.96	74,986.04	8,260.38
32	956,853.53	12,833,049.15	61,524.51	82	11,626.93	61,494.07	7,336.65
33	888,805.69	11,876,195.62	60,233.91	83	9,924.54	49,867.14	6,445.43
34	825,555.83	10,987,389.93	58,993.74	84	8,377.88	39,942.60	5,591.18
35	766,745.51	10,161,834.10	57,780.34	85	6,984.34	31,564.72	4,782.15
36	712,074.79	9,395,088.59	56,603.50	86	5,742.76	24,580.38	4,027.85
37	661,229.37	8,683,013.79	55,437.71	87	4,651.42	18,837.62	3,337.17
38	613,934.56	8,021,784.43	54,275.18	88	3,706.77	14,186.21	2,717.03
39	569,936.85	7,407,849.87	53,110.12	89	2,902.73	10,479.44	2,171.60
40	528,996.82	6,837,913.02	51,933.12	90	2,230.54	7,576.71	1,701.93
41	490,899.19	6,308,916.20	50,742.25	91	1,679.13	5,346.17	1,306.14
42	455,444.84	5,818,017.01	49,536.68	92	1,235.73	3,667.04	979.89
43	422,453.67	5,362,572.17	48,320.73	93	886.49	2,431.31	716.87
44	391,765.85	4,940,118.50	47,106.42	94	617.24	1,544.82	509.46
45	363,223.42	4,548,352.65	45,896.49	95	413.80	927.58	349.09
46	336,679.38	4,185,129.23	44,693.62	96	262.79	513.78	226.94
47	312,000.00	3,848,449.85	43,503.50	97	152.60	250.99	135.09
48	289,057.13	3,536,449.85	42,328.07	98	74.53	98.39	67.67
49	267,726.05	3,247,392.72	41,163.77	99	23.86	23.86	22.20

Table IV	Commutation Columns at 7.75% based on 1980 CSO Mortality Table Females						
Age	D_x	N_x	M_x	Age	D_x	N_x	M_x
0	10,000,000.00	136,661,347.07	170,529.56	50	220,712.03	2,595,699.86	34,014.36
1	9,253,921.11	126,661,347.07	143,708.22	51	203,821.15	2,374,987.84	32,998.36
2	8,580,853.89	117,407,425.95	136,236.25	52	188,156.71	2,171,166.68	31,993.91
3	7,957,218.59	108,826,572.06	129,785.33	53	173,628.04	1,983,009.97	30,998.55
4	7,379,055.53	100,869,353.47	123,951.22	54	160,148.70	1,809,381.93	30,007.54
5	6,843,038.06	93,490,297.94	118,677.88	55	147,647.44	1,649,233.23	29,025.09
6	6,346,020.43	86,647,259.88	113,850.93	56	136,056.26	1,501,585.79	28,053.57
7	5,885,278.64	80,301,239.45	109,551.44	57	125,314.44	1,365,529.53	27,097.70
8	5,458,043.07	74,415,960.82	105,618.98	58	115,367.21	1,240,215.09	26,163.81
9	5,061,923.29	68,957,917.74	102,073.06	59	106,162.46	1,124,847.88	25,256.93
10	4,694,599.06	63,895,994.45	98,831.48	60	97,645.81	1,018,685.42	24,376.10
11	4,353,973.82	59,201,395.39	95,868.81	61	89,764.37	921,039.61	23,517.90
12	4,038,023.01	54,847,421.57	93,080.85	62	82,464.09	831,275.24	22,673.99
13	3,744,887.08	50,809,398.56	90,382.77	63	75,694.00	748,811.15	21,835.20
14	3,472,926.53	47,064,511.49	87,776.06	64	69,405.26	673,117.15	20,990.80
15	3,220,555.17	43,591,584.95	85,197.55	65	63,559.76	603,711.89	20,137.32
16	2,986,373.75	40,371,029.78	82,656.99	66	58,127.54	540,152.13	19,276.69
17	2,769,082.10	37,384,656.04	80,162.53	67	53,083.52	482,024.60	18,413.54
18	2,567,472.46	34,615,573.93	77,721.20	68	48,406.75	428,941.08	17,554.84
19	2,380,469.84	32,048,101.47	75,385.98	69	44,078.67	380,534.32	16,708.45
20	2,206,999.42	29,667,631.63	73,132.64	70	40,075.39	336,455.65	15,875.56
21	2,046,108.62	27,460,632.22	70,981.94	71	36,370.60	296,380.26	15,053.23
22	1,896,908.90	25,414,523.60	68,950.12	72	32,936.74	260,009.66	14,235.35
23	1,758,553.38	23,517,614.70	67,031.21	73	29,746.39	227,072.92	13,414.00
24	1,630,256.53	21,759,061.32	65,219.64	74	26,775.62	197,326.53	12,582.76
25	1,511,274.35	20,128,804.79	63,494.89	75	24,006.61	170,550.91	11,739.60
26	1,400,947.83	18,617,530.43	61,867.92	76	21,427.93	146,544.31	10,887.62
27	1,298,636.33	17,216,582.60	60,320.65	77	19,032.18	125,116.38	10,033.09
28	1,203,760.57	15,917,946.27	58,850.28	78	16,814.73	106,084.20	9,184.54
29	1,115,771.53	14,714,185.70	57,442.63	79	14,771.21	89,269.47	8,350.44
30	1,034,172.68	13,598,414.17	56,096.49	80	12,895.17	74,498.26	7,536.82
31	958,493.34	12,564,241.49	54,800.80	81	11,177.92	61,603.10	6,747.07
32	888,307.63	11,605,748.15	53,555.45	82	9,610.42	50,425.17	5,983.55
33	823,220.06	10,717,440.51	52,360.07	83	8,184.24	40,814.75	5,248.61
34	762,863.35	9,894,220.45	51,214.08	84	6,892.77	32,630.51	4,545.79
35	706,875.18	9,131,357.10	50,095.44	85	5,732.93	25,737.74	3,881.72
36	654,950.21	8,424,481.92	49,013.00	86	4,702.86	20,004.81	3,264.00
37	606,772.65	7,769,531.70	47,943.22	87	3,800.31	15,301.95	2,699.70
38	562,065.76	7,162,759.06	46,878.91	88	3,021.48	11,501.64	2,194.22
39	520,574.59	6,600,693.30	45,814.75	89	2,360.60	8,480.16	1,750.65
40	482,059.31	6,080,118.71	44,742.19	90	1,809.74	6,119.57	1,369.59
41	446,304.14	5,598,059.40	43,659.50	91	1,359.19	4,309.83	1,049.21
42	413,109.87	5,151,755.25	42,565.99	92	997.96	2,950.63	785.73
43	382,296.27	4,738,645.38	41,465.63	93	714.26	1,952.68	573.81
44	353,702.99	4,356,349.11	40,369.29	94	496.16	1,238.42	407.09
45	327,172.79	4,002,646.12	39,279.45	95	331.86	742.25	278.47
46	302,559.68	3,675,473.32	38,198.49	96	210.26	410.39	180.74
47	279,730.81	3,372,913.64	37,131.45	97	121.82	200.13	107.42
48	258,559.55	3,093,182.83	36,080.04	98	59.36	78.32	53.72
49	238,923.42	2,834,623.28	35,041.00	99	18.96	18.96	17.59

Table IV Commutation Columns at 8% based on 1980 CSO Mortality Table Females

Age	D_x	N_x	M_x	Age	D_x	N_x	M_x
0	10,000,000.00	132,826,031.43	161,034.71	50	196,563.18	2,262,779.37	28,949.89
1	9,232,500.00	122,826,031.43	134,275.45	51	181,100.21	2,066,216.19	28,047.15
2	8,541,173.70	113,593,531.43	126,838.04	52	166,794.96	1,885,115.98	27,156.74
3	7,902,087.94	105,052,357.74	120,431.81	53	153,559.47	1,718,321.02	26,276.43
4	7,310,967.82	97,150,269.80	114,651.53	54	141,310.26	1,564,761.55	25,401.99
5	6,764,202.06	89,839,301.98	109,438.95	55	129,977.95	1,423,451.29	24,537.12
6	6,258,389.75	83,075,099.93	104,678.65	56	119,496.68	1,293,473.34	23,683.84
7	5,790,575.02	76,816,710.17	100,448.34	57	109,807.49	1,173,976.66	22,846.25
8	5,357,783.32	71,026,135.15	96,588.12	58	100,857.16	1,064,169.18	22,029.82
9	4,957,437.75	65,668,351.84	93,115.39	59	92,595.28	963,312.01	21,238.83
10	4,587,052.82	60,710,914.09	89,948.07	60	84,969.88	870,716.73	20,472.35
11	4,244,383.03	56,123,861.27	87,059.98	61	77,930.76	785,746.85	19,727.29
12	3,927,272.80	51,879,478.23	84,348.48	62	71,427.15	707,816.09	18,996.33
13	3,633,745.69	47,952,205.43	81,730.48	63	65,411.40	636,388.94	18,271.48
14	3,362,055.85	44,318,459.74	79,206.98	64	59,838.11	570,977.54	17,543.48
15	3,110,524.26	40,956,403.89	76,716.57	65	54,671.53	511,139.43	16,809.35
16	2,877,666.97	37,845,879.63	74,268.48	66	49,883.22	456,467.90	16,070.78
17	2,662,108.36	34,968,212.66	71,870.39	67	45,449.15	406,584.68	15,331.77
18	2,462,573.57	32,306,104.29	69,528.81	68	41,349.05	361,135.53	14,598.27
19	2,277,926.08	29,843,530.73	67,294.18	69	37,564.85	319,786.48	13,876.96
20	2,107,039.54	27,565,604.64	65,142.90	70	34,074.10	282,221.63	13,168.79
21	1,948,914.00	25,458,565.10	63,094.36	71	30,852.52	248,147.54	12,471.22
22	1,802,619.18	23,509,651.10	61,163.54	72	27,874.97	217,295.02	11,779.04
23	1,667,272.52	21,707,031.93	59,344.23	73	25,116.64	189,420.05	11,085.52
24	1,542,057.29	20,039,759.41	57,630.67	74	22,555.90	164,303.41	10,385.28
25	1,426,203.16	18,497,702.11	56,003.01	75	20,176.46	141,747.51	9,676.65
26	1,319,026.65	17,071,498.95	54,471.17	76	17,967.51	121,571.05	8,962.25
27	1,219,867.55	15,752,472.30	53,017.75	77	15,921.71	103,603.54	8,247.37
28	1,128,129.01	14,532,604.75	51,639.77	78	14,034.10	87,681.83	7,539.15
29	1,043,247.74	13,404,475.73	50,323.61	79	12,299.98	73,647.72	6,844.59
30	964,714.40	12,361,227.99	49,067.88	80	10,712.94	61,347.74	6,168.67
31	892,048.20	11,396,513.59	47,862.01	81	9,264.81	50,634.80	5,514.08
32	824,814.22	10,504,465.39	46,705.68	82	7,947.15	41,369.99	4,882.71
33	762,609.51	9,679,651.16	45,598.31	83	6,752.13	33,422.84	4,276.37
34	705,060.76	8,917,041.65	44,539.16	84	5,673.48	26,670.71	3,697.87
35	651,802.54	8,211,980.89	43,507.66	85	4,707.89	20,997.23	3,152.54
36	602,525.08	7,560,178.35	42,511.87	86	3,853.06	16,289.34	2,646.44
37	556,911.72	6,957,653.27	41,530.00	87	3,106.38	12,436.28	2,185.18
38	514,684.41	6,400,741.55	40,555.41	88	2,464.05	9,329.90	1,772.95
39	475,587.44	5,886,057.14	39,583.21	89	1,920.64	6,865.85	1,412.06
40	439,381.14	5,410,469.70	38,605.61	90	1,469.04	4,945.21	1,102.73
41	405,849.84	4,971,088.56	37,621.06	91	1,100.76	3,476.17	843.26
42	374,794.80	4,565,238.72	36,628.97	92	806.34	2,375.41	630.38
43	346,036.23	4,190,443.93	35,632.97	93	575.78	1,569.08	459.55
44	319,413.86	3,844,407.70	34,642.92	94	399.04	993.30	325.46
45	294,771.66	3,524,993.84	33,661.01	95	266.28	594.26	222.26
46	271,965.07	3,230,222.17	32,689.35	96	168.32	327.98	144.02
47	250,862.59	2,958,257.10	31,732.43	97	97.29	159.66	85.47
48	231,339.45	2,707,394.51	30,791.71	98	47.30	62.37	42.68
49	213,275.70	2,476,055.06	29,864.21	99	15.07	15.07	13.96

Solutions

Chapter 1

Exercises 1.1

1.1.1 $3^{3.6}$ must lie between $3^3 = 27$ and $3^4 = 81$.

1.1.2 $9^{1.5} = 9^{3/2} = (9^{1/2})^3 = 3^3 = 27$.

1.1.3 Use the Log Rule: $n = \log(2)/\log(1.09) = 8.0432...$

1.1.4 Use the Log Rule: $n = \log(.11)/\log(.89) = 18.9410...$

Exercises 1.2

1.2.1 January 1, 2002 is 226 years. At the annual interest rate of 2.5%, \$4,000 becomes $\$4{,}000(1.025)^{226} \approx \$1{,}060{,}849$.

1.2.2 The assurance means that $\$15{,}000(1 + i)^{2.5} = \$30{,}000$. Thus, $i = (\$30{,}000/\$15{,}000)^{1/2.5} - 1 = .3195079... = 31.95079...\%$.

1.2.3 If the initial value of the work was V, it has increased in value in n years to $1.5V$ at the rate of 12% per year. This means that $V(1.12)^n = 1.5V$. The V's cancel giving $(1.12)^n = 1.5$ and finally, $n = \log(1.5)/\log(1.12) = 3.57778...$ years. That's almost 3 years and 7 months.

1.2.4 Faramarz is looking for P so that $P(1.06)^4 = \$10{,}000$ giving $P = \$10{,}000(1.06)^{-4} \approx \$7{,}920.94$.

1.2.5 *a*) José Luis's \$500 grows to $\$500(1 + i)^3 = \$750$ in three years at annual interest rate i. This gives $i = (\$750/\$500)^{1/3} - 1 = .14471... \approx 14.471\%$.

b) In five years José Luis has $\$500(1.14471...)^5 \approx \$982.78$.

1.2.6 *a*) Sukie's investment grew from \$11,000 to \$14,500 in three years at annual rate i. Thus $\$11{,}000(1 + i)^3 = \$14{,}500$ so $i = (\$14{,}500/\$11{,}000)^{1/3} - 1 = .096457... = 9.6457...\%$

b) One year from now is two years since it reached \$14,500, so in one year Sukie's investment will be worth $\$14{,}500 \cdot (1.096457...)^2 = \$17{,}432.17$.

1.2.7 Sita's money will be worth $\$6{,}000(1 + i)^{10} = \$24{,}000$ in 10 years. This means that i must be $i = (\$24{,}000/\$6{,}000)^{1/10} - 1 = .148698\ldots = 14.8698\ldots\%$.

1.2.8 Barbro will have $\$12{,}000(1.0784)^5 \approx \$17{,}501.72$. That's \$1.72 more than she needs.

1.2.9 Joanna's money has been exposed to interest for 1,760 days = 1,760/365.25 years = 4.8186... years. Her \$500 has grown to $\$500(1 + i)^{4.8186\ldots} = \$698.18$. This gives

$$i = (\$698.18/\$500)^{1/4.8186\ldots} - 1 = .071744\ldots\ 7.1744\ldots\%.$$

1.2.10 Shimon borrowed $P = \$4{,}130.35(1.085)^{-3} \approx \$3{,}233.68$.

Exercises 1.3

1.3.4 Here is a time diagram for the situation.

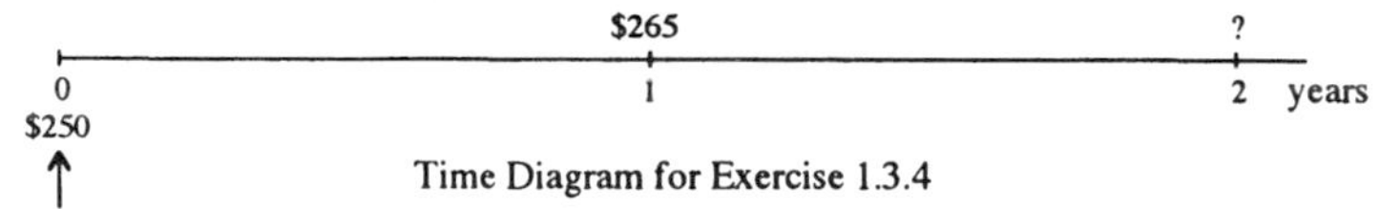

Time Diagram for Exercise 1.3.4

We first find the interest rate that Samantha's sister was charging. That rate satisfies $\$250(1 + i) = \$265$, so $i = 6\%$. To repay her loan in two years, Samantha should pay $P = \$250(1.06)^2 = \$280.90$.

1.3.5 Below is a time diagram for Joel's situation. Again we need the interest rate being charged.

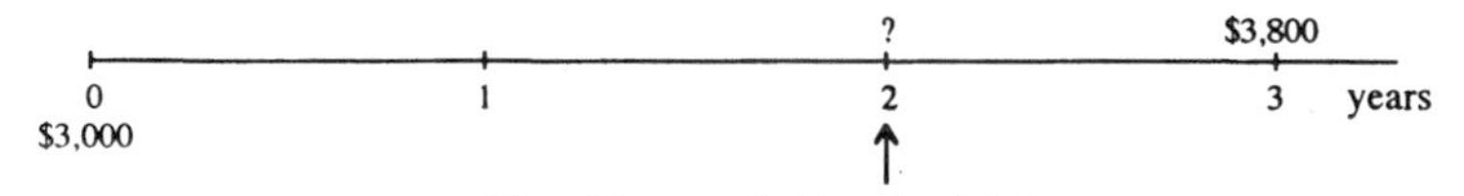

Time Diagram for Exercise 1.3.5

We have $\$3{,}000(1 + i)^3 = \$3{,}800$ which yields $i = (\$3{,}800/\$3{,}000)^{1/3} - 1 = .08198\ldots$ But Joel will pay after two years so he should pay $P = \$3{,}000(1.08198\ldots)^2 = \$3{,}512.07$.

1.3.6 Below is a time diagram for Freddie's transaction. The interest rate must satisfy $\$10{,}000(1 + i)^{10} = \$21{,}560.19$ from which we get $i = .0798545\ldots$

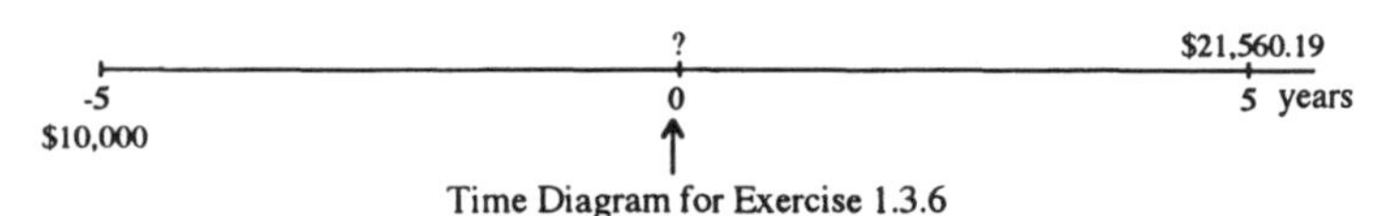

Time Diagram for Exercise 1.3.6

Freddie's balance at time 0 is $\$10{,}000(1.0798545\ldots)^5 \approx$ $14,683.39.

1.3.7 This is an interesting problem because we don't know i or n, and we don't need to. We are just going to need $(1 + i)^n$. From time n years to time $2n$ years, n years have passed so that we have the equation of value $\$12{,}094.47(1 + i)^n = \$18{,}298.76$ which gives $(1 + i)^n = 1.51298\ldots$ From the time diagram below you see that $P(1 + i)^n = P(1.51298\ldots) \approx \$12{,}094.47$, so $P \approx$ $7,993.78.

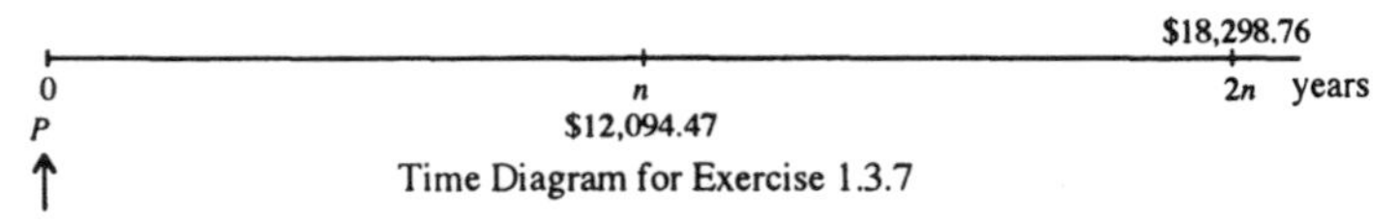

Time Diagram for Exercise 1.3.7

1.3.8 We can find the interest rate from the equivalence of $3,200 in 3 years and $4,000 in 4 years. From $\$3{,}200(1 + i) = \$4{,}000$ we solve for i to get $i = 25\%$. Then, the amount borrowed is $L = \$4{,}000(1.25)^{-4} = \$1{,}638.40$.

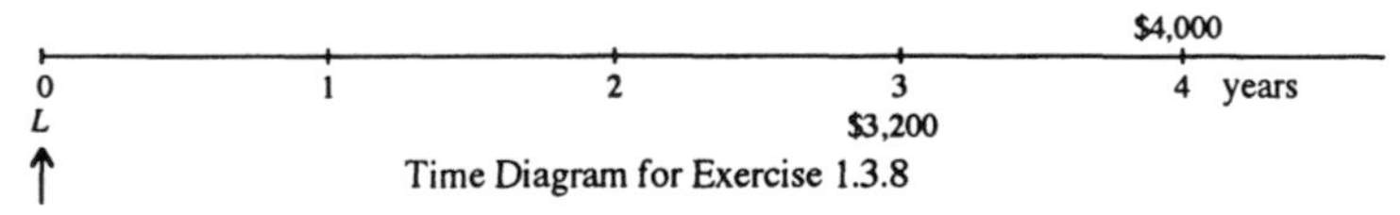

Time Diagram for Exercise 1.3.8

1.3.9 The time diagram below will serve for the whole problem. Move the arrow from -6, to 0 to +4 for parts a), b), and c) respectively.

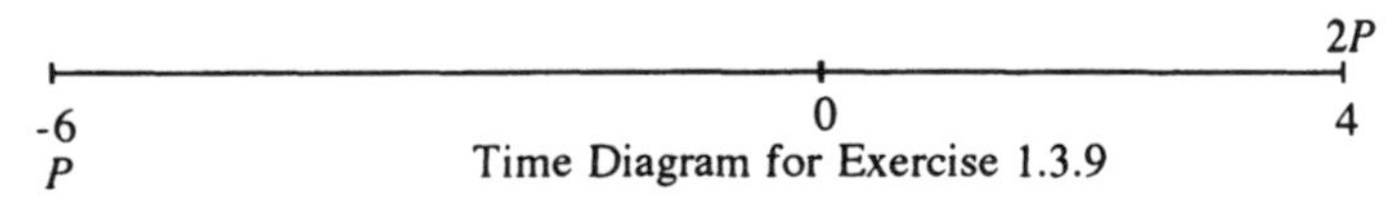

Time Diagram for Exercise 1.3.9

a) With the focal date at -6, we equate P with the value of $2P$ discounted 10 years. That is, $P = 2P(1 + i)^{-10}$.

b) Now, move the arrow to 0. We equate the value of P accumulated 6 years with the value of $2P$ discounted 4 years. We get $P(1 + i)^6 = 2P(1 + i)^{-4}$.

c) If the focal date is four years from now, we equate the value of P accumulated 10 years with $2P$. This gives $P(1 + i)^{10} = 2P$.

d) For each of the three equations of value, divide both sides by P and solve for $1 + i$. In each case we get $1 + i = 2^{1/10}$. Solving for i gives $i = .07177... \approx 7.177\%$.

1.3.10 We could use either Formula 1.1 or 1.2 for this. I put the arrow at 2 years in the time diagram below, so Formula 1.1 it is. The

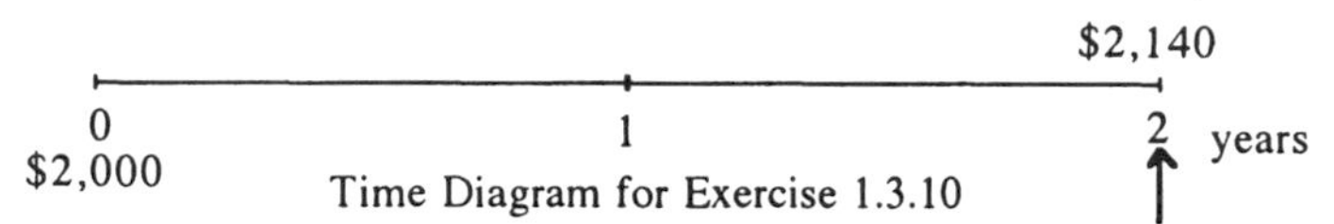

Time Diagram for Exercise 1.3.10

equation of value is $\$2000(1 + i)^2 = \$2140$, or $(1 + i)^2 = 2140/2100 = 1.07$, so $1 + i = (1.07)^{1/2} = 1.034408...$, and $i = 1.034408... - 1 = .034408... \approx 3.4408\%$.

1.3.11 Since each \$1 is to become \$2, we need to find n so that $(1.06)^n = 2$. Using the Log Rule, we find $n = \log(2)/\log(1.06) \approx 11.89566...$ years.

Ewan must pay a 50% higher rate the second year. This amount is 1.5(5.5%) = 8.25%. With the help of the time diagram below, we can write \$4,220(1.0825) = \$4,568.15. Note that this may also have been written \$4,000· (1.055)·(1.0825) = \$4,568.15.

Exercises 1.4

1.4.1 We are looking for δ for which $\$2,499(1 - \delta)^3 = \$945$. The solution is $\delta = 1 - (\$945/\$2,499)^{1/3} = .27686... = 27.686\%$.

1.4.2 Here the equation of value is $\$3,040(1 - .35)^n = \$750$ and it is n that we don't know. Solving for n we get $n = \log(\$750/\$3,040)/\log(1 - .35) = 3.249$ years $\approx$ 3 years, 3 months.

1.4.3 This is best shown in a table:

Year No.	*% Lost*	*Rate of Depreciation*	*Current Value*
0	-	-	\$1,499.00
1	32	1 - .32 = .68	\$1,019.32
2	16	1 - .16 = .84	\$856.23
3	8	1 - .08 = .92	\$787.73
4	4	1 - .04 = .96	**\$756.22**

1.4.4 What we know is that $\$134,500(1 - \delta)^5 = \$10,400$ which tells us that $1 - \delta = (\$10,400/\$134,500)^{1/5} = .59932\ldots$ What we then want is $\$134,500(1 - \delta)^3 = \$134,500(.59932\ldots)^3 \approx$ $28,954.02.

1.4.5 *a*) In 6 years the population has gone from 21,139,656 to 12,171,974 by equal annual percentage decreases. We can write $21,139,656(1 - \delta)^6 = 12,171,974$ giving $1 - \delta =$ $12,171,974/21,139,656)^{1/6} = .91210\ldots$ It's another 4 years to the present so the current population is $12,171,974 \cdot (.91210\ldots)^4 \approx 8,424,350$.

b) The annual rate of decrease is $\delta = 1 - .91210\ldots =$ $087897\ldots = 8.7897\ldots\%$.

1.4.6 Given $D = B(1 - \delta)^n$ the solution for B is $B = D/(1 - \delta)^n = D(1 - \delta)^{-n}$. To find the solution for δ we get in turn $(1 - \delta)^n = D/B$, $1 - \delta = (D/B)^{1/n}$, and finally, $\delta = 1 - (D/B)^{1/n}$. To find the solution for n we get in turn $(1 - \delta)^n = D/B$, and $n = \log(D/B)/\log(1 - \delta)$.

1.4.7 *a*) From $\$985(1 - \delta)^{10} = \$150$ we have in turn $(1 - \delta)^{10} =$ $\$150/\$985$; $(1 - \delta) = (\$150/\$985)^{1/10}$; and $\delta = 1 -$ $(\$150/\$985)^{1/10} = .17155\ldots = 17.155\ldots\%$.

b) In 5 years, the saw will be worth $\$985(1 - .17155\ldots)^5 =$ $384.38.

1.4.8 If the machine loses 40% each year, it retains 60% of its value each year. Thus, after 4 years, the value of Lasse's machine is $\$345(.6)^4 \approx \$44.71$.

1.4.9 In 6 years the car will be worth $\$31,280(1 - .15)^6 \approx \$11,797.24$.

1.4.10 After n years the printer is worth $\$2345(1 - .2)^n$. The values for the first 5 years are given in the table below:

Year	*Value*
1	$1,876.00
2	$1,500.80
3	$1,200.64
4	$960.51
5	$768.41

Miscellaneous Exercises—Chapter 1

1M1. Start with $1{,}590(1-\delta)^2 = 950$, giving $1-\delta = (950/1{,}590)^{1/2} = 0.77297...$ What we want is t such that $1{,}590(1-\delta)^t = 500$, or $(1-\delta)^t = 0.31446...$ This calls for the Log Rule, which yields $t = \log 0.31446.../\log(1-\delta) = \log 0.31446.../\log 0.77297... \approx 4.4925$, or, about 4½ years.

1M2. We can write $\$30{,}000 = \$22{,}500(1.08)^n$ and solve for n. We find $n = \log(1.\overline{3})/\log(1.08) \approx 3.738$ years before the \$30,000 payment. The earlier date is 5 - 3.738 = 1.262 years after time zero.

1M3. If true the interest rate i satisfies $P(1+i)^{0.5} = 2P$. The P's cancel so $i = 2^{1/0.5} - 1 = 3 = 300\%$.

1M4. At time 0, the \$30,000 payment is worth $\$30{,}000(1.08)^{-5} \approx$ \$20,417.50.

1M5. First, we need to find out what interest rate Frema was to earn the first year. This amount is $i = (4{,}220/4{,}000)^{1/1} - 1 = .055 = 5.5\%$. This is from Formula 1.3.

Ewan must pay a 50% higher rate the second year. This amount is 1.5(5.5%) = 8.25%. We may therefore write \$4,220(1.0825) = \$4,568.15. Note that this may also have been written \$4,000·(1.055)·(1.0825) = \$4,568.15.

Chapter 2

Exercises 2.1

2.1.1 The entries in the second column are found from $\$1{,}000\cdot(1 + .08/m)^{5m}$. The missing entries are in bold.

Number of Compoundings per year, m	*Accumulated value of \$1,000 at 8% after 5 years, with compounding m times per year*
1	**\$1,469.33**
2	\$1,480.24
4	**\$1,485.95**
12	\$1,489.85
365	**\$1,491.76**
8,760 (hourly)	**\$1,491.82**

2.1.2 *a*) At $j_1 = 6\%$ the deposit was $\$20{,}000(1.06)^{-21} \approx$

$5,883.11.

b) At $j_2 = 8\%$ the deposit was $\$20{,}000(1 + .08/2)^{-42} \approx$ $3,851.50.

c) At $j_{12} = 12\%$ we get $\$20{,}000(1 + .12/12)^{-252} \approx \$1{,}629.46$ for the deposit.

2.1.3 We'll write t for the number of years. Then $4t$ is the number of quarters. The equation of value is $\$45{,}000\left(1+\frac{.114}{4}\right)^{4t}$ = $135,000 which must be solved for t. We get $4t$ = log(3)/log(1.0285) = 39.095... quarters, or t = 9.774... years.

2.1.4 Mimi's account, 145 years later, is worth $\$200\left(1+\frac{.0275}{4}\right)^{4\times145} \approx$ $10,637.84.

2.1.5 Compounding daily Helen has $\$10{,}000\left(1+\frac{.05}{365}\right)^{365} \approx$ $10,512.67 in one year. With hourly compounding Helen would have $\$10{,}000\left(1+\frac{.05}{365\times24}\right)^{365\times24}$ = $10,512.71 in one year.

2.1.6 If j_2 is the nominal annual rate compounded semi-annually then we can write the equation of value $P\left(1+\frac{j_2}{2}\right)^{5\times2} = 2P$. Canceling the P's and solving for j_2 we get $j_2 = 2(2^{1/10} - 1) = .1435469... =$ 14.35469...%. To find when Valerie's money will triple, write n for the number of years and $2n$ for the number of half-years. Then $P(1+\frac{.1435469...}{2})^{2n} = 3P$. The solution is $2n$ = 15.84962..., so n = 7.9248... years. That's 7.9248... years from the time of the investment, or 2.9248 years from now.

2.1.7 Astrid's accumulated value after 2 years at 10% is $\$10{,}000\left(1+\frac{.1}{m}\right)^{2m}$, where m is the number of compoundings per year. Here are the results for m = 1, 12, 365, 8760 in a table:

Part	*Number m*	*Accumulation*
a)	1	\$12,100.00
b)	12	\$12,203.91
c)	365	\$12,213.69
d)	8,760	\$12,214.01

2.1.8 Jesper's \$5,000 becomes $\$5{,}000\left(1+\frac{.096}{12}\right)^{3\times 12} \approx \$6{,}661.15$, so the interest he earned was \$6,661.15 - \$5,000 = \$1,661.15.

2.1.9 Nat invested P where $P\left(1+\frac{.12}{6}\right)^{10\times 6} \approx \$121{,}719.74$, so $P \approx$ \$37,098.02.

2.1.10 The equation of value for Björn's investment is $\$15{,}000\left(1+\frac{.085}{2}\right)^{2\times n} = \$30{,}000$, where n is measured in years. Solving for n gives $n \approx 8.326\ldots$ years.

2.1.11 We have $\$131{,}119.78 = \$100{,}000\left(1+\frac{j_{12}}{12}\right)^{3\times 12}$, which implies that $j_{12} \approx 9.06544\%$.

Exercises 2.2

2.2.1 Two and one-half years is 10 quarters, so two and one-half years earlier \$10,000 is worth $\$10{,}000(1.015)^{-10} \approx \$8{,}616.67$.

2.2.2 The interest rate per quarter is $i = .05/4 = .0125$. From the time diagram below we can write the equation of value $\$50(1.0125)^3 + \$75(1.0125)^2 = \$60(1.0125) + P$. This gives $P \approx \$68.04$

\$60 P

Feb. 1 May 1 Aug. 1 Nov. 1 quarters

\$50 \$75

Time Diagram for Exercise 2.2.2

2.2.3 Using the time diagram below, we can write $\$10{,}000 = P(1.0615)^{-10} + P(1.0615)^{-11} + P(1.0615)^{-12} + P(1.0615)^{-13} + P(1.0615)^{-14} = P(2.4517479\ldots)$ so that $P \approx \$4{,}078.72$.

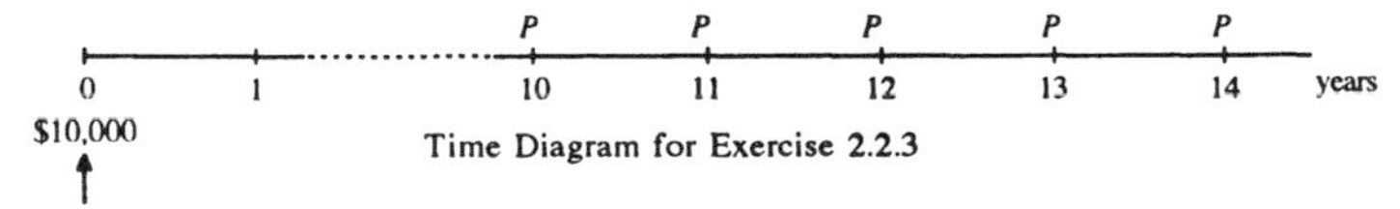

Time Diagram for Exercise 2.2.3

2.2.4 The interest rate per month is $i = .06/12 = .005$.

a) The amount of Stanley's loan is $L =$
$\$3,000(1.005)^{-15} + \$3,000(1.005)^{-27} + \$3,000(1.005)^{-39} \approx$ $7,875.48.

b) We need to find the time when a single payment of $9,000 would be equivalent to the loan of $7,875.48 which in turn is equivalent to the three equal payments of $3,000. From the time diagram we write the equation of value

$$\$7,875.48(1.005)^t = \$9,000.$$

Using the Log Rule, we find $t =$ $\log(9,000/7,875.48)/\log(1.005) \approx 26.761$ months.

2.2.5 The period is the month and the interest rate per month is $i = .091/12 = .00758\overline{3}$. I put the focal date at time 0, but the solution would be unchanged wherever I put it. Using the time diagram as a guide we can write an equation of value. $\$10,000(1.00758\overline{3})^{12} + \$8,000(1.00758\overline{3})^{6} = P(1.00758\overline{3})^{-6} + P(1.00758\overline{3})^{-7} + P(1.00758\overline{3})^{-8} + P(1.00758\overline{3})^{-9} = P(3.77979\ldots)$ or $\$19,319.90 = P(3.77979\ldots)$ and finally, $P =$ $5,111.36.

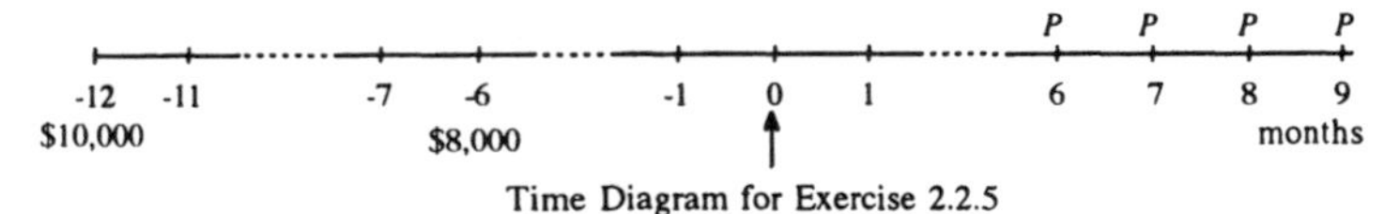

Time Diagram for Exercise 2.2.5

2.2.6 The period is the month and the monthly interest rate is $i = .0375/12 = 003125$. Below is a Time Diagram for this:

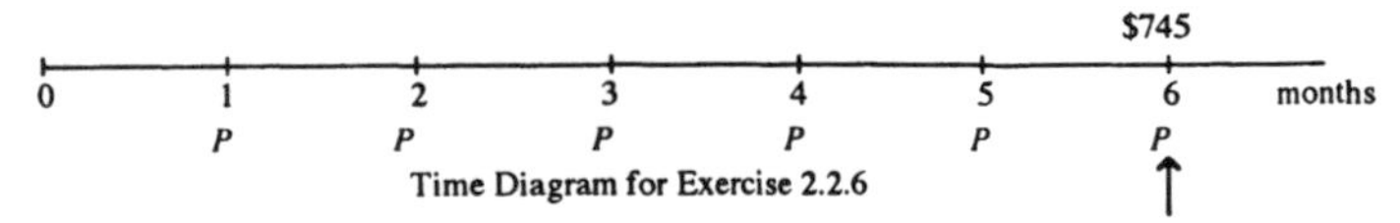

Time Diagram for Exercise 2.2.6

The equation of value with focal date at time 6 months is $\$745 = P(1.003125)^5 + P(1.003125)^4 + P(1.003125)^3 + P(1.003125)^2 + P(1.003125)^1 + P = P(6.047070\ldots)$ so $P = \$745/6.047070\ldots) = \$123.20$.

2.2.7 From the Time Diagram below we can write

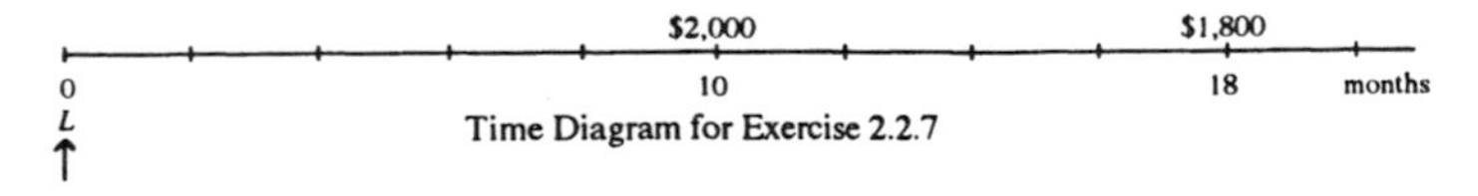

Time Diagram for Exercise 2.2.7

$L = \$2{,}000(1.004\overline{3})^{-10} + \$1{,}800(1.004\overline{3})^{-18} = \$3{,}580.58.$

2.2.8. See the time diagram below and ignore the \$4,300 for part *a*).

a) The equation of value for the annual payment is $\$4{,}000 = P(1.04)^{-1} + P(1.04)^{-2} + P(1.04)^{-3} = P(2.775091\ldots)$ from which we get $P = \$1{,}441.39$.

b) We don't know where to put the \$4,300 payment. Call the time t. What we do know is that $\$4{,}000(1.04)^t = \$4{,}300$. The solution for t is $t = \log(\$4{,}300/\$4{,}000)/\log(1.04) = 1.8439\ldots$ years.

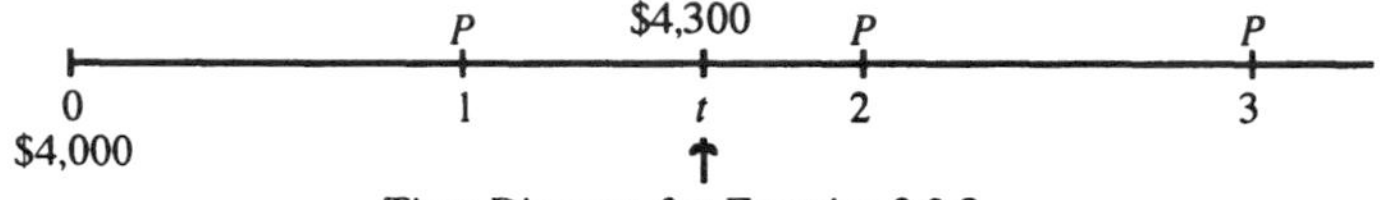

Time Diagram for Exercise 2.2.8

Exercises 2.3

2.3.1 Here's a story to go along with the equation of value:

$$\$23{,}000 = \$7{,}177.45(1.095)^{-1} + \$7{,}177.45(1.095)^{-2} + \$7{,}177.45(1.095)^{-3} + \$7{,}177.45(1.095)^{-4}.$$

Gustavo borrowed \$23,000 at annual interest rate 9.5% to make bail. He will repay the loan with 4 annual payments of \$7,177.45.

2.3.2 The equation of value in Formula 2.1 is

$$D_1(1+i)^{d-u_1} + D_2(1+i)^{d-u_2} + \ldots + D_r(1+i)^{d-u_r} = W_1(1+i)^{d-v_1} + W_2(1+i)^{d-v_2} + \ldots + W_s(1+i)^{d-v_s}.$$

We can factor out $(1+i)^d$ from both sides as follows:

$$(1+i)^d\{D_1(1+i)^{-u_1} + D_2(1+i)^{-u_2} + \ldots + D_r(1+i)^{-u_r}\} = (1+i)^d\{W_1(1+i)^{-v_1} + W_2(1+i)^{-v_2} + \ldots + W_s(1+i)^{-v_s}\}.$$

These common factors can then be cancelled to give

$$\{D_1(1+i)^{-u_1} + D_2(1+i)^{-u_2} + \ldots + D_r(1+i)^{-u_r}\} = \{W_1(1+i)^{-v_1} + W_2(1+i)^{-v_2} + \ldots + W_s(1+i)^{-v_s}\}.$$

2.3.3 The period is the month and the interest rate per month is i = .09/12 = .0075. There are three arrows in the time diagram—one for each of parts a), b), and c). The equations of value with focal dates at times 0, 6 and 12 months are:

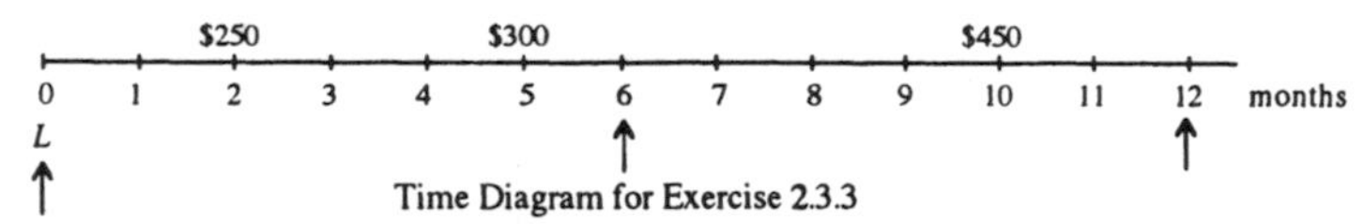

Time Diagram for Exercise 2.3.3

a) $L = \$250(1.0075)^{-2} + \$300(1.0075)^{-5} + \$450(1.0075)^{-10}$;
b) $L(1.0075)^{6} = \$250(1.0075)^{4} + \$300(1.0075) + \$450(1.0075)^{-4}$; and
c) $L(1.0075)^{12} = \$250(1.0075)^{10} + \$300(1.0075)^{7} + \$450(1.0075)^{2}$.

In the first time diagram we solve for L directly: L = \$952.89. In the second we calculate L(1.04585...) = \$996.58417... so that L = \$952.89. In the third equation we get L(1.09380...) = \$1,042.27978... so L = \$952.89 as expected.

2.3.4 The monthly interest rate is i = .06/12 = .005, so we can write $P = \$600(1.005) + \$800(1.005)^{-1} = \$1{,}399.02$ for the equivalent payment in 7 months.

2.3.5 In the time diagram below, I have put the deposits under the time line, and the withdrawals, including the final withdrawal, P, above the time line. I have also included the number of

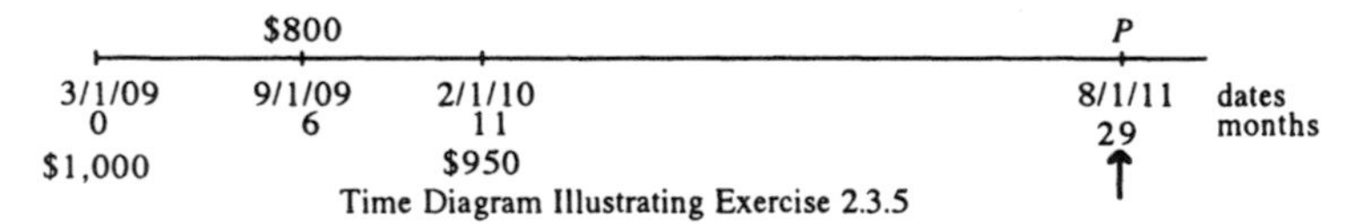

Time Diagram Illustrating Exercise 2.3.5

months measured from 3/1/09 under the time line. The equation of value sets the value of the deposits equal to the value of the withdrawals, all at 8/1/2011. With $i = j_{12}/12 = .051/12 = .00425$, we can write the equation of value

$$\$1{,}000(1.00425)^{29} + \$950(1.00425)^{18} = \$800(1.00425)^{23} + P.$$

Solving this equation for P gives P = \$1,274.27.

2.3.6 Let L denote the amount of Siomak's loan. Then L is the sum of the discounted values of all of Siomak's payments with $i = .05/4 = .0125$. That is,

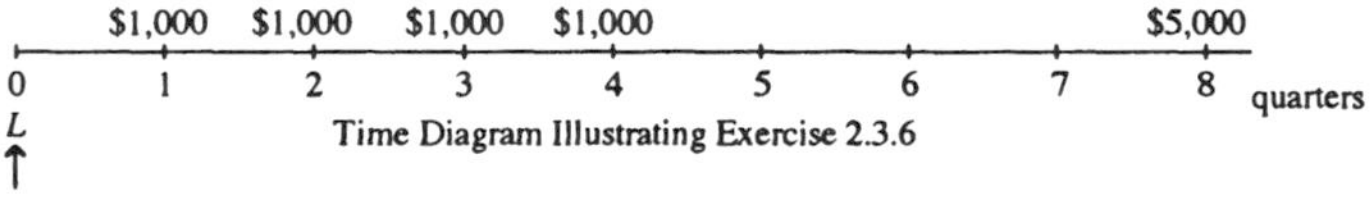

	\$1,000	\$1,000	\$1,000	\$1,000				\$5,000	
0	1	2	3	4	5	6	7	8	quarters
L ↑									

Time Diagram Illustrating Exercise 2.3.6

$$L = \$1{,}000(1.0125)^{-1} + \$1{,}000(1.0125)^{-2} + \$1{,}000(1.0125)^{-3} + \$1{,}000(1.0125)^{-4} + \$5000(1.0125)^{-8} = \$8{,}405.05.$$

Exercises 2.4

2.4.1 Steve's bank's nominal rate compounded semi-annually is (using Formula 2.3a)

$$j_2 = 2\left\{\left(1 + \frac{.0475}{365}\right)^{365/2} - 1\right\} = 4.8065\ldots\%.$$

2.4.2 Lance's certificate is worth $\$5(1.0325)^{500} = \$44{,}055{,}503.45$ in 1992.

2.4.3 We have the equation of value $\$75{,}000(1 + i)^2 = \$76{,}000$ which solved for i gives $i = (\$76{,}000/\$75{,}000)^{1/2} - 1 = .664\ldots\%$.

2.4.4 The EOV is $\$15{,}000 = \$6{,}900(1 + i)^{-1} + \$6{,}900(1 + i)^{-2} + \$6{,}900(1 + i)^{-3}$.
Solution by Solver: Enter the equation

0 = 6900((1 + I)^-1 + (1 + I)^-2 + (1 + I)^-3) – 15000

into the EQUATION EDITOR with a starting value of I = .05. Next press ALPHA then [SOLVE] to find the solution I = .18010334667...= 18.0103...%. Here is what your final screen should look like:

```
EQUATION SOLVER
Eqn: 0 = 6900((1 + I)^-1 + (1 + I)^-2 + (1 + I)^-3) – 15000
I = .18010334667...
Bound = {-1E99, 1E99}
Left – rt = 0
```

Solution by Guess-and-Check: Set $LHS(i) = \$6{,}900(1 + i)^{-1} + \$6{,}900(1 + i)^{-2} + \$6{,}900(1 + i)^{-3}$ and make a table with values of i to make $LHS(i)$ close to \$15,000.

i	*LHS(i)*	*Comments*
.05	\$18,790.41	This guess is far too low.
.10	\$17,159.28	Still far too low.
.20	\$14,534.72	Now we need to back up.
.15	\$15,754.25	The solution is between .15 and .20.
.175	\$15,123.47	We are still a little too low.
.18	\$15,002.48	Very close. We can stop here.

With Guess-and-Check, we got reasonably close in six steps.

2.4.5 The EOV is $\$78{,}000 = \$21{,}000 + \$21{,}000(1 + i)^{-1} + \$21{,}000(1 + i)^{-2} + \$21{,}000(1 + i)^{-3}$.

Solution by Solver: Enter the equation

0 = 21000(1 + (1 + I)^-1 + (1 + I)^-2 + (1 + I)^-3 – 78000

into the EQUATION EDITOR with a starting value of I = .05. Next press ALPHA then [SOLVE] to find the solution I = .05176121920... ≈ 5.1761%. Here is what your final screen should look like:

```
EQUATION SOLVER
Eqn: 0 = 21000(1 + (1 + I)^-1 + (1 + I)^-2 + (1 + I)^-3) – 78000
I = .05176121920...
Bound = {-1E99, 1E99}
Left – rt = 0
```

Solution by Guess-and-Check: The value of i making the two sides equal is the effective rate. Set $LHS(i) = \$21{,}000 + \$21{,}000(1 + i)^{-1} + \$21{,}000(1 + i)^{-2} + \$21{,}000(1 + i)^{-3}$ and we make a table to get $LHS(i)$ close to \$78,000.

i	*LHS(i)*	*Comments*
.05	\$78,188.21	This is fairly close on the first guess.
.055	\$77656.60	The solution is close to .05
.051	\$78,081.22	
.0515	\$78,027.85	.0515 is still too low, but close.
.0517	\$78,006.52	
.0518	\$77,995.87	
.05175	\$78,001.20	This is good enough.

The effective rate is about 5.175%.

From here forward, whenever problems requiring a solver or Guess-and-Check arise, we will provide brief solutions.

2.4.6 Formula 2.2 gives the effective rate $j_1 = (1 + .0413/365)^{365} - 1 =$ 4.216...%.

2.4.7 Formula 2.3a gives $j_{12} = 12\{(1 + .0678/4)^{4/12} - 1\} = 6.742...\%$.

2.4.8 We write the equation of value

$$(1+i)^5 = \left(1+\frac{.04}{2}\right)^2\left(1+\frac{.06}{4}\right)^4\left(1+\frac{.08}{6}\right)^6\left(1+\frac{.10}{12}\right)^{24} = 1.4590745\ldots$$

From $(1 + i)^5 = 1.4590745...$, we get $1 + i = (1.4590745...)^{1/5}$, and $i = (1.4590745...)^{1/5} - 1 = .0784884... = 7.84884...\%$.

2.4.9 From $\left(1+\frac{j_n}{n}\right)^n - 1 = \left(1+\frac{j_m}{m}\right)^m - 1$, we get $\left(1+\frac{j_n}{n}\right)^n = \left(1+\frac{j_m}{m}\right)^m$.

Now we raise both sides to the power $1/n$ to get Formula 2.3*b*:

$$1+\frac{j_n}{n} = \left(1+\frac{j_m}{m}\right)^{m/n}, \text{ and } \frac{j_n}{n} = \left(1+\frac{j_m}{m}\right)^{m/n} - 1.$$

Miscellaneous Exercises – Chapter 2

2M1 The equation of value for this problem is $\$5,000(1.06)^n = \$6,000$. The solution is $n = \log(\$6,000/\$5,000)/\log(1.06) =$ 3.1290 years which is 3 years, 1 month and about 17 days.

2M2. The number of hairs on Seymour's head n years after Seymour was 30 is $985,912(1 - .14)^n = 985,912(.86)^n$. He will have 10,000 hairs left when $985,912(.86)^n = 10,000$. This happens in $n = \log(10,000/985,912)/\log(.86) \approx 30.44$ years after Seymour was 30 so Seymour will be about 60.

2M3. If Jörgen's investment tripled in 4 years, then the interest rate i satisfies the equation of value $(1 + i)^4 = 3$ so that $i = 3^{1/4} - 1 =$ .316074... His investment doubles in n years where n must satisfy $(1.316074...)^n = 2$. This happens in $n = \log(2)/\log(1.316074...) \approx 2.524$ years, or 2 years, 6 months and about 9 days.

2M4. Here is a time diagram:

	$206.03	$206.03	$206.03
$599.99 ↑			months

Time Diagram for Exercise 2M4

This equation of value is:

$$\$599.99 = \$206.03(1 + i)^{-1} + \$206.03(1 + i)^{-2} + \$206.03(1 + i)^{-3}.$$

Solution by Solver: In the EQUATION SOLVER, enter

0 = 206.03((1 + I)^-1 + (1 + I)^-2 + (1 + I)^-3) – 599.99,

and use a starting value of I = .05. The solution provided is I = .01500905485... ≈ 1.501%.

Solution by Guess-and-Check: Set $LHS(i) = \$206.03(1 + i)^{-1} + \$206.03(1 + i)^{-2} + \$206.03(1 + i)^{-3}$ (where $i = j_{12}/12$) and make a table. I got $i = 1.501\%$ in six steps.

a) Since $i = j_{12}/12 \approx .01501$, $j_{12} \approx .1801 = 18.01\%$.

b) The effective rate is $j_1 = (1 + j_{12}/12)^{12} - 1 \approx .195746 = 19.5746\%$.

2M5. Here is a time diagram for this problem:

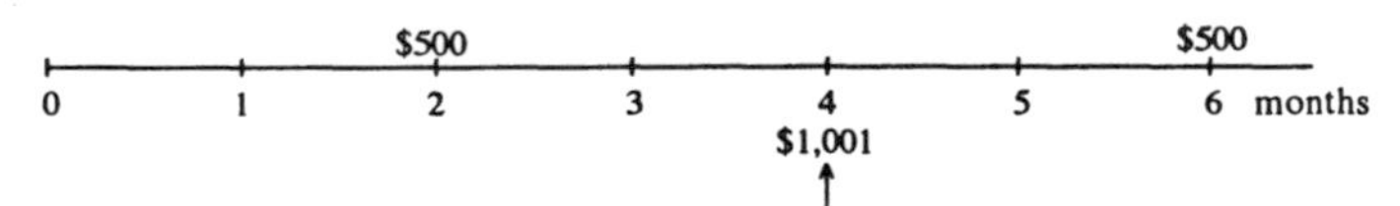

Time Diagram for Exercise 2.M.5.

The equation of value is $\$500(1 + i)^2 + \$500(1 + i)^{-2} = \$1,001$, where $i = j_{12}/12$.

Solution by Solver: In the EQUATION SOLVER, enter

0 = 500((1 + I)^2 + (1 + I)^-2) – 1001,

and use a starting value of I = .05. You should obtain the solution I = .02261064853... ≈ 2.2611%. This makes $j_{12} = 12i = .27132778... \approx 27.133\%$.

Solution by Guess-and-Check: Set $LHS(i) = \$500(1 + i)^2 + \$500(1 + i)^{-2}$ (where $i = j_{12}/12$) and make a Guess-and-Check table. Starting with $i = .05$, I got $i = .0226$ in seven steps. This means that $j_{12} \approx 27.12\%$. As expected, not as accurate as using a solver.

2M6. Study the time diagram below:

$1,300 $1,300 $1,300 $1,300 $1,300 $1,300 *P*

0 1 2 3 4 5 6 7

$6,000 years

Time Diagram for Exercies 2M6

We need to evaluate all payments at the focal date of 7 years. The equation of value for this is

$$P + \$1{,}300\{(1.1)^6 + (1.1)^5 + (1.1)^4 +$$
$$(1.1)^3 + (1.1)^2 + (1.1)\} = \$6{,}000(1.1)^7.$$

We solve this for P to get $P = \$658.98$.

2M7. First, consider the time diagram below. I put d to the right of

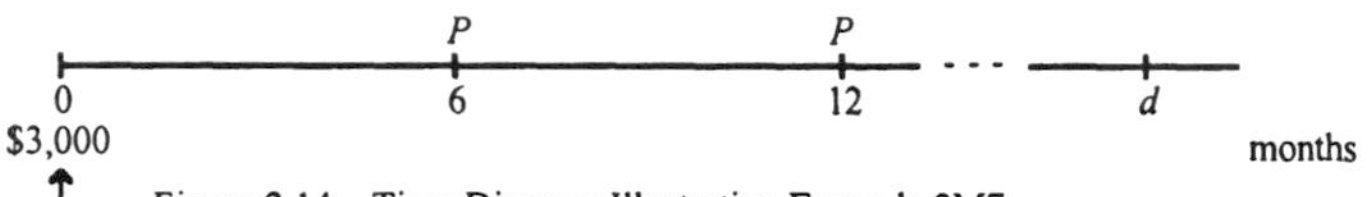

Figure 2.14—Time Diagram Illustrating Example 2M7

12, but the equation of value remains the same regardless of where I put it. The equation of value is

$$\$3{,}000(1.0025)^d = P(1.0025)^{d-6} + P(1.0025)^{d-12}.$$

Since $(1.0025)^d$ is a common factor of both sides of the equation, we can divide both sides by $(1.0025)^d$ to get the equivalent equation

$$\$3{,}000 = P(1.0025)^{-6} + P(1.0025)^{-12}$$

whose solution we already found in Example 2.5 to be 1,534.05.

Chapter 3

Exercises 3.1

3.1.1 Here $a = 2{,}000(.9)^3$, $r = .9$ and $n = 75 - 2 = 73$, so the sum is

$$2{,}000(.9)^3 \frac{1-.9^{73}}{1-.9} \approx 14{,}573.34.$$

3.1.2 We can write $1 + 4 + 16 + 64 + \ldots + 1{,}048{,}576$ in the form

$$1 + 4 + 4^2 + 4^3 + \ldots + 4^{n-1}$$

where $4^{n-1} = 1{,}048{,}576$. From this we see that $a = 1$ and $r = 4$. Using the Log Rule or trial and error, we find $n - 1 = 10$ so that $n = 11$. The sum is

$$G_{11} = 1 \cdot \frac{4^{11} - 1}{4 - 1} = 1{,}398{,}101.$$

3.1.3 The total paid is

$$G_{21} = \$30{,}000 + \$30{,}000(.8) + \$30{,}000(.8)^2 + \ldots$$

$$+ \$30{,}000(.8)^{20} = \$30{,}000 \frac{1 - .8^{21}}{1 - .8} = \$148{,}616.49.$$

3.1.4 The monthly interest rate is $i = .084/12 = .007$. The equation of value is

$$\$2{,}400 = P(1.007)^{-1} + P(1.007)^{-2} + \ldots + P(1.007)^{-24}.$$

The right-hand side is a geometric sum with $a = P(1.007)^{-1}$, $r = (1.007)^{-1}$, and $n = 24$, so it sums to

$$P(1.007)^{-1} \frac{1 - (1.007)^{-24}}{1 - (1.007)^{-1}} = P(22.0216085\ldots),$$

so \$2,400 = $P(22.0216085\ldots)$, and P = \$108.98 is Ladislao's monthly payment.

3.1.5 The first alternative pays \$5,000,000. The second pays

$$G_{30} = \$.01 + \$.01(2) + \$.01(2^2) + \$.01(2^3) + \ldots$$

$$+ \$.01(2^{29}) = .01\,(2^{30} - 1)/(2 - 1) = \$10{,}737{,}418.23,$$

so the second pays more than twice what the first does.

3.1.6 The thing to notice is that each term, except for a common factor of two, is a power of –2/9. The sum is therefore

$$G_{10} = 2 + 2(-2/9) + 2(-2/9)^2 + 2(-2/9)^3 + \ldots + 2(-2/9)^9.$$

The last exponent is 9 because there are 10 terms and the first one is $2(-2/9)^0$. We have $a = 2$, $r = -2/9$, and $n = 10$. The sum is

$$G_{10} = 2\{1 - (-2/9)^{10}\}/\{1 - (-2/9)\} = 1.636\ldots$$

3.1.7 The inputs for Formula 3.1 (a or b) are $a = (1.05)^2$, $r = 1.05$, and $n = 100$. The sum is therefore $G_{100} = 2{,}877.5527355\ldots$

3.1.8 Your parents make 2; your grandparents make 4 more; your great-grandparents make 8 more; ...; your great-great-great-

great-great-great-great-great-great grandparents make 2,048 more for a total of $2 + 2^2 + 2^3 + \ldots + 2^{11} = 2(2^{11}-1)/(2-1) =$ 4,094. If you are not convinced, here is a table:

Ancestors	**Number**
Parents	2
Grandparents	$2^2 = 4$
Great-Grandparents	$2^3 = 8$
$(\text{Great})^2$-grandparents	$2^4 = 16$
$(\text{Great})^3$-grandparents	$2^5 = 32$
$(\text{Great})^4$-grandparents	$2^6 = 64$
$(\text{Great})^5$-grandparents	$2^7 = 128$
$(\text{Great})^6$-grandparents	$2^8 = 256$
$(\text{Great})^7$-grandparents	$2^9 = 512$
$(\text{Great})^8$-grandparents	$2^{10} = 1,024$
$(\text{Great})^9$-grandparents	$2^{11} = 2,048$

3.1.9 Balbino's payments may be written in the form of a geometric progression with $a = \$1,200$, $r = 1.06$, and $n = 10$. The sum of these terms is Balbino's accumulated value:

$$\$1,200\frac{(1.06)^{10}-1}{1.06-1} = \$15,816.95.$$

3.1.10 With the help of the time diagram below, we can write an equation of value using Equation 3.1a. We have \$13,040.16 as the sum, on one side of the

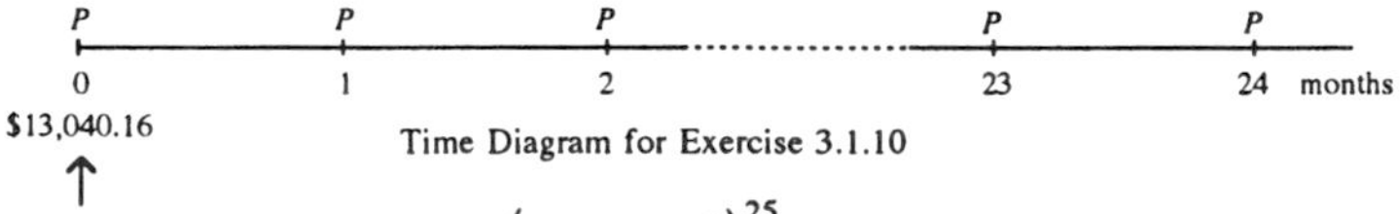

P P P .... P P
0 1 2 23 24 months
\$13,040.16 ↑

Time Diagram for Exercise 3.1.10

equation, and $P\frac{1-\left((1+\frac{.145}{12})^{-1}\right)^{25}}{1-(1+\frac{.145}{12})^{-1}}$ on the other. This is because we have $a = P$, $r = (1 + .145/12)^{-1}$, and $n = 25$ in Equation 3.1a. With a bit of work on our calculators, we get

$$\$13,040.16 = P\frac{1-\left(\left(1+\frac{.145}{12}\right)^{-1}\right)^{25}}{1-\left(1+\frac{.145}{12}\right)^{-1}} = P(21.725634\ldots),$$

so $P = \$13,040.16/21.725634\ldots = \$600.22$.

3.1.11 Below is a time diagram for Simon's series of payments.

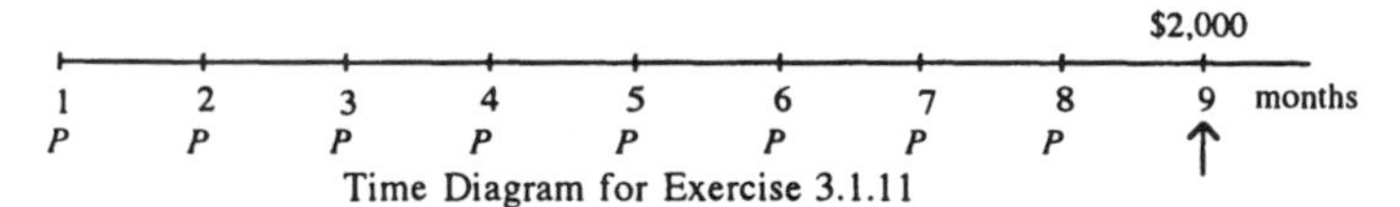

Time Diagram for Exercise 3.1.11

Let's write out the sum of the accumulated values of each payment. With monthly interest rate $i = .04/12 = .00\overline{3}$, we have

$$\$2{,}000 = P(1.00\overline{3})^8 + P(1.00\overline{3})^7 + \ldots + P(1.00\overline{3})^1.$$

Notice that we can look at the right-hand-side in two ways. First, as a geometric sum with $a = P(1.00\overline{3})^8$, $r = (1.00\overline{3})^{-1}$, and $n = 8$. On the other hand, we take the terms of the right-hand-side in reverse order, we could write $a = P(1.00\overline{3})^1$, $r = 1.00\overline{3}$, and $n = 8$. Using either Equation 3.1*a*, or 3.1*b*, the equation of value boils down to $2,000 = $P(8.120938...)$, so $P \approx \$246.28$.

3.1.12 Given $G_n = a + ar + ar^2 + ar^3 + \ldots + ar^{n-1}$ we multiply through r to get

$$rG_n = ar + ar^2 + ar^3 + \ldots + ar^n.$$

Subtracting rG_n from G_n, and canceling like terms, we get

$$\begin{gathered}(1 - r)G_n = G_n - rG_n = \\ a + ar + ar^2 + ar^3 + \ldots + ar^{n-1} - \\ r(a + ar + ar^2 + ar^3 + \ldots + ar^{n-1}) = \\ a + ar + ar^2 + ar^3 + \ldots + ar^{n-1} \\ - (ar + ar^2 + ar^3 + \ldots + ar^{n-1} + ar^n) \\ a - ar^n = a(1 - r^n).\end{gathered}$$

Now dividing both sides by $1 - r$, we have

$$G_n = a\frac{1-r^n}{1-r},$$

provided that $r \neq 1$. This is Formula 3.1 a. To obtain Formula 3.1b, multiply Equation 3.1a by (-1)/(-1).

When $r = 1$, $G_n = a + a\cdot 1 + a\cdot 1^2 + a\cdot 1^3 + \ldots + a\cdot 1^{n-1} = a\cdot n.$

Exercises 3.2

3.2.1 The equation of value is

$$S(P,\ 0.0734/12,\ 60,-1) = P\frac{1-(1.00611\overline{6})^{-60}}{0.00611\overline{6}} =$$

$$P(50.095171...) = \$17,345.40$$

which when solved for P gives P = \$346.25.

3.2.2 We have two annuities going in this problem. One is a 34-year annuity of P (to be determined) deposited monthly; the other is a 25-year annuity of \$2,141.09 payable monthly. The time diagram below shows you the layout

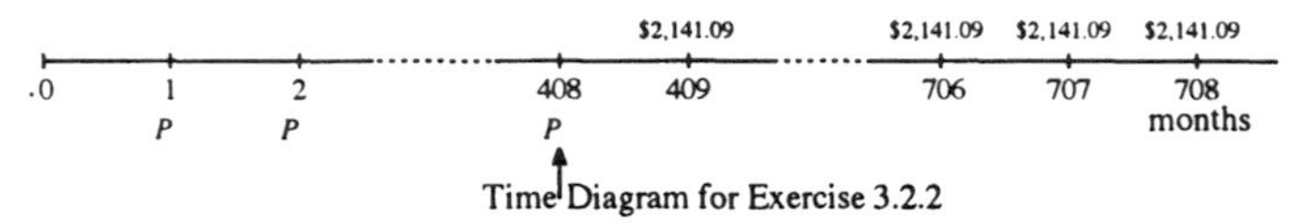

Time Diagram for Exercise 3.2.2

The number line runs from 0 to 408 (34×12) months for the first annuity and from 408 to 708 (25×12 = 300) months for the second annuity. The interest rate per month is i = .082/12 = $0.0068\overline{3}$. The equation of value equates the values of each annuity at the arrow:

$$S(P,\ .0068\overline{3},\ 408,\ 407) = S(\$2,141.09,\ .0068\overline{3},\ 300,\ -1)$$

or

$$P\frac{(1.0068\overline{3})^{408}-1}{.0068\overline{3}} = \$2,141.09\frac{1-(1.0068\overline{3})^{-300}}{.0068\overline{3}}.$$

Notice that both sides are among the special cases of Formula 3.2. Solving for P yields P = \$272,711.48/2,209.05 = \$123.45.

3.2.3 The interest rate per month for Karen's loan is $i = j_{12}/12$ = .0765/12 = .006375 and the equation of value for the loan is

$$S(P,\ .006375,\ 60,\ -1) = P\frac{1-(1.006375)^{-60}}{.006375}$$

$$= P(49.72822...) = \$13,099.95$$

so P = \$263.43.

3.2.4 Using the time diagram below as a guide with the monthly

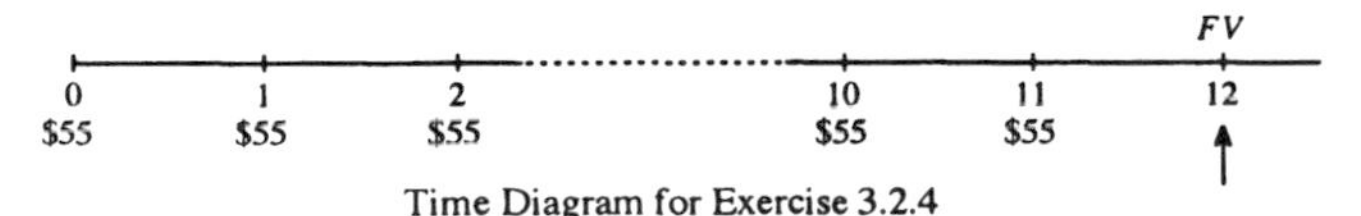

Time Diagram for Exercise 3.2.4

interest rate of $i = j_{12}/12 = .04/12 = .00\overline{3}$, the equation of value is

$$S(\$55,\ .00\overline{3},\ 12,\ 12) = \$55\frac{(1.00\overline{3})^{12} - 1}{.00\overline{3}}(1.00\overline{3}) = \$674.48.$$

3.2.5 This is similar to Example 3.10. What Ron and M. J. owe is the accumulated value of their loan for eight years minus the accumulated value of the eight years of payments of \$1,002.52, all at $j_{12}/12 = .084/12 = .007$. They owe

$$\$125{,}550(1.007)^{96} - S(\$1{,}002.52,\ .007,\ 96,\ 95) =$$

$$\$125{,}550(1.007)^{96} - \$1{,}002.52\frac{(1.007)^{96} - 1}{.007} =$$

$$\$108{,}703.13.$$

3.2.6 The monthly rate for this annuity is $i = .098/12 = .0081\overline{6}$ and its equation of value with focal date 45 years in the future is

$$S(P,\ 0.0081\overline{6},\ 540,\ 539) = P\frac{(1.0081\overline{6})^{540} - 1}{.0081\overline{6}} = \$2{,}000{,}000.$$

Solving for P we get $P = \$204.65$.

3.2.7 Note that this problem is similar to Example 3.9. The period here is the quarter and the quarterly rate is $i = j_4/4 = .0742/4 = .01855$. The 30-years constitute 120 quarters, so Erik has 120 – 35 = 85 payments still to make. What he owes is the present value of these 85 payments of \$889.91:

$$S(\$889.91,\ .01855,\ 85,\ -1) = \$889.91\frac{1 - (1.01855)^{-85}}{.01855}$$

$$= \$37{,}915.80.$$

3.2.8 This is something like the preceding problem except that it is n that we do not know. The monthly interest rate is $i = j_{12}/12 = .00625$. The equation of value is:

$$S(\$250,\ .00625,\ n,\ -1) = \$250\frac{1 - (1.00625)^{-n}}{.00625} = \$27{,}049.30$$

and the solutions is $n = -\dfrac{\log\left(1 - \{\$27{,}049.30 \cdot 0.00625/\$250\}\right)}{\log(1.00625)}$

$= 181.$

3.2.9 With the time diagram below as a guide, and with the arrow at time 0 we write the equation of value:

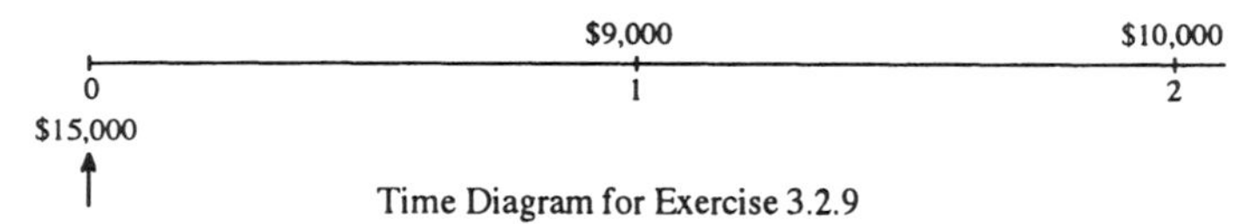

Time Diagram for Exercise 3.2.9

$$\$9{,}000(1+i)^{-1} + \$10{,}000(1+i)^{-2} = \$15{,}000.$$

Solution by Solver: Enter 0 = 9000(1 + I)^-1 + 10000(1 + I)^-2 – 15000 into the EQUATION EDITOR. With starting value I = .05, you should get the solution I = .16986589004... ≈ 16.9866%.

Solving by Guess-and-Check: Define

$$LHS(i) = \$9{,}000(1+i)^{-1} + \$10{,}000(1+i)^{-2}$$

and make a Guess-and-Check table making $LHS(i)$ close to $15,000. After six steps, I got $i \approx .1699 = 16.99\%$

3.2.10 The period is the month, and the monthly interest rate is $i = .075/12 = .00625$. We are looking for n, the number of equal

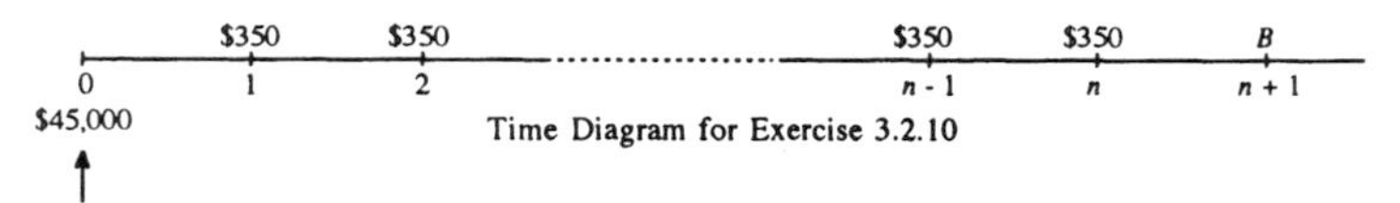

Time Diagram for Exercise 3.2.10

payments and, B, the final smaller payment. The time diagram illustrating the setup appears above. We will find the value n, which must satisfy two conditions: it must be such that the present value of n payments of $350 does not exceed $45,000, but n +1 payments of $350 does. The simplest thing to do is set the present value equal to $45,000. The resulting n will almost certainly not be an integer, but the integer part of the number will be the correct value of n. So set

$$S(\$350,\ .00625,\ n,\ -1) = \$350\frac{1-(1.00625)^{-n}}{.00625} = \$45{,}000$$

and solve for n. The result is $n = 261.2059...$ The means that 261 payments of $350 does not pay off the loan, but 262 payments of $350 is too much. As a check,

$$S(\$350,\ .00625,\ 261,\ -1) = \$44{,}985.88$$

and

$$S(\$350,\ .00625,\ 262,\ -1) = \$45{,}054.29.$$

But now with the help of our time diagram the equation of value is written

$$S(\$350,\ .00625,\ 261,\ -1) + B(1.00625)^{-262}$$
$$= \$44{,}985.88 + B(1.00625)^{-262} = \$45{,}000$$

because adding the final payment will pay off the loan. Solving this for B we get B = (\$45,000 - \$44,985.88)$(1.00625)^{262}$ = \$72.25. So Sandy makes 261 equal payments of \$350 and a final payment of \$72.25 one month after the last payment of \$350.

3.2.11 The equation of value is

$$S(\$6{,}500{,}000,\ i,\ 20,\ 0) = \$6{,}500{,}000\frac{1-(1+i)^{-20}}{i}(1+i)$$
$$= \$49{,}290{,}168.$$

Solution by Solver: In the EQUATION SOLVER, enter

0 = 6500000(1 – (1 + I)^-20)/I*(1 + I) – 49290168,

and use I = .05 as starting value. You should then get I = .13911698029... ≈ 13.912%.
Solving by Guess-and-Check: Set

$$LHS(i) = \$6{,}500{,}000\frac{1-(1+i)^{-20}}{i}(1+i)$$

and make a Guess-and-Check table to make $LHS(i)$ close to \$49,290,168. In my table, I got $LHS(.1391)$ = \$49,294,277 after five steps. This agrees well with the solver method.

3.2.12 The interest rate per month is .06/12 = .005 and the payments are at the beginning of every month for four years. A time diagram and equation of value are:

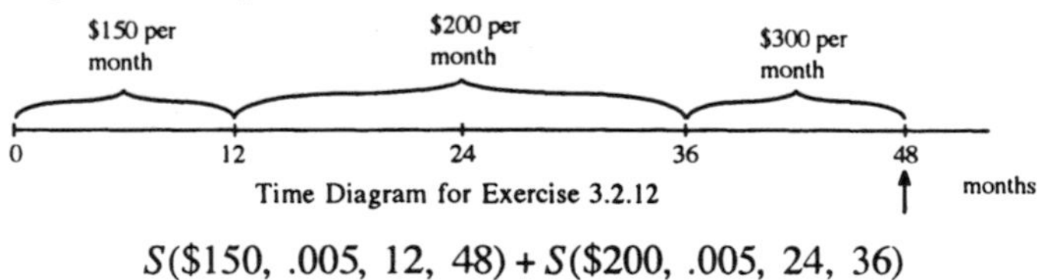

Time Diagram for Exercise 3.2.12

$$S(\$150,\ .005,\ 12,\ 48) + S(\$200,\ .005,\ 24,\ 36)$$
$$+\ S(\$300,\ .005,\ 12,\ 12) = \$2{,}225.33 + \$5{,}427.11$$
$$+ \$3{,}719.17 = \$11{,}371.61.$$

3.2.13 With a monthly interest rate of $i = .075/12 = .00625$ for $240 = 20\times12$ months the price of Betsy's house was

$$S(\$744.39,\ .00625,\ 240,\ -1) = \$744.39\frac{1-(1.00625)^{-240}}{.00625}$$

$= \$92{,}402.72.$

3.2.14 Greg's monthly interest rate for $6\times12 = 72$ months is $i = .0999/12 = .008325$. The equation of value for his loan is:

$$S(P,\ 0.008325,\ 72,\ -1) = P\frac{1-(1.008325)^{-72}}{.008325} =$$

$$P(53.993361\ldots) = \$23{,}130.70$$

so $P = \$428.40$.

3.2.15 Neith's account has a monthly interest rate of $i = .004308\overline{3}$ and there are $14\times12 = 168$ deposits. The accumulated value of her account is:

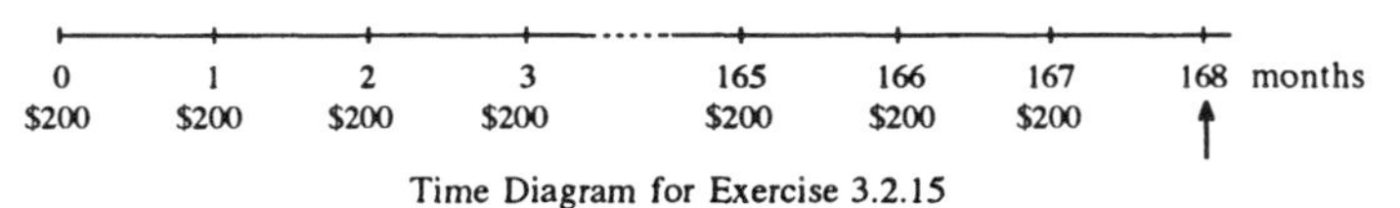

Time Diagram for Exercise 3.2.15

$$S(\$200,\ 0.004308\overline{3},\ 168,\ 168) =$$

$$\$200\frac{(1.004308\overline{3})^{168}-1}{.004308\overline{3}}(1.004308\overline{3}) = \$49{,}374.73.$$

3.2.16 The equation of value is:

$$S(\$1950,\ i,\ 12,\ -1) = \$1950\frac{1-(1+i)^{-12}}{i} = \$20{,}000$$

where $i = j_{12}/12$ is the monthly rate of interest. This is another problem requiring a numerical solution.

Solution by Solver: Enter 0 = 1950(1 − (1 + I)^-12)/I = 20000 into the EQUATION SOLVER. Starting with I = .05, the solution will be I = .02502180807..., so that $j_{12} \approx 12i \approx 30.026\%$.

Solution by Guess-and-Check: Let $LHS(i) = \$1950\frac{1-(1+i)^{-12}}{i}$,

where $i = j_{12}/12$ is the nominal annual rate compounded monthly. I made a Guess-and-Check table and found $i \approx .02502$ in eight steps, making $j_{12} \approx 12(.02502) = .30024 = 30.024\%$

3.2.17 Remember our convention that when nothing is said about whether payments are being made at the end or the beginning of a period, the assumption is that are at the end. Dawson's deposits at the end of every month earn at the rate $i = .0575/12 = .004791\overline{6}$. In 5 years his deposits amount to \$15,000. The equation of value is:

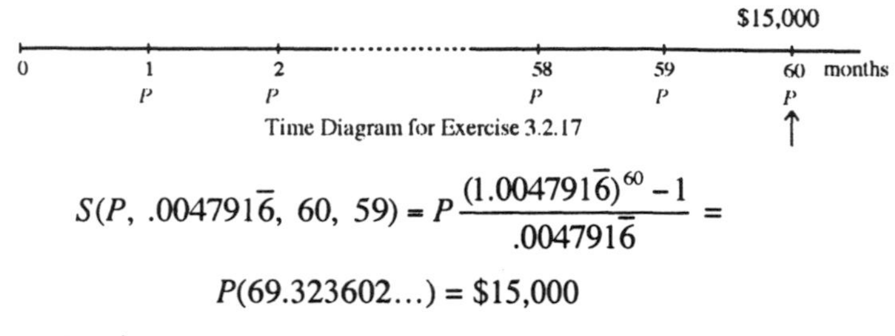

Time Diagram for Exercise 3.2.17

$$S(P, .004791\overline{6}, 60, 59) = P\frac{(1.004791\overline{6})^{60} - 1}{.004791\overline{6}} =$$

$$P(69.323602\ldots) = \$15{,}000$$

so $P = \$216.38$.

3.2.18 Suzanne's payments are worth

$$S(\$645.74, .005, 90, -1) + \$1{,}250(1.005)^{-91} = \$47{,}501.17$$

to Beta Mortgage at $j_{12} = 6\%$.

3.2.19 The monthly interest rate Lloyd's money is earning is $i = .0445/12 = .003708\overline{3}$. His accumulated savings at the end of 14 years are:

$$S(\$150, .003708\overline{3}, 84, 167) =$$

$$\$150\frac{1-(1.003708\overline{3})^{-84}}{.003708\overline{3}}(1.003708\overline{3})^{168} = \$20{,}130.90.$$

Deposits of \$150 every month for 7 years

0 84 168

Time Diagram for Exercise 3.2.19

3.2.20 With interest per month at $i = .089/12 = .00741\overline{6}$ for $5\times12 = 60$ months Antonia can borrow

$$S(\$240, .00741\overline{6}, 60, -1) = \$240\frac{1-(1.00741\overline{6})^{-60}}{.00741\overline{6}}$$

$= \$11{,}588.69$ for her car.

3.2.21 In the time diagram below, we have the focal date at time 0.

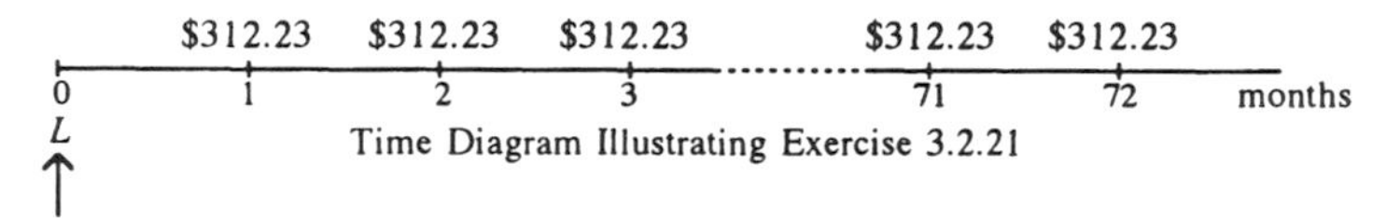

Time Diagram Illustrating Exercise 3.2.21

Using Formula 3.2a, the equation of value for this is

$$L = S(\$312.23, .0425/12, 72, -1) =$$

$$\$312.23\frac{1-(1.003541\overline{6})^{-72}}{.003541\overline{6}} = \$19{,}812.37.$$

3.2.22 Notice that I started the numbering in the time diagram below with 1. I could have run the numbers from 0 to 88 just as well. The point is to show you that the starting point does not matter

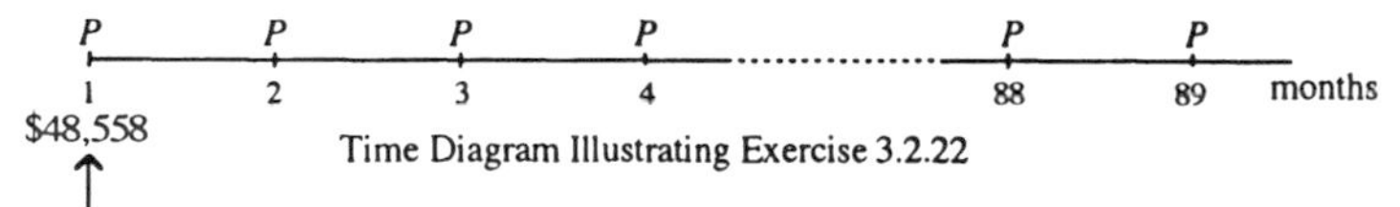

Time Diagram Illustrating Exercise 3.2.22

in calculating d. We just count from the arrow to the time of the first payment. In this case, we have $d = 0$. Using Formula 3.2b, the equation of value can be written

$$\$48{,}558 = S(P, .0712/12, 89, 0) =$$

$$P\frac{1-(1.0059\overline{3})^{-89}}{.0059\overline{3}}(1.0059\overline{3}) = P(69.3983779...),$$

so that $P = \$48{,}558/69.3983779... = \$699.70$.

3.2.23 We want to know Luisa's balance at the end of 3 years = 12 quarters, so the arrow in the time diagram below is placed at the end of the three-year period.

$450 $450 $450 $450 $450

0 1 2 3 11 12 quarters

Time Diagram Illustrating Exercise 3.2.23

From Formula 3.2c, we write

$$S(\$450, .0614/4, 12, 11) = \$450\frac{(1.01535)^{12}-1}{.01535} = \$5{,}880.05.$$

3.2.24 Tran makes 216 deposits of \$220, and we want the accumulated value of all the deposits one month after the last one. From Formula 3.2*d*, we write the accumulated value

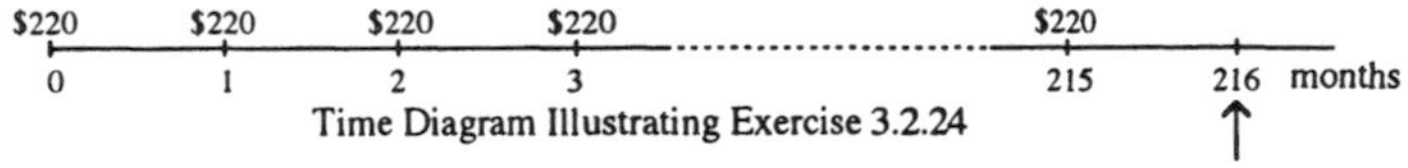

$220 $220 $220 $220 .... $220
0 1 2 3 215 216 months
Time Diagram Illustrating Exercise 3.2.24

$$S(\$220, .07/12, 216, 216) = \$220\frac{(1.00583\overline{3})^{216} - 1}{.00583\overline{3}}(1.00583\overline{3})$$

$$= \$95{,}311.38.$$

3.2.25 We are looking for j_{12}, but first we will find $i = j_{12}/12$. Using Formula 3.2*a* and the time diagram, the equation of value is

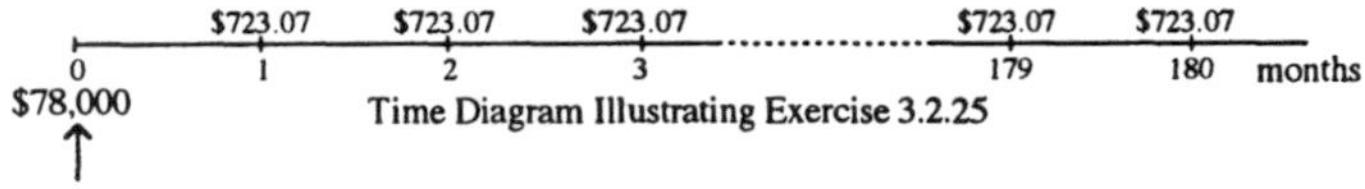

$723.07 $723.07 $723.07 .... $723.07 $723.07
0 1 2 3 179 180 months
$78,000
Time Diagram Illustrating Exercise 3.2.25

$$S(\$723.07, i, 180, -1) = \$723.07\frac{1-(1+i)^{-180}}{i} = \$78{,}000.$$

Solving by Solver: In the EQUATION SOLVER, enter

0 = 723.07(1 – (1 + I)^-180)/I – 78000

We find I = .00625000675... ≈ .00625, so $j_{12} = 12i \approx .075 =$ 7.5%.

Solving by Guess-and-Check: Start with

$$LHS(i) = \$723.07\frac{1-(1+i)^{-180}}{i}.$$

After 7 steps with a Guess-and-Check table, I got *LHS*(.00625) ≈ \$78,000.04. As above $j_{12} \approx 7.5\%$

Exercises 3.3

3.3.1 a) $P = S(\$50, .06, 20, -1) + \$1{,}000(1.06)^{-20} = \$885.30$
b) $P = S(\$60, .05, 30, -1) + \$1{,}000(1.05)^{-30} = \$1{,}153.72$
c) $P = S(\$50, .045, 40, -1) + \$1{,}100(1.045)^{-40} = \$1{,}109.20$

3.3.2 Substituting the information given into Formula 3.4*a*, we have

$$\$5{,}814.90 = S(\$300, .05, 30, -1) + C(1.05)^{-30} =$$

$$\$300\frac{1-(1.05)^{-30}}{.05}+C(1.05)^{-30},$$

which has the solution $C = \$5200.01$.

3.3.3 The inputs for using Formula 3.4*a* are $P = \$10{,}980$, $F = \$10{,}000$, $r = .095/2 = .0475$, $n = 40$, $C = \$10{,}138.13$. What we need to find is Rickie's annual yield rate j_2. We begin by finding $i = j_2/2$. From Formula 3.4*a* we have

$$S(\$475, i, 40, -1) + \$10{,}138.13(1 + i)^{-40} =$$

$$\$475\frac{1-(1+i)^{-40}}{i}+\$10{,}138.13(1+i)^{-40}=\$10{,}980.$$

Solution by Solver: We put

0 = 475(1 – (1 + I)^-40)/I + 40138.13(1 + I)^-40 - 10980

into the EQUATION SOLVER and start with I = .05. The solution is I = .04250000057... ≈ 4.25%, making $j_2 \approx 8.5\%$.

Solution by Guess-and-Check: We set

$$LHS(i) = \$475\frac{1-(1+i)^{-40}}{i}+\$10{,}138.13(1+i)^{-40}$$

and make a table and get $LHS(i)$ as close as we can to \$10,980. In my table, I reached $LHS(.0425) = \$10{,}980.00$ in four steps.

3.3.4 For callable bonds with $i > r$, as we have here, the lowest price is at the latest possible date. We have $i = .07$, $r = .055$, $F = C = \$5{,}000$ and $n = 30$ (the longest possible duration). Substituting these into Formula 3.4*b*, we calculate the price

$$P = S(-\$75, .07, 30, -1) + \$5{,}000 = \$4{,}069.32.$$

3.3.5 The inputs for Formula 3.4*b* are $F = \$5{,}000$, $r = .03$, $C = \$5{,}250$, $i = .035$, and $n = 40$. Bob's purchase price is

$$P = S(-\$33.75, .035, 40, -1) + \$5{,}250 = \$4{,}529.27.$$

3.3.6 This is equivalent to four \$25,000 bonds with coupons at $j_2 = 8\%$ redeemable at par; the first in 5 years, the second in 10 years, the third in 15 years, and the fourth in 20 years. The yield rate is $j_2 = 6\%$. The purchase price for the serial bond is the sum of the prices of the individual bonds. These are:

$$P_1 = S(\$250, .03, 10, -1) + \$25{,}000 = \$27{,}132.55,$$

$$P_2 = S(\$250, .03, 20, -1) + \$25{,}000 = \$28{,}719.37,$$

$$P_3 = S(\$250, .03, 30, -1) + \$25{,}000 = \$29{,}900.11,$$

$$P_4 = S(\$250, .03, 40, -1) + \$25{,}000 = \$30{,}778.69,$$

and their sum is $116,530.72.

3.3.7 We are given Formula 3.4*a*:

$$P = Fr\frac{1-(1+i)^{-n}}{i} + C(1+i)^{-n}.$$

This can be written successively as

$$P = Fr\frac{1-(1+i)^{-n}}{i} - \left(-C(1+i)^{-n}\right),$$

$$P = Fr\frac{1-(1+i)^{-n}}{i} - \left(C - C(1+i)^{-n} - C\right),$$

$$P = Fr\frac{1-(1+i)^{-n}}{i} - C\left(1-(1+i)^{-n}\right) + C,$$

$$P = Fr\frac{1-(1+i)^{-n}}{i} - Ci\frac{1-(1+i)^{-n}}{i} + C,$$

$$P = (Fr - Ci)\frac{1-(1+i)^{-n}}{i} + C.$$

This is Formula 4.5b.

3.3.8 The inputs for Formula 3.4*a* are $F = C = \$50{,}000$, $r = .06$, $i = .065$, and $n = 20$. The price of the bond is

$$P = S(\$3{,}000, .065, 20, -1) + \$50{,}000(1.065)^{-20} = \$47{,}245.37.$$

3.3.9 The only difference between this exercise and Exercise 3.3.5 is that here we are given the yield rate as an effective rate. To find Nadya's purchase price we will convert the yield rate of $j_1 = 7\%$ to the equivalent semiannual rate. When we do this we get $i = j_2/2 = \{(1 + .07)^{1/2} - 1\} = .03440804328$. Also, from $Fr = \$150$, $Ci = \$180.64$ we calculate the purchase price $P = S(\$150 - \$180.64, .0344\ldots, 40, -1) + \$5{,}250 = \$4{,}589.58$.

Miscellaneous Exercises – Chapter 3

3M1. The equation of value with focal date at time 0 is:

$$S(\$25{,}000,\ .066,\ 15,\ -10) = \$25{,}000\frac{1-(1.066)^{-15}}{.066}(1.066)^{-9}$$

$$= \$131{,}398.70.$$

3M2. This exercise is similar to Example 3.13 and Exercise 3.2.10. The period is the quarter and the interest rate per quarter is .0812/4 = .0203. Using the time diagram below, we write the equation of value:

	\$8,200	\$8,200	\$8,200	......	\$8,200	\$8,200	\$8,200	B
0	1	2	3		$n-2$	$n-1$	n	$n+1$
\$190,000 ↑								quarters

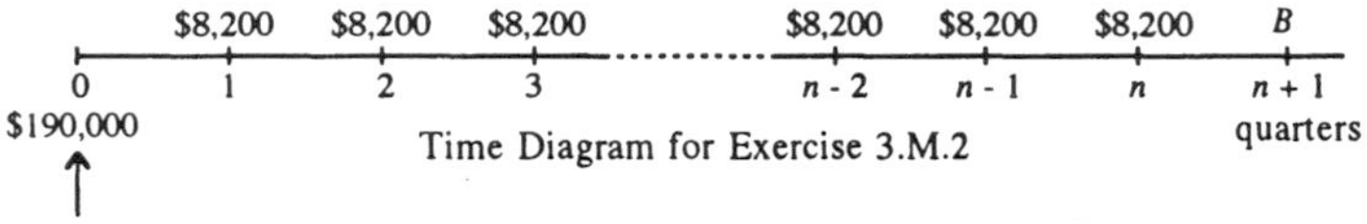

Time Diagram for Exercise 3.M.2

$$S(\$8200,\ .0203,\ n,\ -1) + B(1.0203)^{-(n+1)} =$$

$$\$8{,}200\frac{1-(1.0203)^{-n}}{.0203} + B(1.0203)^{-(n+1)} = \$190{,}000$$

where n is the number of quarters. We need to find n such that

$$\$8{,}200\frac{1-(1.0203)^{-n}}{.0203} < \$190{,}000,$$

but

$$\$8{,}200\frac{1-(1.0203)^{-(n+1)}}{.0203} > \$190{,}000.$$

The simplest way to do this is to find n such that $\$8{,}200\frac{1-(1.0203)^{-n}}{.0203} = 190{,}000$. Then the integer part of that number will be the correct n. Setting $\$8{,}200\frac{1-(1.0203)^{-n}}{.0203} =$ \$190,000 and solving for n we find n = 31.6255..., so the number of equal payments of \$8,200 is n = 31. In fact,

$$\$8{,}200\frac{1-(1.0203)^{-31}}{.0203} = \$187{,}293.56563106,$$

and

$$\$8{,}200\frac{1-(1.0203)^{-32}}{.0203} = \$191{,}604.00.$$

We can rewrite the equation of value now as:

$$S(\$8200,\ .0203,\ 31,\ -1) + B(1.0203)^{-32} =$$

$$\$8,200\frac{1-(1.0203)^{-31}}{.0203} + B(1.0203)^{-32} = \$190,000$$

or

$$\$187,293.56563106 + B(1.0203)^{-32} = \$190,000.$$

The solution is $B = (\$190,000 - \$187,293.56563106)(1.0203)^{32} = \$5,148.61$.

3M3. If DeAnn were getting j_{12} = 7.9% or better, her monthly payments would be $426.42 or less. This is the solution of the equation of value

$$S(P,\ .079/12,\ 60,\ -1) = \$21,080.14$$

for P. DeAnn is currently getting j_{12} = 8.1% as you could verify using a Solver or Guess-and-Check.

3M4. We can write the equation of value

$$S(\$275,\ .0515/12,\ 36,\ 36) =$$

$$\$275\frac{(1.004291\overline{6})^{36}-1}{.004291\overline{6}}(1.004291\overline{6}) = \$10,726.85.$$

3M5. The equation of value may be written

$$\$1,100(1.00\overline{6})^{11} = S(\$100,\ .00\overline{6},\ 10,\ 10) + R.$$

We solve for P to get

$$R = \$1,100(1.00\overline{6})^{11} - S(\$100,\ .00\overline{6},\ 10,\ 10) = \$146.00.$$

Chapter 4

Exercises 4.1

4.1.1 We first need to find $j_{12}/12$ from j_{365}. The formula is

$$\frac{j_{12}}{12} = \left(1+\frac{j_{365}}{365}\right)^{365/12} - 1 = \left(1+\frac{.0475}{365}\right)^{365/12} - 1 = .00396591932.$$

After six years of monthly deposits, Maria's accumulated savings are:

$$S(\$150,\ .00396591943,\ 72,\ 71) = \$150\frac{(1.00396591943)^{72}-1}{.00396591943} =$$

$12,471.41.

4.1.2 We need the interest rate per month:

$$i = \frac{j_{12}}{12} = (1 + j_1)^{1/12} - 1 = (1.1)^{1/12} - 1 = .00797414043.$$

The equation of value for the loan is:

$$S(P,\ .00797414043,\ 60,\ -1) = P\frac{1-(1.00797414043)^{-60}}{.00797414043}$$
$$= \$25{,}000.$$

Solving for P gives $525.89.

4.1.3 We have the effective rate but need the monthly rate:

$$i = \frac{j_{12}}{12} = (1 + j_1)^{1/12} - 1 = (1.068)^{1/12} - 1 = .00549736708.$$

With the time diagram below as a guide we can write the equation of value:

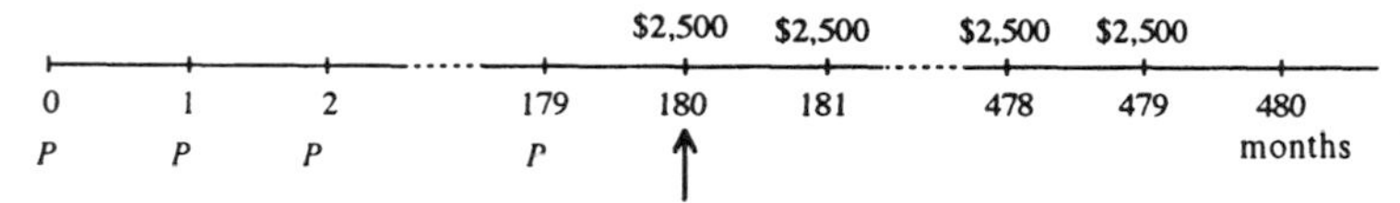

Time Diagram for Exercise 4.1.3

$$S(P,\ .00549736708,\ 180,\ 180) =$$
$$= S(\$2{,}500,\ .00549736708,\ 300,\ 0).$$

I have equated the two annuities at the focal date of 180. Let's write the equation out:

$$P\frac{(1.00549736708)^{180}-1}{.00549736708}(1.00549736708) =$$
$$\$2{,}500\frac{1-(1.00549736708)^{-300}}{.00549736708}(1.00549736708).$$

The solution is $P = \$1{,}198.87$.

4.1.4 We need the weekly interest rate. This is

$$i = \frac{j_{52}}{52} = \left(1 + \frac{.054}{12}\right)^{12/52} - 1 = .00103666896.$$

Evelyn's savings after two years are $S(\$100, .00103666896, 104, 103) = \$10,975.33$.

4.1.5 The monthly interest rate is

$$i = \frac{j_{12}}{12} = (1 + j_1)^{1/12} - 1 = (1.09)^{1/12} - 1 = .00720732332.$$

The equation of value is

$$S(P, .00720732332, 24, -1) = P\frac{1-(1.00720732332)^{-24}}{.00720732332} =$$

$$= P(21.966547\ldots) = \$3,300 - \$800 = \$2,500,$$

from which we get $P = \$2,500/21.966547\ldots = \$113.81$.

4.1.6 We are given $j_4 = 5\%$ but we need $i = j_{12}/12$. We have

$$i = \frac{j_{12}}{12} = \left(1 + \frac{j_4}{4}\right)^{4/12} - 1 = (1.0125)^{1/3} - 1 = .00414942512.$$

Our equation of value is

$$S(P, .00414942512, 60, 59) = P\frac{(1.00414942512)^{60} - 1}{.00414942512} =$$

$$P(67.9701972567) = \$25,000,$$

so $P = \$367.81$.

4.1.7 We'll need the monthly rate

$$i = \frac{j_{12}}{12} = (1 + j_1)^{1/12} - 1 = (1.086)^{1/12} - 1 = .00689878956.$$

Let P be the purchase price for the restaurant. Then

$$P - \$30,000 = S(\$1100, .00689878956, 96, -1) =$$

$$\$1,100\{1-(1.00689878956)^{-96}\}/000689878956 = \$77,038.02,$$

so Pablo paid $P = \$77,038.02 + \$30,000 = \$107,038.02$.

4.1.8 We need to convert the monthly interest rate to the equivalent effective rate. That's $j_1 = \{(1 + .072/12)^{12} - 1\} =$ 7.442416772%. Then, Madhu's accumulated savings at the end of 20 years are $S(\$5{,}000, .07442416772, 20, 19) = \$215{,}156.86$.

4.1.9 We need the monthly interest rate equivalent to the quarterly rate of 8%. That is, $i = j_{12}/12 = \{(1 + .08/4)^{4/12} - 1\} =$ .662270956%. Wai-Han's monthly payments P satisfy $S(P, .00662270956, 360, 359) = \$385{,}000 - \$150{,}000$. Solving for P gives us $P = \$1{,}715.71$.

4.1.10 We need the quarterly interest rate equivalent to the monthly rate of 7.5%. That is, $i = j_4/4 = \{(1 + .075/12)^{12/4} - 1\} =$ 1.886743164%. Yuri's quarterly deposits P satisfy $S(P, .01886743164, 12, 11) = \$30{,}000$. Solving for P gives us $P =$ \$2,251.07.

Exercises 4.2

4.2.1 There are two annuities here. The contributions Tage and Berit make to their retirement fund *and* their increasing annuity. The inputs for the increasing annuity are $k = .0025$, $P = 25{,}000$ Crowns, $i = .0725/12 = .006041666...$, $n = 300$ and $d = 0$. With the help of the time diagram below, we can write the equation of value:

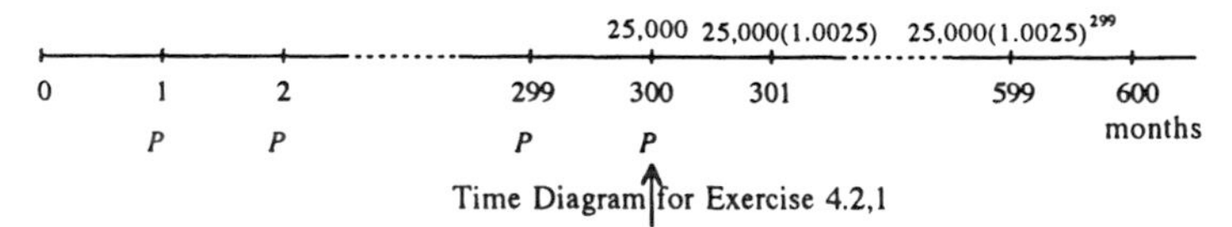

Time Diagram for Exercise 4.2,1

$$S(P, .006041\bar{6}, 300, 299) = S^{(.0025)}(25000, .006041\bar{6}, 300, 0)$$

Writing this out using Formula 4.1, we get

$$P\frac{(1.006041\bar{6})^{300} - 1}{.006041\bar{6}} =$$

$$25{,}000\frac{1 - (0.9964796021)^{300}}{.00354 1\bar{6}}(1.00604 1\bar{6}).$$

Carrying out the calculations and solving we get $P = 5{,}500.35$ Crowns.

4.2.2 For Formula 4.1 we have P to be determined, $i = .005$, $k = .006$, $n = 360$, and $d = -1$. Setting

$$S^{(.006)}(P, .005, 360, -1) = \$750{,}000$$

and solving for P yields $P = \$1{,}742.12$. The second payment is then $\$1{,}742.12(1.006) = \$1{,}752.57$ and the last is $\$1{,}742.12(1.006)^{359} = \$14{,}919.47$.

4.2.3 First look at the time diagram below to see the setup.

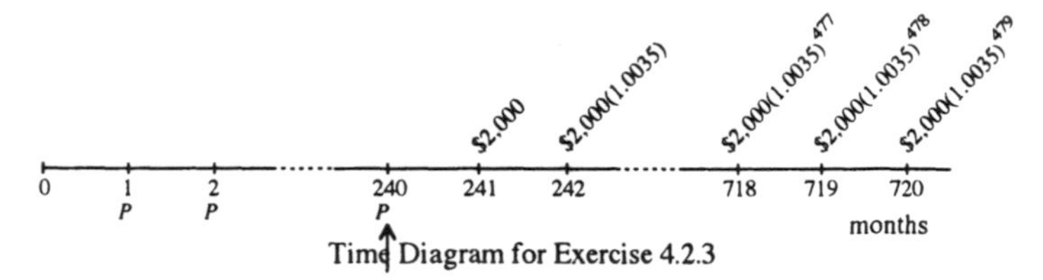

Time Diagram for Exercise 4.2.3

The annuity of P that Federico pays into his fund has the value

$$S(P, .0045, 240, 239) =$$

$$P\frac{(1.0045)^{240}-1}{.0045} = P(430.567557658),$$

at the focal date of 240 months, and this must equal the value of the increasing annuity at the same focal date.

$$S^{(.0035)}(\$2{,}000, .0045, 480, -1) = \$760{,}064.34.$$

Equating these two annuities and solving for P we get $P = \$1{,}765.26$.

4.2.4 The monthly rate of interest is $i = .06/12 = .005$ and, from Formula 4.1, we can write the equation of value as

$$S^{(k)}(\$600, .005, 72, 71) =$$

$$\$600\frac{1-\left(\left(1+k\right)/1.005\right)^{72}}{.005-k}(1.005)^{72} = \$60{,}000$$

Solution by Solver: In the EQUATION EDITOR, enter

0 = 600(1 – ((1 + K)/1.005)^72)/(.005 – K)*1.005^72 – 60000.

With a starting value of K = .05, I found K = .00427199894... ≈ 0.4272%.

Solving by Guess-and-Check: Make a Guess-and-Check table with $LHS(k)$ = $\$600\frac{1-((1+k)/1.005)^{72}}{.005-k}(1.005)^{72}$ and substitute values of k to make $LHS(k)$ close to \$60,000. It took me eleven steps to get $LHS(.004272) \approx \$100.000$. This agrees well with the solver. Therefore, $k \approx .004272 = .4272\%$ is the percentage increase of Mool's deposits.

4.2.5 We begin with the sum, which we are calling $S^{(k)}(P, i, n, d)$,

$$P(1+i)^d + P(1+k)(1+i)^{d-1} + P(1+k)^2(1+i)^{d-2} + \ldots + P(1+k)^{n-1}(1+i)^{d-n+1}.$$

This is an n-term geometric sum with first term $a = P(1 + i)^d$, and common ratio $r = (1 + k)/(1 + i)$. From Formula 3.1a, we have

$$S^{(k)}(P, i, n, d) = P(1+i)^d\frac{1-((1+k)/(1+i))^n}{1-((1+k)/(1+i))} =$$

$$P(1+i)^d\frac{1-((1+k)/(1+i))^n}{(i-k)/(1+i)} = P\frac{1-((1+k)/(1+i))^n}{i-k}(1+i)^{d+1}.$$

4.2.6 We'll need $k = -.005$, $i = .087/12 = .00725$, and $d = 0$. Using Formula 4.1, the present value of Mario's insurance payments is

$$S^{(-.005)}(\$24, .00725, 240, 0) =$$

$$\$24\frac{1-(0.987838173)^{240}}{0.01225}(1.00725)$$

$$\approx \$1,868.72$$

4.2.7 We have $k = .0045$, $i = j_{12}/12 = .005$, $n = 120$, and $d = 119$.

a) Bibi Titi's last deposit will be $\$100(1.0045)^{119} = \$170.63$.

b) With the help of Formula 4.1, at the end of 10 years, Bibi Titi will have

$$S^{(.0045)}(\$100, .005, 120, 119) = \$21,093.47.$$

c) Setting $\$21,093.47 = S(P, .005, 120, 119)$, we find $P =$ \$128.71.

4.2.8 We are given $i = .109/12 = .00908\overline{3}$, $k = .007$, $n = 300$, and $d =$ -1. To find the first payment P, set

$$S^{(.007)}(P,\ .00908\overline{3},\ 300,\ -1) = \$125{,}000$$

and solve to find $P = \$563.60$. Zubeida's final payment will be $\$563.60(1.007)^{299} = \$4{,}537.11$.

4.2.9 The time diagram below tells most of the story. We

P P(1.0065) P(1.0065)^2 .... P(1.0065)^{359}

0 1 2 359 360 months

Time Diagram for Exercise 4.2.9 $2,500,000 ↑

calculate $i = .08/12 = .00\overline{6}$. The equation of value is

$$S^{(.006)}(P,\ .00\overline{6},\ 360,\ 360) = \$2{,}500{,}000.$$

Solving for P gives $P = \$654.08$. Dudley's last payment will be $\$654.08(1.0065)^{359} = \$6{,}695.45$

4.2.10 Here we set $S(P,\ .00\overline{6},\ 360,\ 360) = \$2{,}500{,}000$ and get $P =$ \$1,666.34.

4.2.11 If $i = k$, $S^{(k)}(P, i, n, d)$ is the sum of a geometric progression with $a = P(1 + i)^d$, $r = 1$, and n terms. Since $r = 1$, all n terms are equal to the first, $P(1 + i)^d$. thus, Their sum is $nP(1 + i)^d$.

4.2.12 Here we have $k = .0035$ and $i = j_{12}/12 = .042/12 = .0035$, so $i = k$ and we may use the version for $S^{(k)}$ from Exercise 4.2.11. We have $S^{(k)}(P, i, n, d) = S^{(.0035)}(\$125,\ .0035\ 120,\ 119) = (120)(\$125)(1.0035)^{119} = \$22{,}733.12$.

Exercises 4.3

4.3.1 The inputs to Formula 3.2 for Bev's deposits are $P = \$10{,}000$, $i = .08$, $n = 10$, and $d = 0$. The inputs for Formula 4.3 for the perpetuity are P to be determined, $i = .08$ and $d = 0 - 10 = -10$. Using the time diagram below, we can write the equation of value:

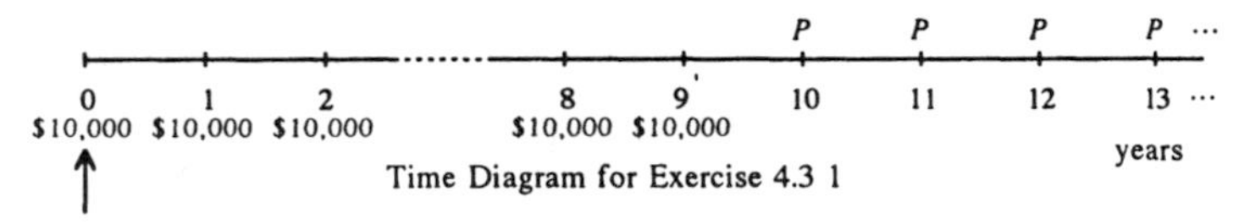

Time Diagram for Exercise 4.3 1

$$S(\$10000,\ .08,\ 10,\ 0) = S(P,\ .08,\ \infty,\ -10),$$

which, when written out becomes

$$\$10,000\frac{1-(1.08)^{-10}}{.08}(1.08) = P\frac{1.08^{-9}}{.08}.$$

Solving this for P yields P = $11,589.25.

4.3.2 Using Formula 4.3, we see that Professor Ümläüt deposited

$$S(\$20000,\ .064,\ \infty,\ 0) = \$20,000\frac{1.064}{.064} = \$332,500.$$

4.3.3 Suzie's deposits amount to S($30000, .086, 10, 10) = $485,634.691356 at the end of 10 years. This money is to finance a perpetuity providing equal annual payments beginning at the end of the 10th year. Therefore

$$\$485,634.691356 = S(P,\ .086,\ \infty,\ 0) =$$

$$P\frac{1.086}{.086} = P(12.6279069767),$$

so P = $38,457.26

4.3.4 We have $176,891 = $\$9,413\frac{1+i}{i}$. This can be solved algebraically for i as follows:

$$\$9,413\frac{1+i}{i} = \$9,413\left(\frac{1}{i}+1\right),$$

so

$$\frac{1}{i}+1 = \$176,891/\$9,413 = 18.7922022735,$$

and

$$\frac{1}{i} = 17.7922022735,$$

and finally,

$$i = 1/17.7922022735 = .056204\ldots = 5.6204\ldots\%.$$

4.3.5 The present value of a perpetuity of $8,000 at the end of every year at 10% effective is worth

$$S(\$8000,\ .10,\ \infty,\ -1) = \$8,000\frac{1}{.1}(1.1)^{0} = \$80,000.$$

4.3.6 First we must find the effective rate equivalent to j_{12} = 12%. This is $j_1 = \{(1 + .12/12)^{12} - 1\}$ = 12.682503013%. The inputs for Formula 4.3 are i = .12682503013, d = -4, and P to be determined. We set

$$P\frac{(1.12682503013)^{-4+1}}{.12682503013} = \$150,000$$

and solve to get P = \$27,218.59.

4.3.7 In this problem we have an annuity and a perpetuity. Equating them at 20 years, we have

$$S(P, .09, \infty, 0) = P\frac{(1.09)^{0+1}}{.09} =$$

$$S(\$5000, .09, 20, 20) = \$5,000\frac{(1.09)^{20}-1}{.09}(1.09),$$

so $P = \$278,822.65/12.\overline{1}$ = \$23,022.05

4.3.8 We will need to convert j_2 = 10% into the equivalent effective rate. This is $j_1 = \{(1 + .10/2)^2 - 1\}$ = .1025 = 10.25%. The present value of Vandana's perpetuity is

$$S(\$45000, .1025, \infty, -2) =$$

$$\$45,000\frac{(1.1025)^{-2+1}}{.1025} = \$398,208.06$$

4.3.9 We have $G = a + ar + ar^2 + ar^3 + \ldots$ If we multiply both sides of the equation by r, we get $rG = ar + ar^2 + ar^3 + \ldots$, and subtracting the latter from the former gives us $(1 - r)G = a$, all other terms canceling each other out. Dividing now by $1 - r$ gives Formula 4.2. Our derivation requires that $-1 < r < 1$.

Exercises 4.4

4.4.1 The present value of Bev's deposits is

$$S(\$10000, .08, 10, 0) =$$

$$\$10,000\frac{1-(1.08)^{-10}}{.08}(1.08) = \$72,468.88,$$

and using Formula 4.4, we can write the present value of the increasing perpetuity as

$$S^{(.03)}(P,\ .08,\ \infty,\ -10) = P\frac{(1.08)^{-9}}{.08-.03}.$$

Setting the two equal and solving for P yields P = \$7,243.28 as the first scholarship award. The next two increase by 3% each and are \$7,460.58 and \$7,684.40.

4.4.2 We use Formula 4.4 for Jan-Ove's fund. The inputs are P = 100,000 Crowns,

$i = \left(1+\frac{.075}{12}\right)^{12} - 1 = .07763259886$, k = .025, and d = 0. Jan-Ove must provide

$$S^{(.025)}(100000,\ .077632...,\ \infty,\ 0) = 100{,}000\frac{(1.077632...)}{.077632...-.025} =$$

2,047,462.26 Crowns.

4.4.3 If you use Formula 4.4, Lois's single deposit looks like this:

$$S^{(.0325)}(\$120000,\ .078,\ \infty,\ 0) = \$120{,}000\frac{(1.04406779661)}{.04406779661}.$$

If you use Formula 4.4 the equation of value looks like this:

$$S^{(.0325)}(\$120000,\ .078,\ \infty,\ 0) = \$120{,}000\frac{(1.078)}{.078-.0325}.$$

In either case, the amount of the deposit is \$2,843,076.92.

4.4.4 From Formula 4.4, we can write

$$\$2{,}500{,}000 = S^{(k)}(\$150000,\ .082,\ \infty,\ 0) = \$150{,}000\frac{(1.082)}{.082-k}.$$

This equation can be solved algebraically for k. After a little rearranging we get

$$.082 - k = .06492,$$

so

$$k = .01708 = 1.708\%.$$

4.4.5 This is an increasing perpetuity with P = \$300, i = 5.3%, k = 3.5%, and d = -1. We find

$$S^{(.035)}(\$300, .053, \infty, -1) = \$300\frac{(1.053)^{-1+1}}{.053-.035} = \$16{,}666.78.$$

4.4.6 Inputs for Zubeida's perpetuity are P = \$85,000, i = .085, k = .05, and d = 0, so her outlay is

$$S^{(.05)}(\$85000, .085, \infty, 0) = \$85{,}000\frac{(1.085)^{0+1}}{.085-.05} = \$2{,}635{,}000.$$

4.4.7 This is similar to Exercise 4.3.5 except that the dividends increase annually. For a yield of 10%, the investor should invest

$$S^{(.03)}(\$8000, .10, \infty, -1) = \$8{,}000\frac{(1.1)^{-1+1}}{.1-.03} = \$114{,}285.71.$$

4.4.8 This is an increasing perpetuity with P to be determined, i = .11, k = .06, and d = 0. Set

$$P\frac{(1.11)^{0+1}}{.11-.06} = \$400{,}000,$$

and solve for P to get P = \$18,018.02.

Exercises 4.5

4.5.1 The price is that of an increasing perpetuity with semi-annual dividends of \$3, $i = j_2/2$ = .045, k = .02, and d = -1. From Formula 4.4, we find the share price to be $S^{(0.02)}(\$3, .045, \infty, -1) = \$3(1.045)^0/(.045 - .02) = \$120$.

4.5.2 The price of \$120 pays for a semi-annual dividend of P in perpetuity, where the semi-annual interest rate is $i = j_2/2$ = .04. Using Formula 4.3, we find $\$120 = S(P, .04, \infty, -1) = P/.04$, so $P = \$120(.04) = \$4.80$.

4.5.3 We have an increasing perpetuity with semi-annual dividends of \$1.50, beginning in 3½ years, with subsequent dividends increasing by 2.5%. The semi-annual rate of return is to be 8%/2 = 4%. From Formula 4.4, the price should be $S(\$1.50, .04, \infty, -8) = \$1.50(1.04)^{-8+1}/(.04 - .025) \approx \$75.99$ per share.

4.5.4 The equation of value we get is $\$180 = S(\$4.50, j_2/2, \infty, -1)$, from which we find $j_2/2$ = .025. But, we want j_1, the effective rate. $1 + j_1 = (1 + j_2/2)^2 = (1.025)^2$ yields j_1 = 5.0625%.

4.5.5 $S = P_1(1 + j_2/2)^{-1} + P_2(1 + j_2/2)^{-2} + P_3(1 + j_2/2)^{-3} + \ldots$

4.5.6 The biennial interest rate is $i = (1.10)^2 - 1 = 1.21 - 1 = .21 =$ 21%. The price is the present value of a perpetuity of \$12 with interest rate 21% per period. The price is $S(\$12, .21, \infty, -1) =$ $\$12(1.21)^{-1+1}/.21 \approx \$57.14$.

Miscellaneous Exercises – Chapter 4

4M1. We can use Formula 3.2 with inputs P (to be determined), $i =$ .08, $n = 20$, and $d = 0$:

$$S(P, .08, 20, 0) = R\frac{1-(1.08)^{-20}}{.08}(1.08) = R(10.6035992)$$

$$= \$2,650,899.80,$$

which gives P = \$250,000. Chauncey won the '\$5,000,000' lottery.

4M2. We are estimating the present value of an annuity increasing by 2.912621359% annually. The inputs we need for Formula 4.1 are $P = \$35,000$, $i = .06$, $k = .02912621359$, $n = 39$, and $d = -1$. The estimated present value of loss of income is

$$S^{(.02912621359)}(\$35,000, .06, 39, -1) =$$

$$\$35,000\frac{1-\left(\frac{1.02912621359}{1.06}\right)^{39}}{.06-.02912621359}(1.02912621359)^{-1+1} = \$775,694.49.$$

4M3. We need $i = \frac{j_4}{4} = \left(1+\frac{j_{12}}{12}\right)^{12/4} - 1 = .015075125$. Then the accumulated value of the 5-year annuity of \$300 per quarter is

$$S(\$300, .015075125, 20, 19) = \$300\frac{(1.015075125)^{20}-1}{.015075125}$$

$$= \$6,942.23.$$

4M4. The value of the cash option is the present value of Rosie's 20-year annuity of \$6,000,000 at $j_1 = .06$. That is

$$S(\$6,000,000, .06, 20, 0) = \$6,000,000\frac{1-(1.06)^{-20}}{.06}(1.06)$$

$$= \$72,948,698.95.$$

4M5. The details of this exercise are the same as they would be if Tomoko had bought a 5-year bond with F = \$20,000, r = .05, i = .045, n = 10, and C to be determined. Therefore, from

$$P = S(Fr, i, n, -1) + C(1 + i)^{-n},$$

we have

$$\$20{,}000 = S(\$1000, .045, 10, -1) + C(1.045)^{-10},$$

so

$$C = \{\$20{,}000 - S(\$1000, .045, 10, -1)\}(1.045)^{10}$$
$$= \$18{,}771.18.$$

4M6. We know that $S^{(k)}(P, i, n, d) = P\dfrac{1-\left(\dfrac{1+k}{1+i}\right)^n}{i-k}(1+i)^{d+1}$. Set $\left(\dfrac{1+k}{1+i}\right)^n = (1+i^*)^{-n} = \dfrac{1}{(1+i^*)^n}$, and solve this for i^*. After a bit of algebra, we get $i^* = (i - k)/(1 + k)$. We now have

$$S^{(k)}(P, i, n, d) = P\frac{1-(1+i^*)^{-n}}{i^*}\frac{(1+i)^{d+1}}{1+k}$$

using the fact that $i - k = i^*(1 + k)$. Next, write

$$\frac{(1+i)^{d+1}}{1+k} = \left(\frac{1+i}{1+k}\right)^{d+1}(1+k)^d = (1+i^*)^{d+1}(1+k)^d.$$

Finally, we can write

$$S^{(k)}(P, i, n, d) = P\frac{1-(1+i^*)^{-n}}{i^*}(1+i^*)^{d+1}(1+k)^d$$
$$= S(P, i^*, n, d)(1+k)^d,$$

as requested.

4M7. The simplest way to see the desired result is to note that, in the preceding exercise, $\{(1+k)/(1+i)\}^n$ can be replaced by zero when n increases without bound (provided that $k < i$). The result is $S^{(k)}(P, i, \infty, d) = S(P, i^*, \infty, d)\cdot(1 + k)^d$.

4M8. From Formula 4.4, we have

$$90 = 4\frac{(1+i)^{-3+1}}{i-0.02},$$

where $i = j_2/2$. This can be solved by Guess-and-Check, yielding $i \approx 0.0595863$. But, we want j_1, the effective rate. Using Formula 2.2, we find $j_1 = (1.0593863)^2 - 1 \approx 0.122722$, so $j_1 \approx 12.27\%$.

Chapter 5

Exercises 5.2

5.2.1 To amortize a debt is to liquidate the debt by means of periodic payments.

5.2.2 a) From Formula 3.2 we have

$$S(P, .025, 20, -1) = P\frac{1-(1.025)^{-20}}{.025} = \$50,000,$$

so if Pavel could pay to the correct fraction of a penny, his payments would be $P = \$3,207.35643578$. Rounded up to the nearest dollar, Pavel will make his first 19 payments in the amount of \$3,208.

b) If we let B denote Pavel's final payment, then it must satisfy the equation of value

$$S(\$3,208, .025, 19, -1) + B(1.025)^{-20} =$$

$$\$48,052.2834271 + B(1.025)^{-20} = \$50,000$$

The solution for B is $B = (\$50,000 - \$48,052.2834271)\cdot (1.025)^{20} = \$3,191.56$.

5.2.3 a) The amount Akira's monthly payments for the first 10 years is the solution of the equation of value

$$S(P, .105/12, 360, -1) = \$136,000.$$

That solution is \$1,244.05.

b) We need to find what Akira owes on the original loan after 10 years of paying \$1,244.05. We will use Formula 5.2 with inputs $P = \$1,244.05$, $i = .00875$, $n - r = 240$, and $d = -1$. Akira owes

$$S(\$1,244.05, .00875, 240, -1) = \$124,606.88.$$

If Akira now borrows this amount at j_{12} = 7.2% over the next 15 years, his payments will be the solution of the equation of value

$$S(P, .072/12, 180, -1) = \$124{,}606.88$$

for P. That solution is P = \$1,133.98.

5.2.4 This calls for Formula 3.2 with inputs P (to be determined), i = .06, n = 6, and d = –1. The equation of value is

$$\$16{,}000 = S(P, .06, 6, -1) =$$

$$P\frac{1-(1.06)^{-6}}{.06} = P(4.917324326)$$

and its solutions is P = \$3,253.80.

5.2.5 Here the period is the half-year and the interest rate per period is $i = j_2/2$ = .045. Nadia's semi-annual payments must satisfy the equation

$$S(P, .045, 12, -1) = P\frac{1-(1.045)^{-12}}{.045} = \$16{,}000,$$

or P = \$1,754.66.

5.2.6 From the time diagram below we can write the equation of value

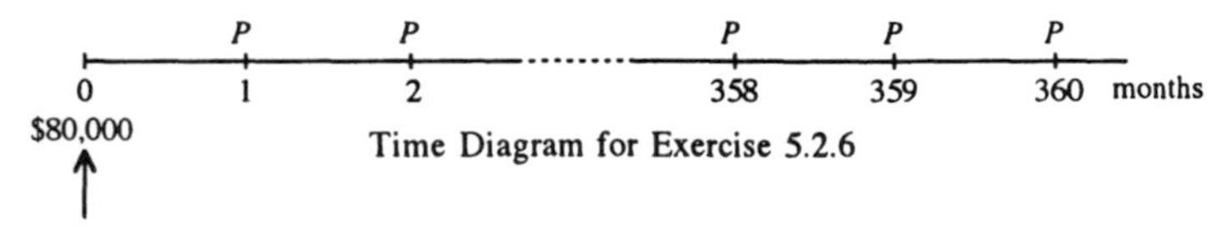

Time Diagram for Exercise 5.2.6

$$S(P, .09/12, 360, -1) = P\frac{1-(1.0075)^{-360}}{.0075} = \$80{,}000,$$

which gives the solution P = \$643.70.

5.2.7 The price of Michiko's house was S(\$715.12, .0799/12, 180, -1) = \$74,875.80.

5.2.8 The equation of value can be written

$$S(\$1854.59, i, 240, -1) = \$1{,}854.59\frac{1-(1+i)^{-240}}{i} = \$230{,}000,$$

where $i = j_{12}/12$.

Solving by Solver: We enter

0 = 1854.59(1 – (1 + I)^-240)/I – 230000.

Using the starting value I = .05, we find at I = .00626022296..., making $j_{12} = 12i \approx .075123 = 7.5123\%$.

Solving by Guess-and-Check: Letting $LHS(i) = \frac{1-(1+i)^{-240}}{i}$, and making a Guess-and-Check table, I found *LHS*(.00626) = \$230,004.68 in six steps.

We find $j_{12} \approx 12(.00626) = 7.512\%$ by Guess-and-Check.

5.2.9 We don't know the amount of the loan, nor do we know its term. But we have all we really need to find what Debra owes at the time she has 100 payments remaining. She owes the present value of 100 payments of \$1,248.13 at $j_{12} = 9.8\%$. Debra owes

$$S(\$1{,}248.13, .098/12, 100, -1) = \$85{,}070.97.$$

5.2.10 Using Formula 5.1, we find Carl-Axel's outstanding balance to be

$$\$15{,}559(1 + .07/12)^{32} - S(\$284.71, .07/12, 32, 31)$$

$$= \$8{,}757.30.$$

5.2.11 We can do this problem by the retrospective method using Formula 5.1 with $i = .12/12 = .01$. After his 70th payment, Sorie owes

$$\$30{,}000\,(1.01)^{70} - S(\$400, .01, 70, 69) =$$

$$\$30{,}000\,(1.01)^{70} - \$400\frac{(1.01)^{70}-1}{.01}$$

$$= \$19{,}932.37.$$

5.2.12 To use the prospective method, we need to know how many payments Nancy has remaining. The term of the loan, n, must satisfy the equation of value

$$S(\$826.15, .0734/12, n, -1) = \$826.15\frac{1-(1.00611\overline{6})^{-n}}{.00611\overline{6}}$$

$$= \$90{,}000.$$

The solution for n is n = 179.999666514 ≈ 180. This discrepancy is going to cause a small disagreement between the two methods. With n = 180, we see that Nancy has 180 – 60 = 120 payments of $826.15 left to make. Using Formula 5.2, we find her outstanding principal to be

$$OP_{120} = S(\$826.15, .00611\overline{6}, 120, -1) = \$70{,}090.92.$$

Now using Formula 5.1 to find Nancy's outstanding principal by the retrospective method, we have

$$OP_{120} = \$90{,}000(1.00611\overline{6})^{60} - S(\$826.15, .00611\overline{6}, 60, 59) =$$

$$\$129{,}760.68 - \$59{,}669.91 = \$70{,}090.77.$$

5.2.13 Andy's monthly payments are the solution of the equation of value

$$S(P, .088/12, 60, -1) = \$24{,}500$$

for P, the solution being P = $506.20. After his 29th payment, Andy owes

$$S(\$506.20, .007\overline{3}, 31, -1) = \$13{,}990.70,$$

and the interest on that amount in the 30th payment is ($13,990.70)($.007\overline{3}$) = $102.60.

5.2.14 Shira is to make 36 bi-monthly payments, so just after her 6th payment, she has 30 to go. We are going to need the amount of her payments. An equation of value for her loan is

$$S(P, .06/6, 36, -1) = P\frac{1-(1.01)^{-36}}{.01} = 60{,}000 \text{ shekels},$$

so P ≈ 1,992.86 bi-monthly. Shira's outstanding principal just after her 6th payment is

$$S(\$1{,}992.86, .01, 30, -1) = 51{,}431.15 \text{ shekels}.$$

5.2.15 We are to show that

$$OP_r = L(1+i)^r - S(P,\ i,\ r,\ r-1) = S(P,\ i,\ n-r,\ -1),$$

where L, P, i, and n satisfy

$$L = S(P,\ i,\ n,\ -1) = P\frac{1-(1+i)^{-n}}{i}.$$

Substituting the right-hand side of the last equation into Formula 5.1, we get

$$OP_r = P\frac{1-(1+i)^{-n}}{i}(1+i)^r - P\frac{(1+i)^r-1}{i} =$$

$$P\frac{(1+i)^r - (1+i)^{-(n-r)} - (1+i)^r + 1}{i} =$$

$$P\frac{1-(1+i)^{-(n-r)}}{i} = S(P,\ i,\ n-r,\ -1).$$

This last expression is the outstanding principal by Formula 5.2. Reversing the steps shows that the two are equivalent.

5.2.16 We are given enough information here to write the equation of value

$$S(\$716.69,\ i,\ 132,\ -1) = \$716.69\frac{1-(1+i)^{-132}}{i} = \$62{,}819.38,$$

where $i = j_{12}/12$.

Solution by Solver: In the EQUATION SOLVER put

0 = 716.69(1 – (1 + I)^-132)/I – 62819.38,

and use I = .05 as the starting value. The solution I get is I = .00665628471..., so $j_{12} = 12i \approx 7.9875\%$.

Solution by Guess-and-Check: We set up a Guess-and-Check table for $LHS(i) = \$716.69\frac{1-(1+i)^{-132}}{i}$. I got $LHS(.00666) \approx$ \$62,806.20 after eight steps, so from Guess-and-Check, I found $j_{12} \approx 12(.00666) = 7.992\%$.

Exercises 5.3

5.3.1 Below is the amortization schedule for Norma's loan.

Year	*Annual Payment*	*Interest Paid*	*Principal Repaid*	*Outstanding Principal*
0	—	—	—	$45,000.00
1	$9,882.32	$3,825.00	$6,057.32	$38,942.68
2	$9,882.32	$3,310.13	$6,572.19	$32,370.49
3	$9,882.32	$2,751.49	$7,130.83	$25,239.66
4	$9,882.32	$2,145.37	$7,736.95	$17,502.72
5	$9,882.32	$1,487.73	$8,394.59	$9,108.13
6	$9,882.32	$774.19	$9,108.13	$0.00

5.3.2 To find these quantities, we will need the amount of Merrick and April's monthly payment and the outstanding principal on their loan after their 74th payment. At that time they had 300 – 74 = 226 payments left to make. First, their monthly payment P satisfies

$$S(P, .071/12, 300, -1) = \$265,000,$$

and the solution is P = $1,889.90. Next, their outstanding principal just after their 74th is

$$S(\$1,889.90, .00591\overline{6}, 226, -1) = \$235,212.59.$$

Therefore, the interest portion of their 75th payment is $\$235,212.59(.00591\overline{6}) = \$1,391.67$. The principal repaid is $1,889.90 - $1,391.67 = $498.23.

5.3.3 Let OP_r denote Sasha's outstanding principal after his r^{th} payment. Then what we are given tell us that $OP_3 - OP_4 =$ $800, so that we can find P from

$$OP_3 - OP_4 = P\frac{1-(1.12)^{-5}}{.12} - P\frac{1-(1.12)^{-4}}{.12} =$$

$$P\frac{(1.12)^{-4}-(1.12)^{-5}}{.12} = P(.56742685572) = \$800.$$

Consequently, P = $1,409.87334656... Finally,

$$L = S(P, .12, 8, -1) = P\frac{1-(1.12)^{-8}}{.12} = P(4.96763976684)$$

$$\approx \$7,003.74.$$

5.3.4 a) The amount of Ken and Amy's monthly payments P are found from the equation of value

$$\$30,800 = S(P, .075/12, 240, -1)$$

to be P = \$248.122703618. From this we find their outstanding balance after their 11th payment to be S(\$248.122703618, .00625, 229, -1) = \$30,168.6669295. The interest paid in their next payment is therefore (\$30,168.6669295)(.00625) = \$188.554168309.

b) The amount of interest paid in the 12th payment can be written as

$$(.00625)\times S(\$248.122703618, .00625, 229, -1).$$

Similarly, the amount of interest paid in their 1st, 2nd , 3rd , ..., 11th payments can be written as

$$(.00625)\times S(\$248.122703618, .00625, 240, -1),$$

$$(.00625)\times S(\$248.122703618, .00625, 239, -1),$$

$$(.00625)\times S(\$248.122703618, .00625, 238, -1), \ldots$$

$$\ldots (.00625)\times S(\$248.122703618, .00625, 230, -1),$$

so that the total amount of interest paid in the first year is the sum of these 11 terms plus the one we already found on part a). We could evaluate each of these and add them up, but we can learn something for solving similar problems by writing everything out. The twelve interest payments can be written as follows:

1st: $(.00625)\cdot\$248.12\ldots\cdot\dfrac{1-(1.00625)^{-240}}{.00625}=$
$\$248.12\ldots(1-(1.00625)^{-240)},$

2nd: $\$248.12\ldots(1-(1.00625)^{-239}),$

3rd: $\$248.12\ldots(1-(1.00625)^{-238}),$

. . .

11th: $\$248.12\ldots(1-(1.00625)^{-230}),$

12th: $\$248.12\ldots(1-(1.00625)^{-229}),$

Adding these, we get $\$248.12\ldots(12-\{(1.00625)^{-229}+1.00625)^{-230}+(1.00625)^{-239}+\ldots+(1.00625)^{-240})\})$.

The term in braces is a geometric sum with $a = (1.00625)^{-229}$, $r = (1.00625)^{-1}$, and n = 12. The sum is \$2,286.57.

5.3.5 and 5.3.6 I suggest you ask your instructor for help with a spreadsheet program.

5.3.7 Jamila's interest rate per month is $i = j_{12}/12 = .084/12 = .007$. Her monthly payment is the solution for P of the equation of value

$$\$121,719.74 = S(P,\ .007,\ 360,\ -1) = P\frac{1-(1.007)^{-360}}{.007}$$

$$= P(131.26156061).$$

The solution is $P = \$927.31$.

a) After her 120th payment Jamila has 240 payments remaining. Her outstanding balance just after her 120th payment is $S(\$927.31, .007, 240, -1) = \$107,638.07$. We find this amount by using the payment value \$927.306817...internally. If we calculate with the rounded value of \$927.31, we get \$107,638.44 as Jamila's outstanding balance.

b) The interest on Jamila's 121st payment is $\$107,638.07 \cdot (.007) = \$753.47$. Therefore, the principal repaid is $\$927.31 - \$753.47 = \$173.84$.

5.3.8 We first find Raza's monthly payment. It is the solution for P in the equation of value

$$\$2,000 = S(P, .01, 6, -1) = P\frac{1-(1.01)^{-6}}{.01} = P(5.7954764746)$$

which yields $P = \$345.096733421$. In the table below, I will retain all possible decimal places internally, but display only two places:

Payment Number	*Payment Amount*	*Interest Paid*	*Principal Repaid*	*Outstanding Principal*
0				\$2,000.00
1	\$345.10	\$20.00	\$325.10	\$1,674.90
2	\$345.10	\$16.75	\$328.35	\$1,346.56
3	\$345.10	\$13.47	\$331.63	\$1,014.92
4	\$345.10	\$10.15	\$334.95	\$679.98
5	\$345.10	\$6.80	\$338.30	\$341.68
6	\$345.10	\$3.42	\$341.68	\$0.00

Amortization Schedule for Exercise 5.3.8

5.3.9 Lori is has been making monthly payments of \$600.13 on her house for the past twelve years. Her original loan was for \$87,500. Her current outstanding balance is \$62,927.85. What nominal annual interest rate compounded monthly is she being charged? *Hint*: Use Guess-and-Check.
From Formula 5.1, we have

$$\$62{,}927.85 = \$87{,}500(1 + i)^{144} - \$600.13\{(1+i)^{144} - 1\}/i,$$

and we must solve this for i. Put $f(i) = \$87{,}500(1 + i)^{144} - \$600.13\{(1+i)^{144} - 1\}/i$ and use Guess-and-Check to make this sufficiently close to \$62,927.85. I got $i = j_{12}/12 \approx 0.005583$, so $j_{12} \approx 0.067 = 6.7\%$.

Exercises 5.4

5.4.1 We can solve this using formula 3.2 with inputs P (to be determined), $i = .06$, $n = 15$, $d = 14$. We are also given that $S(P, .06, 15, 14) = \$40{,}000$. The solution $P = \$1{,}718.51$.

5.4.2 a) The annual interest payment is \$50,000(.11) = \$5,500.
b) The annual sinking fund deposit, P, must satisfy

$$S(P, .08, 10, 9) = \$50{,}000,$$

yielding $P = \$3{,}451.47$.
c) Total annual outlay is \$3,451.47 + \$5,500 = \$8,951.47.
d) If she amortized her loan at 11% annually, her annual outlay would be P', where P' satisfies $S(P', .11, 10, -1) = \$50{,}000$. This value is $P' = \$8{,}940.07$.

5.4.3 Following Table 5.5, here is a sinking fund schedule:

Deposit Number	*Deposit Amount*	*Interest Earned at 8%*	*Increase In Fund*	*Amount Accumulated*
0	-	-	-	0.000
1	\$3,451.47	\$0.00	\$3,451.47	\$3,451.47
2	\$3,451.47	\$276.12	\$3,727.59	\$7,179.07
3	\$3,451.47	\$574.33	\$4,025.80	\$11,204.87
4	\$3,451.47	\$896.39	\$4,347.86	\$15,552.73
5	\$3,451.47	\$1,244.22	\$4,695.69	\$20,248.42
6	\$3,451.47	\$1,619.87	\$5,071.35	\$25,319.77
7	\$3,451.47	\$2,025.58	\$5,477.06	\$30,796.83
8	\$3,451.47	\$2,463.75	\$5,915.22	\$36,712.05
9	\$3,451.47	\$2,936.96	\$6,388.44	\$43,100.49
10	\$3,451.47	\$3,448.04	\$6,899.51	\$50,000.00

5.4.4 We can use Formula 3.2 with inputs $i = j_{12}/12 = .005$, $n = 15\times12 = 180$, and $d = 180 - 1 = 179$ to find P:

$$S(P, .005, 180, 179) = P\frac{1-(1.005)^{-180}}{.005} = \$100{,}000,$$

so $P = \$343.86$.

5.4.5 *Hint:* From $i = i' = j_4/4 = .025$ you should get $P = \$2{,}225.43497437\ldots$ Ask your instructor for help with the spreatsheet program if necessary.

5.4.6 Koji's monthly payment P satisfies the equation of value

$$\$2{,}000 = S(P, .085/12, 6, 5) = P\frac{(1.00708\overline{3})^6 - 1}{.00708\overline{3}}$$

$$= P(6.10725881506),$$

so $P = \$327.479162184$. His interest payments are $(.10/12)\cdot(\$2{,}000) = \$16.67$. Below is a Sinking Fund Schedule showing the details.

Deposit Number	*Interest Payment*	*Deposit in Fund*	*Interest Earned in at 8.5%*	*Increase in Fund*	*Amount Accumulated*
0					$0.00
1	$16.67	$327.48	$0.00	$327.48	$327.48
2	$16.67	$327.48	$2.32	$329.80	$657.28
3	$16.67	$327.48	$4.66	$332.13	$989.41
4	$16.67	$327.48	$7.01	$334.49	$1,323.90
5	$16.67	$327.48	$9.38	$336.86	$1,660.76
6	$16.67	$327.48	$11.76	$339.24	$2,000.00

5.4.7 We have the monthly rate $i = .0999885/12 = .008332375$. Nan is going to have to make n monthly payments, where

$$\$5{,}000 = S(\$293.43, .008332375, n, n-1) =$$
$$\$293.43\frac{(1.008332375)^n - 1}{.008332375}.$$

The solution is $n = \log(1.14198232969)/\log(1.008332375) \approx 16$ months (almost exactly).

5.4.8 By amortization, STNX will pay P monthly, where P satisfies

$$\$2{,}200{,}000 = S(P, .0075, 120, -1)$$

$$= P\frac{1-(1.0075)^{-120}}{.0075} = P(78.9416926689),$$

so that P = \$27,868.67. On the other hand by the sinking fund method, STNX's monthly interest payments would be \$2,200,000(.095/12) = \$17,416.67, and their sinking fund deposits would be \$10,453.76 which is the solution for P' of the equation

$$\$2{,}200{,}000 = S(P', .1047/12, 120, 119).$$

The total outlay for the company by this method would be \$17,416.67 + \$10,453.76 = \$27,870.43, just \$1.76 more per month than by amortization.

Exercises 5.5

5.5.1 With the help of the time diagram below,

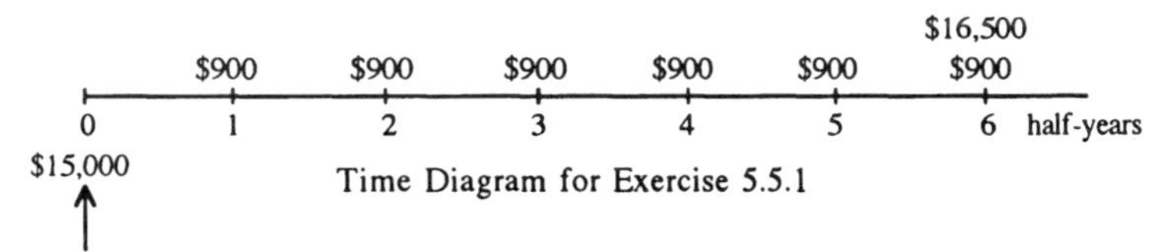

Time Diagram for Exercise 5.5.1

we have the equation of value for Gordy's transaction:

$$\$900\frac{1-(1+i)^{-6}}{i} + \$16{,}500(1+i)^{-6} = \$15{,}000.$$

where $i = j_2/2$.

Solution by Solver: In the EQUATION SOLVER, enter

0 = 900(1 – (1 + I)^-6)/I + 16500(1 + I)^-6 – 15000,

with starting value I = .05. I got I = .07384475289..., so $j_2 = 2i \approx$ 14.76895%.

Solution by Guess-and-Check: We set

$$LHS(i) = \$900\frac{1-(1+i)^{-6}}{i} + \$16{,}500(1+i)^{-6},$$

and make a Guess-and-Check table. After eleven steps, I found *LHS*(.073845) = \$14,999.98. This agrees quite well with the solver method above.

5.5.2 Below is a time diagram for Rose's transaction.

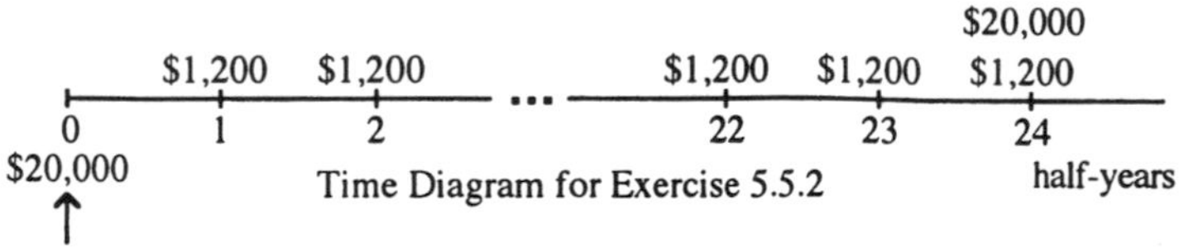

$20,000
$1,200 $1,200 ... $1,200 $1,200 $1,200
0 1 2 22 23 24
$20,000 ↑
half-years

Time Diagram for Exercise 5.5.2

The equation of value for Rose' transaction is

$$\$20{,}000 = \$1{,}200\frac{1-(1+i)^{-24}}{i} + \$20{,}000(1+i)^{-24}$$

Although we could use a Solver or Guess-and-Check, it is not necessary in this case. Because of the equality of the investment and its return, this particular problem can be solved algebraically. The equation of value can be rewritten as follows:

$$\$20{,}000\left(1-(1+i)^{-24}\right) = \$1{,}200\frac{1-(1+i)^{-24}}{i}.$$

Now cancel $\{1 - (1 + i)^{-24}\}$ from both sides and solve for i to get $i = \$1{,}200/\$20{,}000 = .06$. That is, $i = .06$ is the rate per half-year, so that the nominal annual yield compounded semi-annually is $j_2 = 12\%$.

5.5.3 Looking at the time diagram below, we can write the equation of value

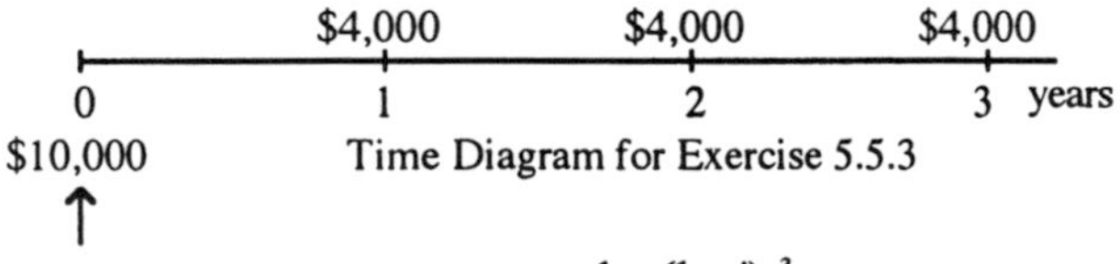

$4,000 $4,000 $4,000
0 1 2 3 years
$10,000 ↑

Time Diagram for Exercise 5.5.3

$$S(\$4{,}000, i, 3, -1) = \$4{,}000\frac{1-(1+i)^{-3}}{i} = \$10{,}000.$$

Solution by Solver: In the EQUATION SOLVER, put

0 = 4000(1 – (1 + I)^-3)/I - 10000

Starting as usual with I = .05, the solution for I = i is I = .09701025740... ≈ 9.701% effective.

Solution by Guess-and-Check: Define

$$LHS(i) = \$4,000\frac{1-(1+i)^{-3}}{i}.$$

In my Guess-and-Check Table, I got $LHS(.097) = \$10,000.18$ in six steps, so $i \approx 9.7\%$ effective.

5.5.4 To realize a yield rate of 10% Imre would have to pay P where P satisfies

$$\$10,000 = S(P, .10, 3, -1) =$$

$$P\frac{1-(1.1)^{-3}}{.1} = P(2.48685199098).$$

Thus, $P = \$4,021.15$.

5.5.5 The equation of value for this exercise is $\$5,000(1 + i)^4 = \$7,320.50$. The solution is $i = .10 = 10\%$. Damien's yield on his money was 10%.

5.5.6 A payment of $\$5,000(1.12)^4 = \$7,867.60$ would make Damien's yield rate 12%.

5.5.7 Let the yield rate per half-year be $i = j_2/2$. Then we can write the equation of value using Formula 4.5a:

$$\$300\frac{1-(1+i)^{-20}}{i} + \$10,000(1 + i)^{-20} = \$9,850.$$

Solution by Solver: Enter

0 = 300(1 – (1 + I)^-20)/I + 10000(1 + I)^-20 - 9850

in the EQUATION EDITOR with starting value I = .05. The solver produced I = .03101774651..., so $j_2 = 2i \approx 6.20355\%$ is the nominal annual yield rate compounded semi-annually.

Solution by Guess-and-Check: Set

$$LHS(i) = \$300\frac{1-(1+i)^{-20}}{i} + \$10,000(1 + i)^{-20}.$$

In my Guess-and-Check table, I got $LHS(.03102) = \$9,849.67$ for an approximate value of $j_2 = 2(.03102) = 6.204\%$.

5.5.8 A purchase price of

$$P = S(\$300, .065/2, 20, -1) + \$10,000(1 + .065/2)^{-20} =$$

$$= \$9,636.52.$$

would yield $j_2 = 6.5\%$.

Miscellaneous Exercises – Chapter 5

5M1. *a)*

Qtr.	*Interest Paid*	*Deposit in Fund*	*Interest Earned*	*Increase In Fund*	*Amount Accumulated*
0	—	—	—	—	$0.00
1	$80	$977.78	$0.00	$977.78	$977.78
2	$80	$977.78	$14.67	$992.45	$1,970.22
3	$80	$977.78	$29.55	$1,007.33	$2,977.56
4	$80	$977.78	$44.66	$1,022.44	$4.000.00

b) Leah's total amount paid by the sinking fund method is 4×($80 + $977.78) = $4,231.12. By amortization at j_4 = 8%, she would pay P per quarter, where P satisfies

$$\$4,000 = P\frac{1-(1.02)^{-4}}{.02}.$$

The solution is P = $1,050.50, so Leah's total outlay would be $4,201.98.

5M2. Victor is going to need to make deposits of P into the sinking fund, where

$$\$10,000 = S(P, .025, 4, 3).$$

This yields P = $2,408.18. The sinking fund schedule is

a)

Deposit No.	*Deposit*	*Interest Earned*	*Increase in Fund*	*Amount Accumulated*
0				$0.00
1	$2,408.18	$0.00	$2,408.18	$2,408.18
2	$2,408.18	$60.20	$2,468.38	$4,876.56
3	$2,408.18	$121.91	$2,530.09	$7,406.65
4	$2,408.18	$185.17	$2,593.35	$10,000.00

b) By the above sinking fund method, Victor's total outlay is 4×($300 + $2,408.18) = $10,832.72. By amortization at j_2 = 6%, Victor's semi-annual payments would be P = $2,690.27 from the equation of value

$$\$10,000 = S(P, .03, 4, -1).$$

Victor's total outlay by amortization would therefore be 4×$2,690.27 = $10,761.08.

5M3. a) From the equation of value

$$\$28{,}000 = S(P, .0325, 16, 15),$$

we arrive at Joshua's semi-annual payments of $P =$ \$1,361.92.

b) Joshua's semi-annual interest payments are .045×\$28,000 = \$1,260, so his total semi-annual outlay is \$1,361.92 + \$1,260 = \$2,621.92.

c) Had Joshua amortized the loan at 8% over 8 years, his semi-annual outlay would be the solution of the equation of value

$$\$28{,}000 = S(P, .045, 16, -1),$$

namely, $P =$ \$2,492.43.

5M4. Tractor A's costs have a present value of

$$S(\$500, .06, 9, -1) + \$99{,}500 = \$102{,}900.85.$$

If Tractor B's price is P_B, then its total costs have a present value of

$$S(\$350, .06, 10, -1) + P_B,$$

and for this to equal the costs for Tractor A, Tractor B's price must be \$102,900.85 - S(\$350, .06, 10, -1) = \$102,900 - \$2,576.03 = \$100,324.82.

5M5. Michael's payments are the solution of the equation

$$S(P, .0725/12, 360, -1) = \$195{,}000,$$

giving $P =$ \$1,330.24. Just after Michael's 120th payment,

$$\begin{aligned} \textit{Seller's Equity} &= S(\$1{,}330.24, .0725/12, 240, -1) \\ &= \$168{,}305.18, \end{aligned}$$

$$\textit{Buyer's Equity} = \$195{,}000 - \textit{Seller's Equity} = \$26{,}694.82.$$

5M6. Per and Ulla's Balloon payment, B, is found from the equation of value

$$\$125{,}000 = S(\$925, .08/12, 299, -1) + B(1 + .08/12)^{-300},$$

which has the solution

$$\begin{aligned} B &= \{\$125{,}000 - S(\$925, .08/12, 299, -1)\}(1 + .08/12)^{300} \\ &= \$38{,}747.58. \end{aligned}$$

5M7. We can use Formula 4.1 with inputs P (first payment to be determined), $i = .072/12 = .006$, $n = 180$, $d = 180 - 1 = 179$:

$$S^{(.002)}(P, .006, 180, 179) =$$

$$P\frac{1-(1.00399201597)^{-180}}{.00399201597}(1.00399201597)(1.006)^{179} =$$

$$P(127.96248532) = \$40{,}000,$$

from which we have P = \$106.50. The last payment is $P(1.002)^{179}$ = \$152.29.

5M8. Fredrik paid a fee of \$108,000(.03) = \$3,240 to the mortgage company. His monthly payments are P = \$819.30 (rounded), which is the solution for P in the equation of value

$$S(P, .078/12, 300, -1) = \$108{,}000$$

In effect, Fredrik has only borrowed \$108,000 - \$3,240 = \$104,760, but is paying \$819.30 per month for 25 years. Therefore, his true interest rate $i = j_{12}/12$ per month satisfies the equation:

$$S(\$819.30, i, 300, -1) =$$

$$\$819.30\frac{1-(1+i)^{-300}}{i} = \$104{,}760,$$

Solution by Solver: The equation to enter into the EQUATION SOLVER is

0 = 819.30(1 – (1 + I)^-300)/I – 104760.

Starting with I = .05, I found I = .00679534661..., which means that $j_{12} = 12i \approx 8.1544\%$ is Fredrik's actual nominal monthly rate of interest.

Solution by Guess-and-Check: Setting

$$LHS(i) = \$819.30\frac{1-(1+i)^{-300}}{i},$$

I found $LHS(.006795) \approx \$104{,}763.71$. On this basis, $j_{12} \approx 12(.006795) = .08154 = 8.154\%$.

Chapter 6

Exercises 6.1

6.1.1 a) Yes. From property *e*), the probability is 1 - .31205 = .68795.

b) No. We do not have enough information to answer this.

6.1.2 The probability is 1 by property *b*).

6.1.3 Since the woman will either die between 70 and 90 or not, the probability of not dying in the interval is 1 – the probability of dying in the interval = 1 - .61313 = .38687 by property *e*).

6.1.4 No. Probabilities must be between 0 and 1 by property *a*).

6.1.5 The horse finishing both second and third is impossible, so the probability is 0 by property *c*).

6.1.6 The event 'at least one birthday match' is complementary to the event 'all have different birthdays,' so the probability is 1 - .2937 = .7063 by property *e*).

Exercises 6.2

6.2.1 The probability that a woman now aged 50 will die at age x is d_x/l_{50}, $x \geq 50$, so the age at which she is most likely to die is the age $x \geq 50$ where d_x is largest. A quick inspection of the table shows that this x is 85. The probability that (50) will die at 85 is $d_{85}/l_{50} = 379{,}033/9{,}219{,}130 \approx .04111$.

6.2.2 The probability that a 35-year-old woman will survive to age 70 and then die within the next two years is $(l_{70} - l_{72})/l_{35} \approx .03540$ using Table II.

6.2.3 The probability that a 20-year-old male will die between the ages of 75 and 85 is $(l_{75} - l_{85})/l_{20} = .32360$ using Table I.

6.2.4 See the Figure below.

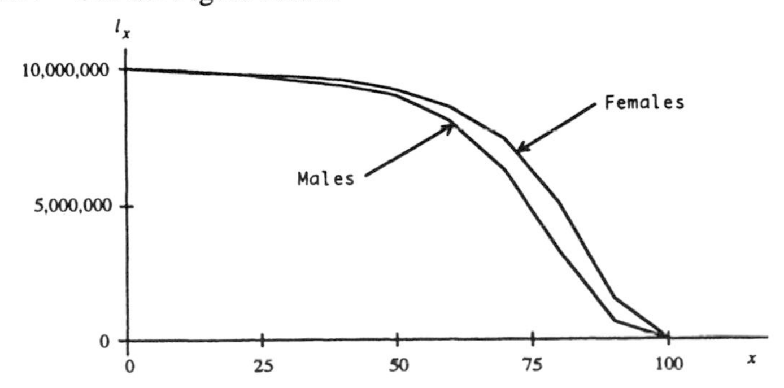

6.2.5 The man dies before age 64 with probability d_{63}/l_{63} = .02106, and survives to age 64 with probability 1 - .02106 = .97894. There is only a 2.106% chance that the insurance company will have to make a payment, so the man should pay \$100,000·(.02106) = \$2,106.

6.2.6. A 10-year-old boy dies at age 10 with probability d_{10}/l_{10} = .00073. A 51-year-old male dies at age 51 with probability 10(.00073) = .0073.

6.2.7. A 10-year-old girl dies at age 10 with probability d_{10}/l_{10} = .00068. A 54-year-old woman dies at age 54 with probability .00661, and a 55-year-old woman dies at that age with probability .00709, and 10(.00068) = .0068 is about 40% of the way from .00661 to .00709, so the age we seek is roughly 54.4.

6.2.8. Using Table I, this is $(l_{70} - l_{80})/l_{28}$ = (6,274,160 - 3,274,541)/9,612,750 = .31205.

6.2.9. For males, we want ${}_{45}q_{35}$ = $(l_{35}$ - $l_{80})/l_{35}$ = (9,491,711 - 3,274,541)/9,491,711 = 0.65501

6.2.10. For females, we want ${}_{40}p_{40}$ = l_{40}/l_{40} = 5,056,025/9,545,345 = 0.52968.

6.2.11. For males, we want $(l_{60}$ - $l_{70})/l_{20}$ = (8,084,266 - 6,274,160)/ 9,754,159 = 0.18557. I would expect the probability to be lower for females. 20-year-old females should be less likely to die as early as between 60 and 70 than males.

6.2.12. For females, we want $(l_{60}$ - $l_{70})/l_{20}$ = (8,603,801 - 7,448,816)/9,820,821 = 0.11761. This *is* lower than the corresponding probability for males.

Exercises 6.3

6.3.1 Since ${}_{t|}q_{40} = d_{40+t}/l_{40}$, the sum equals $(d_{40} + d_{41} + ... + d_{99})/l_{40}$. Using Example 6.3, the last numerator is l_{40}, so that the last ratio is 1. In words, a 40-year-old is certain to die at one of the ages 40, 41, 42, ..., or 99.

6.3.2 $_{40|}q_0$ is the probability that a newborn will die at 40 whereas $_{0|}q_{40}$ is the probability that a 40-year-old will die at 40. For males,

$$_{40|}q_0 = .00283 < .00302 = {}_{0|}q_{40}.$$

For females,

$$_{40|}q_0 = .00231 < .00242 = {}_{0|}q_{40}.$$

6.3.3 a) The probability that (80) dies at age x is d_x / l_{80}, $x \geq 80$. d_x is highest when $x = 85$. Thus, age 85 is the most likely age at death for a woman aged 80.

b) The probability that (85) dies at age x is d_x / l_{85}, $x \geq 85$. d_x is highest when $x = 85$. Thus, age 85 is also the most likely age at death for a woman aged 85.

c) The two probabilities are not equal but they give the most likely ages at death for the two ages. The fact that a) and b) have the same answer owes to the fact that d_x peaks at $x = 85$.

6.3.4 A 40-year-old is not likely to die at any particular age, but 78 is the most likely of all ages. Since the probability of his dying at 78 is so low, you might rather say that 78 is his least unlikely age at death.

6.3.5 No. We know that $_{a|}q_x = \dfrac{d_{x+a}}{l_x}$ and $_{b|}q_x = \dfrac{d_{x+b}}{l_x}$. Just because we know that $a < b$ does not tell us which is the larger of $_{a|}q_x$ and $_{b|}q_x$. This is because the d_x columns of Tables I & II neither increase nor decrease steadily. They fluctuate.

6.3.6 The probability distribution of ages at death of 95-year-old women is composed of the numbers $\{_{0|}q_{95}, {}_{1|}q_{95}, {}_{2|}q_{95}, {}_{3|}q_{95}, {}_{4|}q_{95}\}$, where $_{t|}q_{95} = d_{95+t}/l_{95}$. From Table II, we find {0.31732, 0.25651, 0.20242, 0.14675, 0.07700}. Note that these five numbers sum to 1, as they should.

Exercises 6.4

6.4.1 From Table I for males, *AAD*(71) is 81.39.

6.4.2 From Table II for females, *AAD*(76) is 85.71.

6.4.3 I'll show Formula 6.5. *AAD*(91) = 91 + .5 + (l_{92} + l_{93} + ... + l_{99})/l_{91} = 91.5 + 1,225,875/502,572 = 93.9392027411 ≈ 93.94.

6.4.4 We are looking for the age at which half the original cohort has died and half survived. Table II shows that l_{80} = 5,056,025 and l_{81} = 4,722,378, so the median is somewhere between 80 and 81, but closer to 80. In fact, 5,000,000 is about 17% of the way from 80 to 81 because (5,056,025 – 5,000,000)/(5,056,025 – 4,722,378) = 56,025/333,647 ≈ .168 = 16.8%.

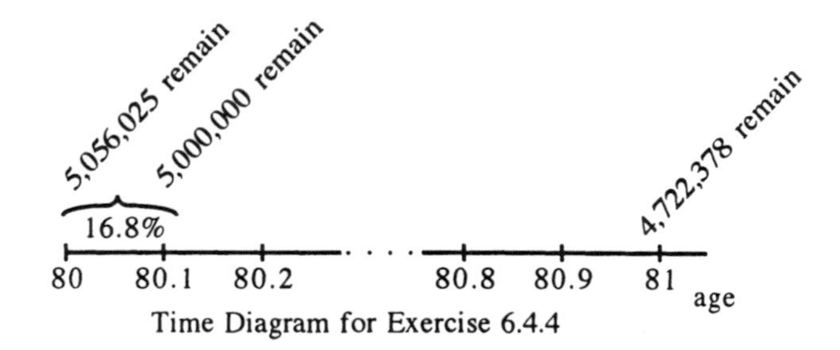

Time Diagram for Exercise 6.4.4

6.4.5 Since l_{69} = 7,603,626 and l_{70} = 7,448,816, Q_1 is between 69 and 70. By the same reasoning as in the preceding problem, Q_1 is about 67% of the way from 69 to 70, so Q_1 = 69.67. Similarly, since l_{87} = 2,512,591 and l_{88} = 2,152.486, and since 2,500,000 is about 3% of the way from 87 to 88, Q_3 = 87.03.

6.4.6 Formula 6.4 is:

$$AAD(x) = (x + .5)(_{0|}q_x) + (x + 1.5)(_{1|}q_x) +$$

$$(x + 2.5)(_{2|}q_x) + \ldots + (99.5)(_{99-x|}q_x).$$

This can first be written as

$$x \cdot {}_{0|}q_x + (x + 1)(_{1|}q_x) + (x + 2)(_{2|}q_x) + \ldots + (99)\,(_{99-x|}q_x) +$$

$$(.5)(_{0|}q_x + {}_{1|}q_x + {}_{2|}q_x + \ldots + {}_{99-x|}q_x) =$$

$$.5 + x \cdot {}_{0|}q_x + (x + 1)(_{1|}q_x) + (x + 2)(_{2|}q_x) + \ldots + (99)(_{99-x|}q_x),$$

because, as we have seen before, ${}_{0|}q_x + {}_{1|}q_x + {}_{2|}q_x + \ldots + {}_{99-x|}q_x = 1$.

Now ${}_{t|}q_x = \dfrac{d_{x+t}}{l_x}$, so we rewrite once again

$$AAD(x) = .5 + \frac{x \cdot d_x + (x+1) \cdot d_{x+1} + (x+2) \cdot d_{x+2} + \ldots + 99 \cdot d_{99}}{l_x}.$$

Using the facts that

a) $d_y = l_y - l_{y+1}$

and

b) $l_y = d_y + d_{y+1} + d_{y+2} + \ldots + d_{99}$,

we are going to rewrite

$$x \cdot d_x + (x+1) \cdot d_{x+1} + (x+2) \cdot d_{x+2} + \ldots + 99 \cdot d_{99}.$$

We get

$$\begin{aligned}
x \cdot d_x + (x+1) \cdot d_{x+1} + (x+2) \cdot d_{x+2} + \ldots + 99 \cdot d_{99} &= \\
x \cdot (d_x + d_{x+1} + d_{x+2} + \ldots + d_{99}) & \\
+ d_{x+1} + 2d_{x+2} + 3d_{x+3} + \ldots + (99-x)d_{99} &= \\
x \cdot l_x + d_{x+1} + d_{x+2} + d_{x+3} + \ldots + d_{99} &+ \\
d_{x+2} + d_{x+3} + \ldots + d_{99} &+ \\
+ d_{x+3} + \ldots + d_{99} &+ \\
\ldots & \\
d_{98} + d_{99} &+ \\
+ d_{99} &
\end{aligned}$$

$$= x \cdot l_x + l_{x+1} + l_{x+2} + l_{x+3} + \ldots l_{99}.$$

Finally we can write

$$AAD(x) = .5 + \frac{x \cdot l_x + l_{x+1} + l_{x+2} + \ldots + l_{99}}{l_x} =$$

$$x+.5+\frac{l_{x+1}+l_{x+2}+\ldots+l_{99}}{l_x},$$

which is Formula 6.5.

6.4.7 $AAD(21) = 77.10$ for females and 72.47 for males. The difference is 4.63 years.

6.4.8 A search of Tables I and II shows that the male partner should be around 25. $AAD(30) = 77.65$ for females while $AAD(25) = 72.84$ for males, a difference of 4.81 years. Thus, a 25-year-old male and a 30-year-old female have about the same average remaining life. Of course, this just an average. In reality there would be very large variations from the averages.

6.4.9 We can use Formula 6.4 or 6.5. If you did Exercise 6.3.6, you calculated the probability distribution for the ages at death. These are {0.31732, 0.25651, 0.20242, 0.14675, 0.07700}. Formula 6.4 gives

$$AAD(95) = (95.5)(0.31732) + (96.5)(0.25651) +$$

$$(97.5)(0.20242) + (98.5)(0.14675) + (99.5)(0.07700)$$

$$\approx 96.91.$$

Using Formula 6.5, we get

$$AAD(95) = 95.5 + \frac{l_{96}+l_{97}+l_{98}+l_{99}}{l_{95}} = 95.5 + \frac{561{,}947}{398{,}655}$$

$$\approx 96.91,$$

the same as before. Note that this is also the value given in Table II.

Exercises 6.5

6.5.1 Possible causes for higher mortality in males between ages 11 and 28 are: higher risk of death during military service, reckless behavior such as driving too fast, drinking and driving, riding motorcycles, and similarly open vehicles without proper protective gear, and general feeling of immortality. Maybe you can think of some more, like homicide and suicide.

6.5.2 Using the approximation $\mu(x) \approx q_x$, we see from Table I that this is lowest at $x = 10$. That is the lowest q_x is $q_{10} = .00073$.

6.5.3 Using the approximation $\mu(x) \approx q_x$, we see from Table II that this is lowest at $x = 10$. That is the lowest q_x is $q_{10} = .00068$.

6.5.4 Formula 6.6 says that $\mu(x) \approx \dfrac{l_x - l_{x+t}}{t \cdot l_x}$ for small t. Note that in calculating the approximation, we can ignore the factor of 10,000,000 because it cancels out. This means that Formula 6.6 becomes

$$\mu(x) \approx \frac{(.933)^x - (.933)^{x+t}}{t \cdot (.933)^x} = \frac{1-(.933)^t}{t}.$$

The right-hand side does not depend on the age x. We will get the same force of mortality for any age. If we evaluate this for $t = .1, .01$, and $.001$ in succession we get

$$\mu(x) \approx .06911 \text{ when } t = .1$$
$$\mu(x) \approx .06933 \text{ when } t = .01$$
$$\mu(x) \approx .06935 \text{ when } t = .001$$

Remark: The fact that the force of mortality in the above problem is the same for any age stems from the fact that the survival function is an exponential function—a function of the form $l_x = l_0 \cdot c^x$, where c is a constant between 0 and 1. An interesting consequence of having such a survival function is that the beings in the population it models do not age. This is not to say that they do not die. They just do not die from aging. While human populations do not have constant forces of mortality, certain electronic components have been found to have nearly constant forces of mortality.

6.5.5 When $t = 1$, $\dfrac{1 - {}_tp_x}{t} = 1 - {}_1p_x = {}_1q_x = \dfrac{d_x}{l_x} = q_x$ as claimed.

Miscellaneous Exercises – Chapter 6

6M1. The missing entries for l_x and d_x are found from the relationship $l_x - d_x = l_{x+1}$. Then $q_x = d_x/l_x$ and I would use Formula 6.5 to find $AAD(x)$.

x	l_x	d_x	q_x	$AAD(x)$
0	1,000	**600**	**0.6000**	**1.132**
1	**400**	255	**0.6375**	**2.080**
2	145	**87**	0.6000	**3.100**
3	**58**	**29**	0.5000	**4.000**
4	**29**	**29**	1.0000	**4.500**

For example $AAD(2) = 2 + .5 + \dfrac{l_3 + l_4}{l_2} = 2.5 + \dfrac{58 + 29}{145} = 3.10.$

6M2. Below are the results in a table:

x	l_x	d_x	q_x	$AAD(x)$
0	10,000	2,000	0.20	2.0104
1	8,000	3,200	0.40	2.3880
2	4,800	2,880	0.60	2.9800
3	1,920	1,536	0.80	3.7000
4	384	384	1.00	4.5000

6M3. On planet Q0317B, 66.5153846154; on Q0317C, 77.3159113673

6M4. $n = .47722181601$.

Chapter 7

Exercises 7.2

7.2.1 a) In the time diagram below, I've put the arrow at age 60,

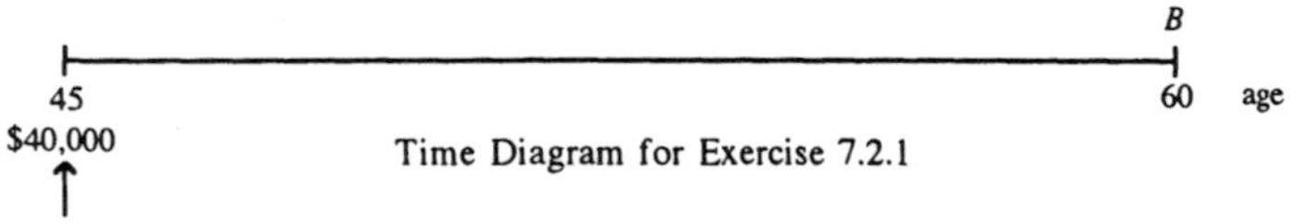

Time Diagram for Exercise 7.2.1

but I could just as well have put it at age 45 or anywhere else for that matter. The equation of value is

$$\$40{,}000 = B \cdot {}_{15}E_{45} = B \cdot \frac{D_{60}}{D_{45}} =$$

$$B \cdot \frac{613{,}356.44}{1{,}298{,}140.38} = B(.47248853009),$$

with $i = 4.5\%$, from which we arrive at $B = \$84{,}658.14$.

b) Here we change the interest rate to 6% and get $B =$

$104,836.46.

7.2.2 a) This calls for Formula 1.2:

$$P = \$50{,}000(1.0375)^{-25} = \$19{,}918.99.$$

b) This is the present value of a 25-year pure endowment of $50,000 to a man aged 20. Referring to Formula 7.2, $P =$

$$\$50{,}000\ {}_{25}E_{20} = \$50{,}000\frac{D_{45}}{D_{20}} = \$50{,}000\frac{1{,}757{,}148.70}{4{,}671{,}192.05}$$

$= \$18{,}808.35$.

7.2.3 a) Johanna will receive a 30-year pure endowment of B, where

$$B\cdot {}_{30}E_{30} = B\frac{D_{60}}{D_{30}} = \$100{,}000,$$

so

$$B = \$100{,}000\frac{D_{30}}{D_{60}} = \$603{,}715.54.$$

b) If Johanna or her estate were sure to receive the payment, the equation of value would be

$$B(1.0575)^{-30} = \$100{,}000,$$

or $B = \$100{,}000(1.0575)^{30} = \$535{,}070.84$

7.2.4 From the preceding problem, we know that $B = \$100{,}000\frac{D_{30}}{D_{60}}$,

so

$$B = \$100{,}000\frac{D_{30}}{D_{60}} = \$100{,}000\cdot\frac{(1.0575)^{-30}10^7(1-\frac{30}{105})}{(1.0575)^{-60}10^7(1-\frac{60}{105})}$$

$= \$891{,}784.73$.

7.2.5 From the preceding two problems, we know that $B = \$100{,}000\frac{D_{30}}{D_{60}}$, so

$$B = \$100{,}000\frac{D_{30}}{D_{60}} = \$100{,}000\cdot\frac{(1.0575)^{-30}10^7\sqrt{1-\frac{30}{105}}}{(1.0575)^{-60}10^7\sqrt{1-\frac{60}{105}}}$$

$= \$690{,}773.48.$

7.2.6 Rong has bought a pure endowment of B with a net single premium of \$50,000. We are looking for B assuming $i = j_1 =$ 4.5%. The equation of value is

$$\$50{,}000 = B \cdot {}_{25}E_{35} = B\frac{D_{60}}{D_{35}}$$

$$= B\frac{613{,}356.44}{2{,}064{,}796.84} = B(.29705413536),$$

so $B = \$168{,}319.49$.

7.2.7 For most ages, it is the case that x-year-old females have a higher probability of surviving for another t years than x-year-old males do. You can observe this by making some calculations in the tables, but we cannot verify this mathematically. Put this information into the annuity formulas and you see that women would pay more than men for the same annuity. This would not be the case if their mortality experiences were combined into a single table.

7.2.8 Each entry in the Table will be of the form $\$100{,}000\frac{D_{x+20}}{D_x}$.

$j_1 \backslash x$	25	45	65
3%	\$52,773.57	\$44,062.67	\$13,162.37
5%	\$35,923.20	\$29,993.65	\$8,959.68
7%	\$24,631.19	\$20,565.52	\$6,143.32

7.2.9 Again, each entry in the Table will be of the form $\$100{,}000\frac{D_{x+20}}{D_x}$, but now we go to Table IV.

$j_1 \backslash x$	25	45	65
3%	\$53,337.82	\$47,863.68	\$22,222.62
5%	\$36,307.26	\$32,581.02	\$15,127.04
7%	\$24,894.53	\$22,339.58	\$10,372.04

Exercises 7.3

7.3.1 The equation of value is

$$\$90{,}000 = P\frac{N_{50}}{D_{29}},$$

from which we find

$$P = \$90{,}000\frac{D_{29}}{N_{50}} = \$25{,}115.31.$$

7.3.2 When Naomi is 55, her pension fund is worth

$$\$43{,}600\frac{N_{55}}{D_{55}} = \$636{,}416.71.$$

7.3.3 For this problem, we calculate $\$50{,}000\dfrac{N_{55}(i)}{N_{32}(i)-N_{55}(i)}$ for i = .02, .03, …, .08. We have already done the calculations for i = .06, .07 and .08. I have done the others and put them all in the table below:

i	0.02	0.03	0.04	0.05	0.06	0.07	0.08
P	$27,266.14	$21,579.54	$17,157.49	$13,698.39	$10,977.92	$8,827.87	$7,121.05

Below is a graph of the annual payments versus the interest rate.

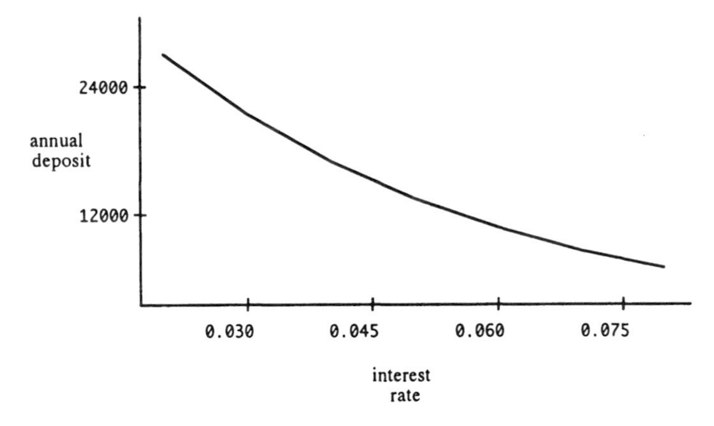

7.3.4 The equation of value for Sven's transaction is

$$\$250{,}000 = P\frac{N_{73}}{D_{72}} = P(7.17267361455),$$

so $P = \$34{,}854.51$.

7.3.5 For $35 \le x \le 65$ the x-th term of the forborne annuity, $\$5{,}000\frac{N_{35} - N_{65}}{D_{65}}$ is

$$\$5{,}000\frac{D_x}{D_{65}} = \$5{,}000(1.06)^{65-x}\frac{l_x}{l_{65}},$$

and each of these terms either exceed or equal $\$5{,}000(1.06)^{65-x}$ since the ratio $\frac{l_x}{l_{65}}$ exceeds or equals 1. On the other hand, for $35 \le x \le 65$, $\$5{,}000(1.06)^{65-x}$ is the x-th term of the 30-year annuity certain of \$5,000 at 6%. This means that term by term the forborne annuity exceeds the annuity certain, so the sum of the terms of the forborne annuity exceed the sum of the terms of the annuity certain.

7.3.6 The time diagram below describes Elena's annuity. Her net

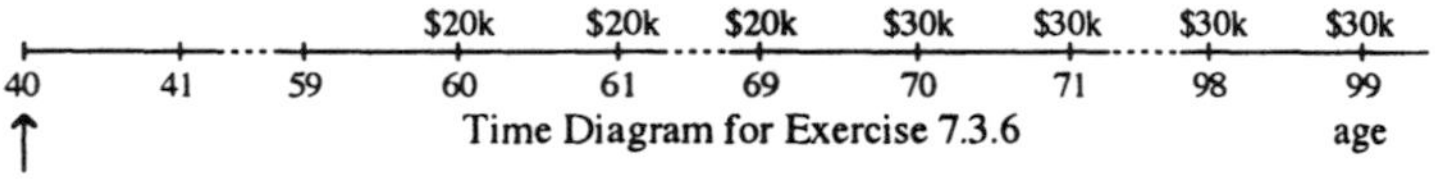

			\$20k	\$20k	\$20k	\$30k	\$30k	\$30k	\$30k	
40	41	59	60	61	69	70	71	98	99	age
↑										

Time Diagram for Exercise 7.3.6

single premium is

$$NSP = \$20{,}000\frac{N_{60} - N_{70}}{D_{40}} + \$30{,}000\frac{N_{70}}{D_{40}} =$$

$$\$10{,}000\frac{2N_{60} + N_{70}}{D_{40}} = \$79{,}644.10.$$

7.3.7 I have added 20 payments of P to the time diagram of the preceding problem.

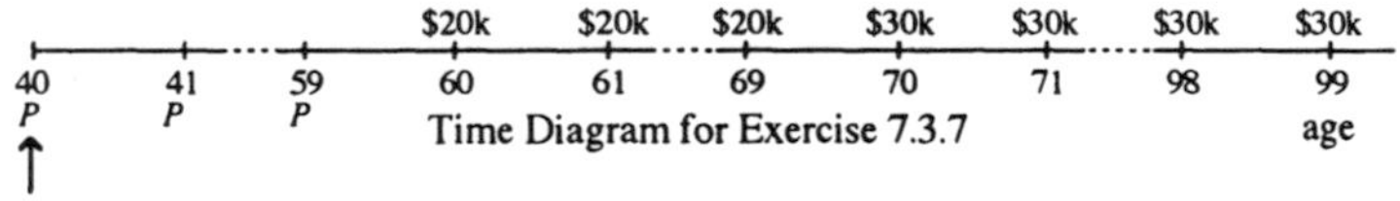

			\$20k	\$20k	\$20k	\$30k	\$30k	\$30k	\$30k	
40	41	59	60	61	69	70	71	98	99	age
P	P	P								
↑										

Time Diagram for Exercise 7.3.7

The equation of value is

$$P\frac{N_{40} - N_{60}}{D_{40}} = \$10{,}000\frac{2N_{60} + N_{70}}{D_{40}},$$

and solving this for P gives

$$P = \$10{,}000\,\frac{2N_{60} + N_{70}}{N_{40} - N_{60}} = \$6{,}740.18.$$

You may have noticed that we could have used the numerical result from the preceding problem to solve this one. The reason that I went through the algebra is to make the next problem easier.

7.3.8 From Exercises 7.3.6 and 7.3.7 we have the formulas:

$$NSP = \$10{,}000\,\frac{2N_{60} + N_{70}}{D_{40}},$$

and

$$NAP = P = \$10{,}000\,\frac{2N_{60} + N_{70}}{N_{40} - N_{60}}$$

Using Table IV with $i = 3\%$, we get $NSP = \$192{,}072.41$ and $NAP = \$12{,}952.04$.

7.3.9 The two are identical except for the first payment of \$50,000 immediately. It is not necessary to calculate the values of the two annuities. Thus, the difference between the two is \$50,000.

7.3.10 We use Formula 7.4 twice in this problem. We see that the value of Rodica's contributions to her pension have present value

$$P\frac{N_{31} - N_{65}}{D_{31}} = P\,\frac{39{,}046{,}202.41 - 3{,}919{,}749.18}{2{,}136{,}271.62}$$

$$= P(16.4428778163).$$

The present value of the pension of \$80,000 annually beginning at age 65 is

$$\$80{,}000\,\frac{N_{65}}{D_{31}} = \$80{,}000\,\frac{3{,}919{,}749.18}{2{,}136{,}271.61} = \$146{,}788.42.$$

Setting the two present values equal, we get P = \$146,788.42/16.4428778163 = \$8,927.17.

7.3.11 With the help of the time diagram

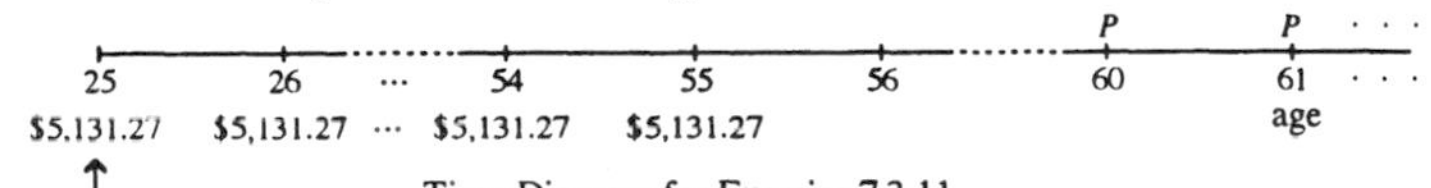

Time Diagram for Exercise 7.3.11

we can write an equation of value

$$\$5{,}131.27\frac{N_{25}-N_{56}}{D_{25}}=P\frac{N_{60}}{D_{25}}.$$

The solution, using Table III, with $i = .055$ is

$$P=\$5{,}131.27\frac{N_{25}-N_{56}}{N_{60}}=\$54{,}836.05.$$

7.3.12 The time diagram helps us write the equation of value

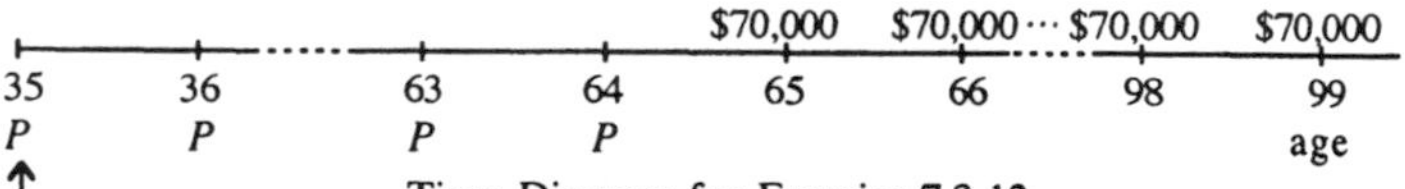

$70,000 $70,000 ··· $70,000 $70,000
35 36 63 64 65 66 98 99
P P P P age
↑

Time Diagram for Exercise 7.3.12

$$P\frac{N_{35}-N_{65}}{D_{35}}=\$70{,}000\frac{N_{65}}{D_{35}},$$

and using Table IV with $i = 4.75\%$ we get

$$P=\$70{,}000\frac{N_{65}}{N_{35}-N_{65}}=\$10{,}767.30.$$

7.3.13 The net single premium is

$$NSP=\$30{,}000\,\frac{N_{35}(i)}{D_{35}(i)},$$

for $i = .04$, .05 and .06. Using Table III, the respective solutions are $587,477.36, $514,357.55 and $456,061.59.

7.3.14 The net single premiums should be higher for women since they are expected to live longer. They receive payments longer, on average. As in the preceding problem, the net single premium is written

$$NSP=\$30{,}000\,\frac{N_{35}(i)}{D_{35}(i)},$$

this time using Table IV for $i = .04$, .05 and .06. The respective solutions are $615,488.24, $534,172.23 and $470,307.71

7.3.15 Use Formula 7.4.

Exercises 7.4

7.4.1 Use Formula 7.5 with inputs

$P = \$43{,}000$,
$x = a = c = 60$,
$i = .0675$,
$k = .0338983$,
$i^* = (i - k)/(1 + k) = .0325$,
gender = *female*,
and $b = 100$.

We calculate

$$\hat{S}^{(.0338...)} = \$43{,}000 \frac{N_{60}(.0325)}{D_{60}(.0325)} \frac{D_{60}(.0675)}{D_{60}(.0675)} = \$655{,}057.20.$$

7.4.2 Use Formula 7.5 with inputs

$P = \$10{,}000$,
$x = 30$,
$i = .06$,
$k = .0291262$,
$i^* = (i - k)/(1 + k) = .03$,
gender = *male*,
$a = 40$,
$b = 60$,
and $c = 30$.

Manu's net single premium is

$$NSP = \hat{S}^{(.0291...)} = \$10{,}000 \frac{N_{40}(.03) - N_{60}(.03)}{D_{40}(.03)} \frac{D_{40}(.06)}{D_{30}(.06)}.$$

$$= \$80{,}203.72.$$

7.4.3 We are given that $i^* = 0$ and $a = c$. When we substitute these into Formula 7.5, we get

$$\hat{S}^{(k)} = P \frac{N_a(0) - N_b(0)}{D_a(0)}.$$

Using the definitions of D and N with $i^* = 0$, we next write

$$\hat{S}^{(k)} = P\frac{l_a + l_{a+1} + \ldots + l_{b-1}}{l_a} =$$

$$P\frac{(l_a + l_{a+1} + \ldots + l_{99}) - (l_b + l_{b+1} + \ldots + l_{99})}{l_a} =$$

$$P\left\{\frac{l_a + l_{a+1} + \ldots + l_{99}}{l_a} - \frac{l_b}{l_a}\left(\frac{l_b + l_{b+1} + \ldots + l_{99}}{l_b}\right)\right\} =$$

$$P\left\{\left(a + \frac{1}{2} + \frac{l_{a+1} + \ldots + l_{99}}{l_a} + \frac{1}{2} - a\right) - \frac{l_b}{l_a}\left(b + \frac{1}{2} + \frac{l_{b+1} + \ldots + l_{99}}{l_b} + \frac{1}{2} - b\right)\right\}$$

$$= P\left\{\left(AAD(a) + \frac{1}{2} - a\right) - \frac{l_b}{l_a}\left(AAD(b) + \frac{1}{2} - b\right)\right\}.$$

This last expression confirms Formula 7.6.

7.4.4 We will use Formula 7.5 with inputs

$P = \$6{,}400$,
$x = 23$,
$i = .055$,
$k = .0343137255$,
$i^* = (i - k)/(1 + k) = .02$,
gender = *male*,
$a = 24$,
$b = 66$,
and $c = 65$.

The result is

$$\hat{S}^{(.03431\ldots)} = \$6{,}400\,\frac{N_{24}(.02) - N_{66}(.02)}{D_{24}(.02)}\,\frac{D_{24}(.055)}{D_{65}(.055)}$$

$$= \$2{,}061{,}709.12.$$

7.4.5 Here we have $i^* = 0$, $a = c = 22$, $b = 66$, and $B = \$32{,}000$. From Formula 7.6 we calculate

$$\hat{S}^{(03)} = \$32{,}000\left\{\frac{1}{2} + AAD(22) - 22 - \frac{l_{66}}{l_{22}}\left(\frac{1}{2} + AAD(66) - 66\right)\right\}$$

$$= \$1{,}310{,}000$$

to the nearest $10,000.

7.4.6 We need to add the expected retirement income to what we found in the preceding exercise. Now we have $i^* = 0$, $a = 66$, $b = 100$, $c = 22$, and $B = \$32{,}000(1.03)^{42}(.75) = \$83{,}056.70$. From Formula 7.6 and the remark following it, we have the expected retirement income of

$$\hat{S}^{(.03)} = \$83{,}056.70\left(\frac{1}{2} + AAD(66) - 66\right)\left(\frac{D_{66}(.03)}{D_{22}(.03)}\right)$$

$$= \$231{,}000.$$

Note that the second term in Formula 7.6 vanishes because $l_{100} = 0$. Combining the last result with the result of Exercise 7.5.5, the widow should be trying to recover $1,541,000.

7.4.7 Magnus's retirement income forms an increasing life annuity. The inputs to Formula 7.5 are $P = \$80{,}000$, $x = 32$, $a = 60$, $b = 100$, $c = 32$, $i = .065$, $k = .03398058$, $i^* = .03$, and $g = male$. The value of Magnus's retirement income when he is 32 is

$$\hat{S}^{(.03398\ldots)} = \$80{,}000\,\frac{N_{60}(.03)}{D_{60}(.03)}\,\frac{D_{60}(.065)}{D_{32}(.065)} =$$

$$\$80{,}000\,\frac{18{,}504{,}969.74}{1{,}372{,}167.45}\cdot\frac{184{,}783.90}{1{,}272{,}483.82} = \$156{,}669.02.$$

7.4.8 Csilla will contribute to an increasing life annuity with inputs $x = 27$, P is to be determined, $g = female$, $k = .033816425$, $i = .07$, $i^* = .035$, $a = 28$, $b = 65$, and $c = 65$. In addition, we are given that $\hat{S}^{(.033816\ldots)} = \$1{,}000{,}000$. The equation of value is

$$\$1{,}000{,}000 = P\,\frac{N_{28}(.035) - N_{65}(.035)}{D_{28}(.035)}\,\frac{D_{28}(.07)}{D_{65}(.07)}$$

$$= P(300.159432802),$$

so $P = \$3{,}331.56$. Csilla's last payment at age 64 will be $P(1.0338\ldots)^{36} = \$11{,}030.91$.

7.4.9 The inputs for Formula 7.5 are $x = 50$, $P = \$50{,}000$, g = female, $k = 3\%$, $i = 5.5\%$, $i^* = (.055 - .03)/(1.03) = 0.02427184466$, $a = 55$, $b = 99$, and $c = 50$. We need to evaluate

$$\hat{S}^{(03)} = \$50{,}000\,\frac{N_{55}(0.02427...)}{D_{55}(0.02427...)}\,\frac{D_{55}(.055)}{D_{50}(.055)}.$$

We do not have a table for $i = .02427...$ The idea here is for you to generate a spreadsheet to find $N_{55}(0.02427...)$ and $D_{55}(0.02427...)$. I will not reproduce that spreadsheet here, but the result we sought is $\hat{S}^{(03)}$ = \$50,000·(18.7964323563)·(.74340248857) = \$698,665.73.

Miscellaneous Exercises – Chapter 7

7M1. a) The present value of the first 15 payments to Arvind is, from Formula 3.2,

$$S(\$75{,}000, .0225, 15, -32) = \$474{,}556.6983.$$

The contingent part of Arvind's annuity has present value

$$\hat{S} = \$75{,}000\,\frac{N_{75}}{D_{28}} = \$104{,}934.55655.$$

The entire premium is \$474,556.6983 + \$104,934.55655 = \$579,491.25.

b) Here, the present value of the first 15 payments to Arvind is, from Formula 3.2,

$$S(\$75{,}000, .0775, 15, -32) = \$64{,}451.1273.$$

The contingent part of Arvind's annuity has present value

$$\hat{S} = \$75{,}000\,\frac{N_{75}}{D_{28}} = \$7{,}003.6656.$$

The entire premium is \$64,451.1273 + \$7,003.6656 = \$71,454.79.

c) Arvind pays less in b) because his premium grows at a faster rate.

7M2. Let B denote the size of Nelly's pure endowment at age 65. Then its present value is

$$B \cdot {}_{40}E_{25} = B\frac{D_{65}}{D_{25}}.$$

The present value of Nelly's \$1,000 payments is

$$\widehat{S} = \$1{,}000\frac{N_{25} - N_{65}}{D_{25}}.$$

Setting these two equal and solving for B, and referring to Table IV with $i = 6.75\%$, we get

$$B = \$1{,}000\frac{N_{25} - N_{65}}{D_{65}} = \$234{,}255.99.$$

7M3. If all payments were certain, Nelly's 'endowment-certain' would be

$$S(\$1{,}000, .0675, 40, 40) = \$199{,}850.08.$$

7M4. We can use Formula 7.5 with inputs

$$P = \$35{,}000,$$

$x = 35$,
$i = .06$,
$k = .0291262$,
$i^* = (i - k)/(1 + k) = .03$,
gender = *male*,
$a = 35$,
$b = 100$,
and $c = 35$.

Our estimate is

$$\widehat{S}^{(.0291\ldots)} = \$35{,}000\frac{N_{35}(.03)}{D_{35}(.03)} = \$35{,}000\frac{76{,}529{,}520.62}{3{,}373{,}196.51}$$

$$= \$794{,}063.79.$$

7M5. The net single premium on all three planets can be written

$$\widehat{S} = \$10{,}000\frac{N_{35}}{D_{35}}.$$

For the Earth-woman, use Table IV with $i = 5\%$ to find $\widehat{S} =$ \$178,057.41. For the other women, we have to take apart $\widehat{S}$:

$$\widehat{S} = \$10{,}000\frac{N_{35}}{D_{35}} = \$10{,}000\frac{D_{35} + D_{35} + D_{35} + \ldots + D_{99}}{D_{35}}.$$

Next we note that

$$D_x = (1.05)^{-x} l_x.$$

Now we can evaluate $\widehat{S}$ using a spreadsheet or other program. When I did this, I got \$148,095.00 for the Q0317B-woman and \$169,532.61 for the Q0317C-woman.

Chapter 8

Exercises 8.2

8.2.1 The formula for the net single premium is

$$I = \$200{,}000\frac{M_{35}(i)}{D_{35}(i)},$$

for $i = .04$, .05 and .06. From Table III we find a) \$49,364.78, b) \$36,711.89, and c)\$27,901.29.

8.2.2 As in the preceding problem, the formula for the net single premium is

$$I = \$200{,}000\frac{M_{35}(i)}{D_{35}(i)},$$

for $i = .04$, .05 and .06, but this time we use Table IV. We find a) \$42,182.50, b) \$30,421.51, and c) \$22,525.39.

8.2.3 In general the net single premiums for life annuities decrease as the interest rate increases. Also, women pay more than men for the same level of annuity because they are expected to receive more payments.

The net single premium for a whole-life insurance policy decreases as the interest rate increases. Women pay less than men for the same level of insurance because the benefit will likely be paid later than for men of the same age at purchase.

8.2.4 We have

$$M_{96} = 37{,}805(1.02375)^{-97} + 29{,}054(1.02375)^{-98}$$

$$+ 20,693(1.02375)^{-99} + 10,757(1.02375)^{-100} = 9,845.91216486\ldots$$

8.2.5 Using Table III we get the following:

		Interest Rate (i)		
		3%	**5%**	**7%**
Age (x)	**30**	\$29,832.26	\$15,045.14	\$8,403.63
	50	\$48,971.53	\$32,524.11	\$22,704.69
	70	\$72,602.97	\$60,078.65	\$50,565.89

Net Single Premiums for Males for Exercise 8.2.5

8.2.6 Dmitrij should pay the net single premium

$$I = 4,365,000 \frac{M_{37}}{D_{37}} \text{ RUR} = 867,706.24 \text{ RUR}.$$

8.2.7 Dmitrij should pay the net annual premium

$$NAP = 4,365,000 \frac{M_{37}}{N_{37}} \text{ RUR} = 51,571.00 \text{ RUR}.$$

8.2.8 Given the lower mortality of women, they tend to pay less than men for the same level of life insurance because the insurance company pays benefits at a later age to women than to men, on average.

Exercises 8.3

8.3.1 Linda will pay

$$I = \$150,000 \frac{M_{36}(.05) - M_{65}(.05)}{D_{36}(.05)} = \$9,794.15.$$

8.3.2 The time diagram below shows the elements of the problem.

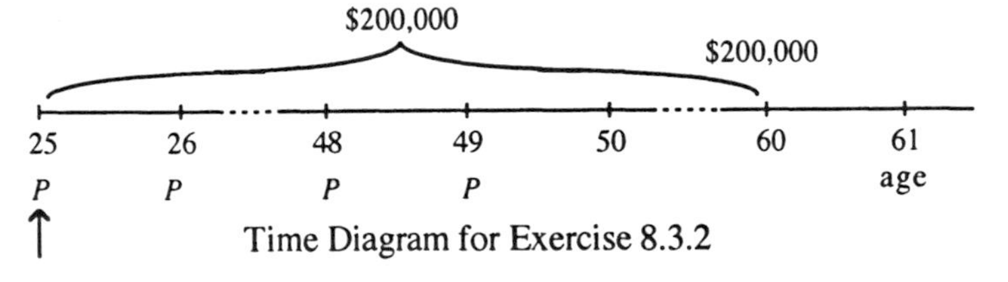

Time Diagram for Exercise 8.3.2

The equation of value is

$$P\frac{N_{25}-N_{50}}{D_{25}} = \$200,000\frac{M_{25}-M_{60}+D_{60}}{D_{25}}.$$

Solving for P gives

$$P = \$200,000\frac{M_{25}-M_{60}+D_{60}}{N_{25}-N_{50}} = \$3,510.17.$$

8.3.3 Adelaida's net single premium looks like this:

$$NSP = I = \$1,000,000\frac{M_{25}(i)-M_{26}(i)}{D_{25}(i)},$$

for i = .02, .04, .06, and .08. Using Table IV, here are the results in a table:

i	2%	4%	6%	8%
NSP(i)	\$1,137.25	\$1,115.38	\$1,094.33	\$1,074.07

8.3.4 Pierre's net single premium looks like Adelaida's, but we use Table III:

$$NSP = I = \$1,000,000\frac{M_{25}(i)-M_{26}(i)}{D_{25}(i)},$$

for i = .02, .04, .06, and .08. Using Table III, here are the results in a table:

i	2%	4%	6%	8%
NSP(i)	\$1,735.34	\$1,701.97	\$1,669.86	\$1,638.93

8.3.5 Eva's equation of value is

$$\$1,000\frac{N_{25}-N_{45}}{D_{25}} = B\frac{M_{25}-M_{55}}{D_{25}},$$

which, when solved for B yields

$$B = \$1,000\frac{N_{25}-N_{45}}{M_{25}-M_{55}} = \$380,463.21.$$

8.3.6 We can write an equation of value:

$$\$1,200\frac{N_{35}-N_{65}}{D_{35}} = \frac{\$100,000(M_{35}-M_{65})+B\cdot D_{65}}{D_{35}}.$$

Solving this equation for B gives Sarafino's endowment:

$$B = \frac{\$1{,}200(N_{35} - N_{65}) - \$100{,}000(M_{35} - M_{65})}{D_{65}} = \$67{,}970.16.$$

8.3.7 A time diagram and equation of value for Elliot's term insurance and life annuity are given below:

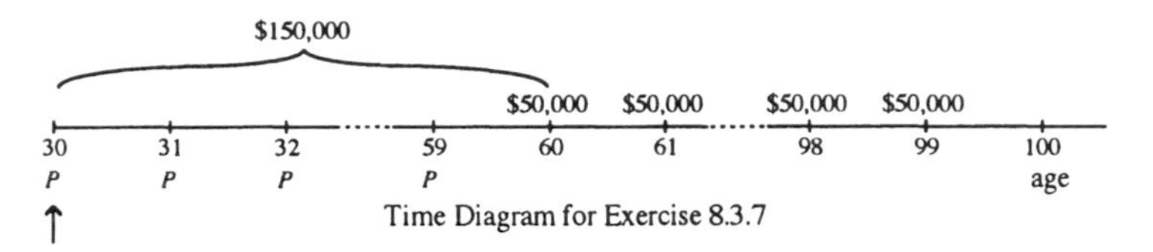

Time Diagram for Exercise 8.3.7

$$P\frac{N_{30} - N_{60}}{D_{30}} = \$150{,}000\frac{M_{30} - M_{60}}{D_{30}} + \$50{,}000\frac{N_{60}}{D_{30}},$$

which implies that Elliot's 30 annual premiums will be

$$P = \frac{\$150{,}000(M_{30} - M_{60}) + \$50{,}000N_{60}}{N_{30} - N_{60}} = \$6{,}866.17.$$

8.3.8 a) Kai's net single premium is

$$NSP = I = \$200{,}000\frac{M_{27}}{D_{27}} = \$26{,}631.42.$$

b) Kai's net annual premium is the solution for P of the equation of value

$$P\frac{N_{27} - N_{60}}{D_{27}} = \$200{,}000\frac{M_{27}}{D_{27}}.$$

Using Table IV, we get

$$P = \$200{,}000\frac{M_{27}}{N_{27} - N_{60}} = \$1{,}536.70.$$

8.3.9 The net single premium is

$$NSP = I = \$100{,}000\frac{M_x - M_{x+1}}{D_x};$$

for $x = 30, 40, 50, 60, 70, 80$, and 90. See the Table below:

	Net Single	*Net Single*
	Premium	***Premium***
Age	***(Males)***	***(Females)***
30	$164.76	$128.57
40	$287.62	$230.48
50	$639.05	$472.38
60	$1,531.43	$901.90
70	$3,762.86	$2,105.71
80	$9,413.34	$6,284.76
90	$21,120.89	$18,166.69

Net Single Premiums at 5% for Males & Females, Ages 30 to 90 In Steps of 10

8.3.10 The equation of value for a $100,000 whole-life policy for (50) is

$$P\frac{N_{50}}{D_{50}} = \$100{,}000\frac{M_{50}}{D_{50}},$$

which has the solution

$$NAP = P = \$100{,}000\frac{M_{50}}{N_{50}}.$$

This is evaluated for interest rates 2% through 8% in the table below:

%	*Net Annual Premium (Males)*	*Net Annual Premium (Females)*
2	$3,104.44	$2,548.81
3	$2,795.21	$2,240.14
4	$2,527.18	$1,977.98
5	$2,295.29	$1,756.27
6	$2,094.87	$1,569.30
7	$1,921.66	$1,411.89
8	$1,771.85	$1,279.40

Net Annual Premiums for 50-Year Old Males & Females, i=2% to 8%

8.3.11 The equation of value for the endowment insurance is

$$\$27{,}500 = B\frac{M_{27} - M_{60} + D_{60}}{D_{27}} = B(.23599997533),$$

so $B = \$116{,}525.44$.

Exercises 8.4

8.4.1 Melinda's equal annual premiums from age 35 to aged 59 have present value

$$\$950\frac{N_{35}(.065) - N_{60}(.065)}{D_{35}(.065)} = \$12{,}028.4731781.$$

Using Formula 8.8 with inputs $a = c = x = 35$, B (to be determined), $g = female$, $k = .0390244$, $i = .065$, $i^* = (i - k)/(1 + k) = .025$, and $b = 100$, the present value of her increasing insurance is

$$I^{(.0390...)} = B\frac{M_{35}(.025)}{D_{35}(.025)}\frac{D_{35}(.065)}{D_{35}(.065)} = B(.3614348\ldots).$$

Setting the two present values equal and solving for B, we find $B = \$33{,}279.78$.

8.4.2 For the net single premium, we want to use Formula 8.8 with inputs $a = c = x = 40$, $B = \$125{,}000$, $g = male$, $k = .03$, $i = .055$, $i^* = (i - k)/(1 + k) = .02427184466$, and $b = 65$. The net single premium is

$$I^{(.03)} = \$125{,}000\frac{M_{40}(.02427\ldots) - M_{65}(.02427\ldots)}{D_{40}(.02427\ldots)}\frac{D_{40}(.055)}{D_{40}(.055)} =$$

$$\$125{,}000\frac{M_{40}(.02427\ldots) - M_{65}(.02427\ldots)}{D_{40}(.02427\ldots)}.$$

$i^* = .02427184466$ is 70.87...% of the way from .0225 to .025, so we will take as our approximation, the amount 70.87...% of the way from $I^{(k)}$ evaluated at .0225 to $I^{(k)}$ evaluated at .025. At $i^* = .0225$, we get $I^{(k)} = \$19{,}028.5873599$. At $i^* = .025$, $I^{(k)} = \$18{,}315.6281145$. The value 70.87...% from the former to the latter is $I^{(k)} = \$18{,}523.29$.

We also need Lonny's net annual premium. Its present value is

$$P\frac{N_{40}(.055)-N_{65}(.055)}{D_{40}(.055)} = P(13.4198825395),$$

giving $P =$ \$18,523.29/13.4198825395 = \$1,380.29.

8.4.3 The time diagram below suggests that we write Peter's net single premium as

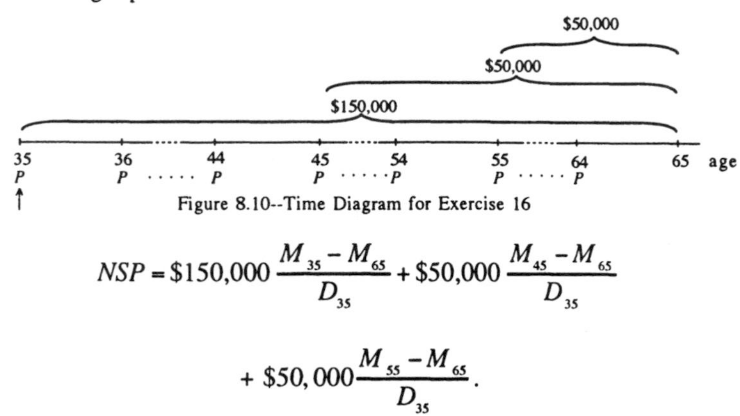

Figure 8.10--Time Diagram for Exercise 16

$$NSP = \$150,000\frac{M_{35}-M_{65}}{D_{35}} + \$50,000\frac{M_{45}-M_{65}}{D_{35}}$$

$$+\ \$50,000\frac{M_{55}-M_{65}}{D_{35}}.$$

We can rewrite this as

$$\frac{\$150,000M_{35} + \$50,000M_{45} + \$50,000M_{55} - \$250,000M_{65}}{D_{35}},$$

We do not need to go on because this last expression matches the expression we found in Example 8.14.

8.4.4 a) Syed's benefit if he died at 79 would be

$$\$50,000(1.04411764706)^{50} = \$432,954.05.$$

b) Noting that $i^* = (.065 - .04411764706)/1.04411765 = .02$, and using Formula 8.8, Syed's net single premium would be

$$NSP = I^{(.0441...)} = \$50,000\frac{M_{30}(.02)}{D_{30}(02)}\frac{D_{30}(.065)}{D_{30}(065)} =$$

$$\$50,000\frac{M_{30}(.02)}{D_{30}(02)} = \$21,805.09.$$

c) Syed's net annual premium P satisfies

$$P\frac{N_{30}(.065)}{D_{30}(.065)} = \$21{,}805.09\,.$$

This leads to

$$P = \$21{,}805.09\frac{D_{30}(.065)}{N_{30}(.065)} = \$1{,}472.62.$$

8.4.5 a) The net single premium for the fixed benefit of \$50,000 is

$$NSP_1 = \$50{,}000\frac{M_{25}(.045)}{D_{25}(.045)} = \$7{,}411.687.$$

The net single premium for the decreasing benefit requires Formula 8.8 with $i^* = (.045 + .027904)/(1 - .027904) = .075$. This is

$$NSP_2 = \$50{,}000\frac{M_{25}(.075) - M_{55}(.075)}{D_{25}(.075)}\frac{D_{25}(.045)}{D_{25}(.045)}$$

$$= \$1{,}546.826.$$

Lars's net single premium is \$7,411.687 + \$1,546.826 = \$8,958.51.

b) We set $\$8{,}958.51 = P\dfrac{N_{25}(.045) - N_{55}(.045)}{D_{25}(.045)}$ and solve for P, getting $P = \$540.40$.

8.4.6 The net single premium for the insurance evaluated at age a is

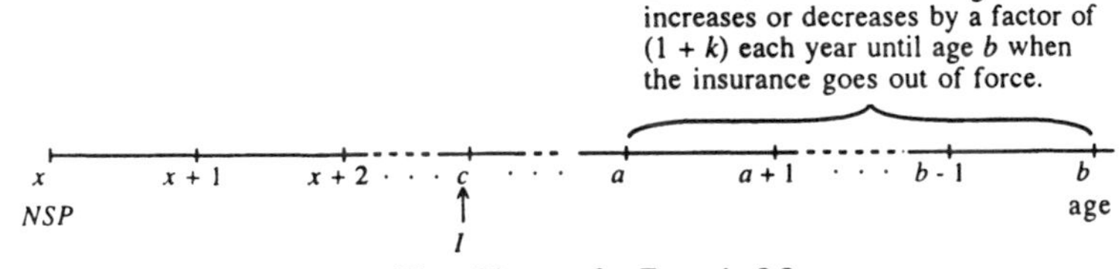

Time Diagram for Formula 8.8

$$B\{{}_{0|}q_a(1+k)(1+i)^{-1} + {}_{1|}q_a(1+k)^2(1+i)^{-2} + \cdots$$

$$+ {}_{b-a-1|}q_a(1+k)^{b-a}(1+i)^{-(b-a)}\}\,.$$

As before I have written the net single premium as a weighted average. In this case it is the weighted average of the benefits,

$B(1+k), B(1+k)^2, B(1+k)^3, \ldots, B(1+k)^{b-a}$ with weights ${}_{t|}q_a$, $t = 0, 1, \ldots, b-a-1$.

Now, replace each factor of $(1+k)(1+i)^{-1}$ by $(1+i^*)^{-1}$, write each ${}_{t|}q_a$ as $\frac{d_{a+t}}{l_a}$, and gather terms to get:

$$B\frac{d_a(1+i^*)^{-1} + d_{a+1}(1+i^*)^{-2} + \ldots + d_{b-1}(1+i^*)^{-(b-a)}}{l_a} =$$

$$B\frac{d_a(1+i^*)^{-(a+1)} + d_{a+1}(1+i^*)^{-(a+2)} + \ldots + d_{b-1}(1+i^*)^{-b}}{(1+i^*)^{-a}l_a} =$$

$$B\frac{M_a(i^*) - M_b(i^*)}{D_a(i^*)}.$$

Here, I have used the definitions of the commutation functions M and D. To evaluate this insurance at age c we multiply by the accumulation/discount factor $D_a(i)/D_c(i)$ to get

$$I^{(k)} = B\frac{M_a(i^*) - M_b(i^*)}{D_a(i^*)}\frac{D_a(i)}{D_c(i)}.$$

8.4.7 From the derivation in the preceding exercise, we see that when $i^* = 0$ we can write

$$I^{(k)} = B\frac{M_a(i^*) - M_b(i^*)}{D_a(i^*)}\frac{D_a(i)}{D_c(i)} =$$

$$B\left(\frac{d_a + d_{a+1} + \ldots + d_{b-1}}{l_a}\right)\left(\frac{D_a(i)}{D_c(i)}\right) = B\left(\frac{l_a - l_b}{l_a}\right)\left(\frac{D_a(i)}{D_c(i)}\right).$$

When $a = c$, the second factor is equal to 1 and need not be written.

8.4.8 We have $i^* = 0$, $a = c = 35$, $b = 55$, and $B = \$250{,}000$. With the aid of the result of the previous exercise, we find Adelle's net single premium to be

$$I^{(04)} = \$250{,}000\frac{l_{35} - l_{55}}{l_{35}} = \$17{,}636.04.$$

Adelle's equal annual payments of P satisfy the equation

$$P\frac{N_{35}(.04)-N_{55}(.04)}{D_{35}(.04)} = P(13.7469115311) = \$17,636.04,$$

so that $P = \$1,282.91$.

Miscellaneous Exercises – Chapter 8

8M1. With the help of the time diagram below, we see that the present value of the insurance at age a is

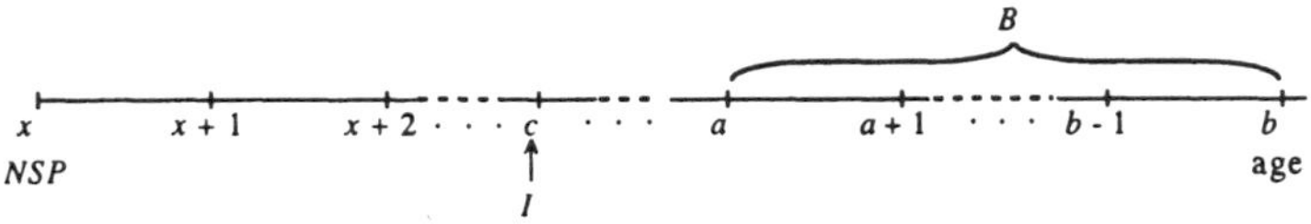

B

x x + 1 x + 2 · · · c · · · a a + 1 · · · b - 1 b

NSP I age

Time Diagram for Formula 8.3

$$B\{{}_{0|}q_a(1+i)^{-1} + {}_{1|}q_a(1+i)^{-2} + \ldots + {}_{b-a-1|}q_a(1+i)^{-(b-a)}\} =$$

$$B\frac{d_a(1+i)^{-1} + d_{a+1}(1+i)^{-2} + \ldots + d_{b-1}(1+i)^{-(b-a)}}{l_a} =$$

$$B\frac{d_a(1+i)^{-(a+1)} + d_{a+1}(1+i)^{-(a+2)} + \ldots + d_{b-1}(1+i)^{-b}}{l_a(1+i)^{-a}} =$$

$$B\frac{M_a - M_b}{D_a}.$$

To find the value the insurance at age c, we accumulate or discount the previous value according to whether $c < a$ or $c > a$. We do this using Formula 8.4. In this case, the factor is D_a/D_c. The value of the insurance at age c is therefore

$$I = B\frac{M_a - M_b}{D_a}\frac{D_a}{D_c} = B\frac{M_a - M_b}{D_c}.$$

If $b = 100$, $M_b = 0$ and this gives the reduction indicated in Formula 8.3.

8M2. The net single premium may be written $NSP =$

$$\$300,000\frac{M_{30} - M_{40}}{D_{30}} + \$150,000\frac{M_{40} - M_{55}}{D_{30}} + \$75,000\frac{M_{55}}{D_{30}} =$$

$$\frac{\$300,000M_{30} - \$150,000M_{40} - \$75,000M_{55}}{D_{30}} =$$

$$\$1,000\frac{300M_{30} - 150M_{40} - \$75M_{55}}{D_{30}}.$$

8M3. With $i^* = (.075 - .043689\ldots)/1.043689\ldots = .03$, the equation of value is

$$\$8,000\frac{N_{30}(.03) - N_{60}(.03)}{D_{30}(.03)}\frac{D_{30}(.075)}{D_{30}(.075)} =$$

$$\$250,000\frac{M_{30}(.03) - M_{60}(.03)}{D_{30}(.03)}\frac{D_{30}(.075)}{D_{30}(.075)} +$$

$$P\frac{N_{60}(.03)}{D_{60}(.03)}\frac{D_{60}(.075)}{D_{30}(.075)}.$$

Solving for P, we get $P = \$102,635.77$.

8M4. (🖳) The net single premium on all three planets takes the form

$$NSP = \$100,000\frac{M_{35}}{D_{35}}.$$

For the earthling, we can use Table IV with $i = 5\%$ to get

$$NSP_{Earth} = \$15,210.76.$$

For the other two we need to use the definitions of M and D with their particular survival functions. To evaluate the premiums you need a spreadsheet program or some other program able to evaluate long sums. I have done this with a spreadsheet and gotten

$$NSP_{Q0317B} = \$32,500.12,$$

and

$$NSP_{Q0317C} = \$21,245.38.$$

Chapter 9

Exercises 9.2

9.2.1 A time diagram will help to determine the annual premium:

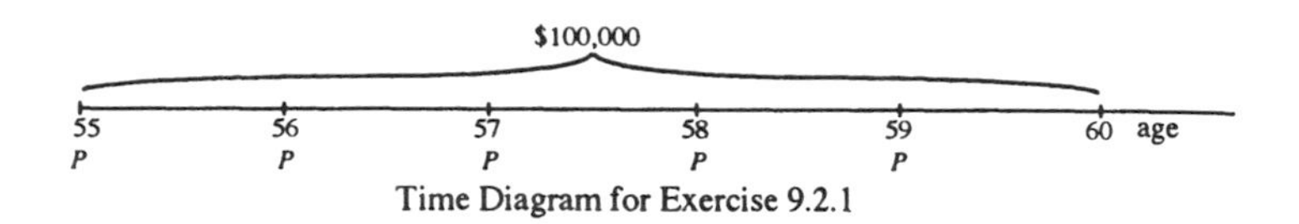

Time Diagram for Exercise 9.2.1

Using Formulas 7.6 and 8.3, the net annual premium satisfies the equation

$$P\frac{N_{55}-N_{60}}{D_{55}} = \$100,000\frac{M_{55}-M_{60}}{D_{55}},$$

so

$$P = \$100,000\frac{M_{55}-M_{60}}{N_{55}-N_{60}} = \$1,193.3912241002$$

$$\approx \$1,193.39.$$

Following the pattern of Table 9.2, and using Table III with $i =$ 4.25%, we get the following table:

Yr	*Premiums*	*Beginning-of Year Fund*	*Fund Plus Interest*	*No. of Deaths*	*End-of-Year fund*	*No. of Survivors*	*Reserve/ Policy*
1	10,276,936,262	10,276,936,262	10,713,706,053	90,163	1,697,406,053	8,521,377	$199.19
2	10,169,336,529	11,866,742,582	12,371,079,142	97,655	2,605,579,142	8,423,722	$309.31
3	10,052,795,909	12,658,375,051	13,196,355,991	105,212	2,675,155,991	8,318,510	$321.59
4	9,927,236,832	12,602,392,822	13,137,994,517	113,049	1,833,094,517	8,205,461	$223.40
5	9,792,325,147	11,625,419,664	12,119,500,000	121,195	0	8,084,266	$0.00

Table for Exercise 9.2.1

Notice that the reserve per policy at the end of the fifth year is $0, as it should be.

9.2.2 The time diagram for this exercise differs from the previous one only by the payment of $100,000 to the insured, if he survives to age 60.

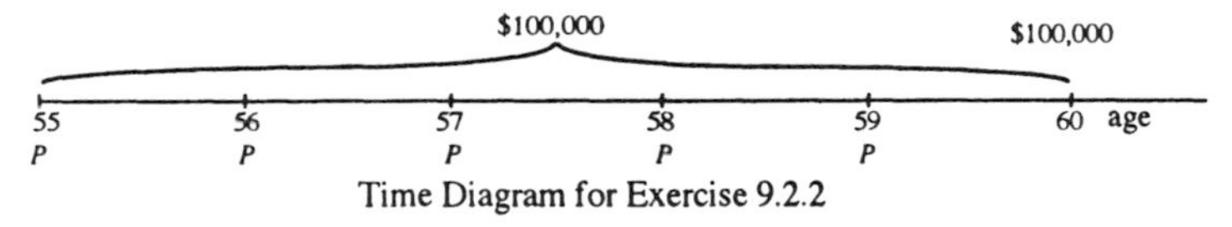

Time Diagram for Exercise 9.2.2

The equation for the net annual premium is now

$$P\frac{N_{55}-N_{60}}{D_{55}} = \$100,000\frac{M_{55}-M_{60}+D_{60}}{D_{55}},$$

so

$$P = \$100,000\frac{M_{55}-M_{60}+D_{60}}{N_{55}-N_{60}} = \$18,103.379725117 \approx \$18,103.38.$$

Yr	Premiums	Beginning-of Year Fund	Fund Plus Interest	No. of Deaths	End-of-Year fund	No. of Survivors	Reserve/ Policy
1	155,897,978,638	155,897,978,638	162,523,642,730	90,163	153,507,342,730	8,521,377	$18,014.38
2	154,265,723,612	307,773,066,342	320,853,421,662	97,655	311,087,921,662	8,423,722	$36,929.98
3	152,497,838,065	463,585,759,726	483,288,154,515	105,212	472,766,954,515	8,318,510	$56,833.13
4	150,593,145,277	623,360,099,792	649,852,904,033	113,049	638,548,004,033	8,205,461	$77,819.88
5	148,546,576,303	787,094,580,336	820,546,100,000	121,195	808,426,600,000	8,084,266	$100,000.00

Table for Exercise 9.2.2

Notice that the reserve per policy in the fifth year is $100,000—sufficient to pay the endowment.

9.2.3 We already know from Example 9.2 that Martina's net annual premium is P = $7,067.94464105. With the help of the time diagram in Figure 9.1, and thinking retrospectively, we can write the tth reserve as

$$R_t = \begin{cases} P\dfrac{N_{30}-N_{30+t}}{D_{30+t}} - \$100,000\dfrac{M_{30}-M_{30+t}}{D_{30+t}}, & 1 \le t \le 5 \\[2ex] P\dfrac{N_{30}-N_{35}}{D_{30+t}} - \$100,000\dfrac{M_{30}-M_{30+t}}{D_{30+t}}, & 6 \le t \le 10 \end{cases}$$

I used a spreadsheet program to generate a table and found a table identical to Table 9.3.

9.2.4 In Example 9.5 we found P = $21,195.7247920833 ≈ $21,195.72 . Using Figure 9.2 as a guide, we can write the prospective reserves as

$$R_t = \begin{cases} \$25,000\dfrac{N_{65}-N_{71}}{D_{60+t}} - P\dfrac{N_{60+t}-N_{65}}{D_{60+t}}, & 1 \le t \le 4 \\[2ex] \$25,000\dfrac{N_{60+t}-N_{71}}{D_{60+t}}, & 5 \le t \le 10 \end{cases}$$

Again, I used a spreadsheet program, and I duplicated Table 9.4.

9.2.5 Here is a time diagram for this policy:

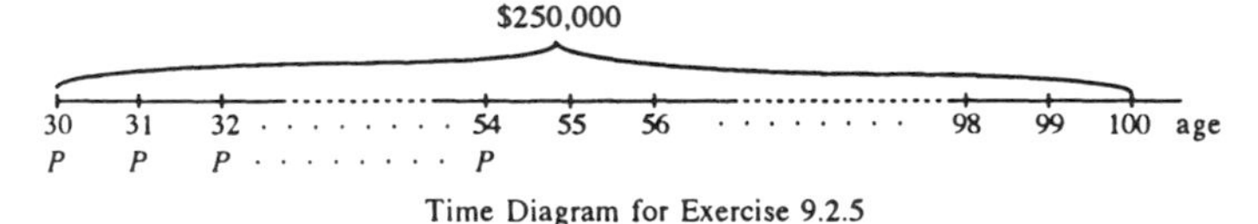

Time Diagram for Exercise 9.2.5

We calculate her annual premium with i = 6% from the equation

$$P\frac{N_{30}-N_{55}}{D_{30}} = \$250,000\frac{M_{30}}{D_{30}},$$

and get

$$P = \$250,000\frac{M_{30}}{N_{30}-N_{55}} = \$1,630.71133885$$

$$\approx \$1,630.71.$$

Now, looking at the time diagram, and thinking retrospectively, we get

$$R_{15} = P\frac{N_{30}-N_{45}}{D_{45}} - \$250,000\frac{M_{30}-M_{45}}{D_{45}}$$

$$= \$29,650.06.$$

9.2.6 Using the time diagram and payment amount found above, we can write the 15th reserve prospectively as

$$R_{15} = \$250,000\frac{M_{45}}{D_{45}} - P\frac{N_{45}-N_{55}}{D_{45}} = \$29,650.06,$$

as we found retrospectively.

9.2.7 Here is a time diagram:

$75,000 $75,000 ... $75,000

41 42 53 54 55 59 60 61 99 100 age

P P .... P P

Time Diagram for Exercise 9.2.7

The equation of value giving us his annual payments is

$$P\frac{N_{40}-N_{55}}{D_{40}} = \$75,000\frac{N_{60}}{D_{40}},$$

so, at i = 5.5%,

$$P = \$75{,}000\frac{N_{60}}{N_{40} - N_{55}} = \$23{,}689.06.$$

With the help of the time diagram, and thinking retrospectively, we write

$$R_{10} = P\frac{N_{40} - N_{50}}{D_{50}} = \$331{,}407.97,$$

and

$$R_{30} = P\frac{N_{40} - N_{55}}{D_{70}} - \$75{,}000\frac{N_{60} - N_{70}}{D_{70}} = \$612{,}034.18.$$

9.2.8 We can use the same time diagram and value of P as in Exercise 9.2.7. This time, we think prospectively, and get

$$R_{10} = \$75{,}000\frac{N_{60}}{D_{50}} - P\frac{N_{50} - N_{55}}{D_{50}} = \$331{,}407.97,$$

and

$$R_{30} = \$75{,}000\frac{N_{70}}{D_{70}} = \$612{,}034.18,$$

just as we got retrospectively.

9.2.9 Below, we write both RR_t and PR_t the tth retrospective and prospective reserves:

$$RR_t = P\frac{N_x - N_{x+t}}{D_{x+t}} - B\frac{M_x - M_{x+t}}{D_{x+t}},$$

and

$$PR_t = B\frac{M_{x+t}}{D_{x+t}} - P\frac{N_{x+t}}{D_{x+t}},$$

where $P = B \cdot \frac{M_x}{N_x}$. This means that we need to show that

$$B \cdot \frac{M_x}{N_x} \cdot \frac{N_x - N_{x+t}}{D_{x+t}} - B\frac{M_x - M_{x+t}}{D_{x+t}} = B\frac{M_{x+t}}{D_{x+t}} - B \cdot \frac{M_x}{N_x} \cdot \frac{N_{x+t}}{D_{x+t}}.$$

Transposing gives

$$B \cdot \frac{M_x}{N_x} \cdot \frac{N_x - N_{x+t}}{D_{x+t}} + B \cdot \frac{M_x}{N_x} \cdot \frac{N_{x+t}}{D_{x+t}} = B\frac{M_{x+t}}{D_{x+t}} + B\frac{M_x - M_{x+t}}{D_{x+t}},$$

or

$$\frac{B\cdot M_x\cdot N_x - B\cdot M_x\cdot N_{x+t} + B\cdot M_x\cdot N_{x+t}}{N_x\cdot D_{x+t}}$$

$$= \frac{B\cdot M_{x+t} + B\cdot M_x - B\cdot M_{x+t}}{D_{x+t}}.$$

Canceling where we can gives

$$\frac{B\cdot M_x}{D_{x+t}} = \frac{B\cdot M_x}{D_{x+t}},$$

Which is certainly true, so $RR_t = PR_t$.

9.2.10 What we need to show is

$$RR_t = VPP_t - VPB_t = VFB_t - VFP_t = PR_t,$$

or, after transposing,

$$VPP_t + VFP_t = VFB_t + VPB_t.$$

What this says in its present form is that the value of all premiums is equal to the value of all benefits t years after issue. Then, discounting t years with interest and survivorship gives

$$(VPP_t + VFP_t)\cdot {}_tE_x = (VFB_t + VPB_t)\cdot {}_tE_x,$$

which states that the value of all premiums is equal to the value of all benefits at issue. This is the basic principle of equations of value.

Exercises 9.3

9.3.1 The time diagrams for the net and gross transactions, respectively, look like this:

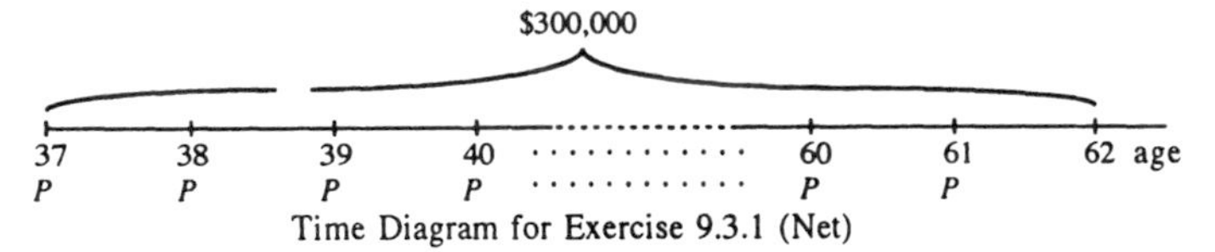

Time Diagram for Exercise 9.3.1 (Net)

$300,000(1.015)

0.50G 0.25G 0.08G 0.08G 0.08G 0.08G

37 38 39 40 .... 60 61 62 age

G G G G .... G G

Time Diagram for Exercise 9.3.1 (Gross)

The equation of value for the net transaction is

$$P\frac{N_{37}-N_{62}}{D_{37}} = \$300,000\frac{M_{37}-M_{62}}{D_{37}},$$

or

$$P = \$300,000\frac{M_{37}-M_{62}}{N_{37}-N_{62}} = \$1,890.56.$$

On the other hand, the equation of value for the gross transaction can be written

$$G\frac{N_{37}-N_{62}}{D_{37}} = \$300,000(1.015)\frac{M_{37}-M_{62}}{D_{37}} + 0.42G +$$

$$0.17G\frac{N_{38}-N_{39}}{D_{37}} + 0.08G\frac{N_{37}-N_{62}}{D_{37}}.$$

The solution for G is

$$G = \frac{\$304,500(M_{37}-M_{62})}{0.92(N_{37}-N_{62}) - 0.42D_{37} - 0.17(N_{38}-N_{39})} = \$2,166.40.$$

9.3.2 a)

$85,000
$250
0.55G 0.07G 0.07G 0.07G
25 26 27 54 55 56 . 98 99 100 age
G G G G

Time Diagram for Exercise 9.3.2a

With the help of the time diagram above, we can write an equation of value:

$$G\frac{N_{25}-N_{55}}{D_{25}} = \$85,000\frac{M_{25}}{D_{25}} + \$250 + 0.48G + 0.07G\frac{N_{25}-N_{55}}{D_{25}}.$$

Solving for G, we get

$$G = \frac{\$85,000M_{25} + \$250D_{25}}{0.93(N_{25}-N_{55}) - 0.48D_{25}} = \$1,208.17.$$

b) Let's look at a time diagram for the net annual premium:

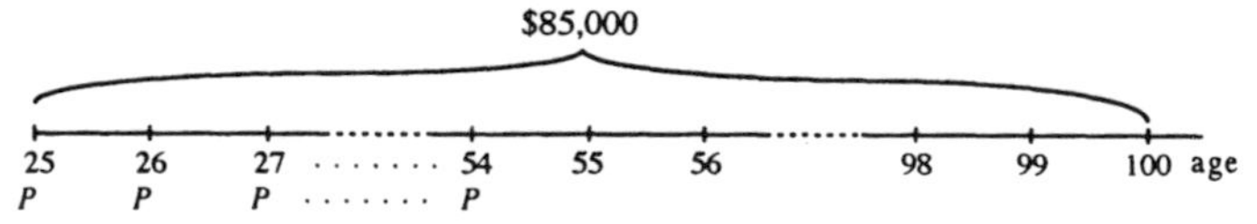

Time Diagram for Exercise 9.3.2b

The equation of value is

$$P\frac{N_{25}-N_{55}}{D_{25}} = \$85{,}000\frac{M_{25}}{D_{25}},$$

so that

$$P = \$85{,}000\frac{M_{25}}{N_{25}-N_{55}} = \$1{,}082.78.$$

9.3.3 Here is a time diagram for Devean's policy:

\$1,000,000(1.025)

0.40G
\$550
0.035G 0.035G 0.035G 0.035G 0.035G 0.035G 0.035G 0.035G 0.035G 0.035G
25 26 27 28 29 30 31 32 33 34 35
G G G G G G G G G G age

Time Diagram Illustrating Exercise 9.3.3

From the diagram we can write the equation of value.

$$G\frac{N_{25}-N_{35}}{D_{25}} = \$1{,}000{,}000(1.025)\frac{M_{25}-M_{35}}{D_{25}} + 0.40G + \$550 +$$

$$+\ 0.035G\frac{N_{25}-N_{35}}{D_{25}}.$$

We get

$$G = \frac{\$1{,}025{,}000(M_{25}-M_{35}) + \$550D_{25}}{0.965(N_{25}-N_{35}) - 0.40D_{25}} = \$1{,}982.97.$$

9.3.4 Here is a time diagram:

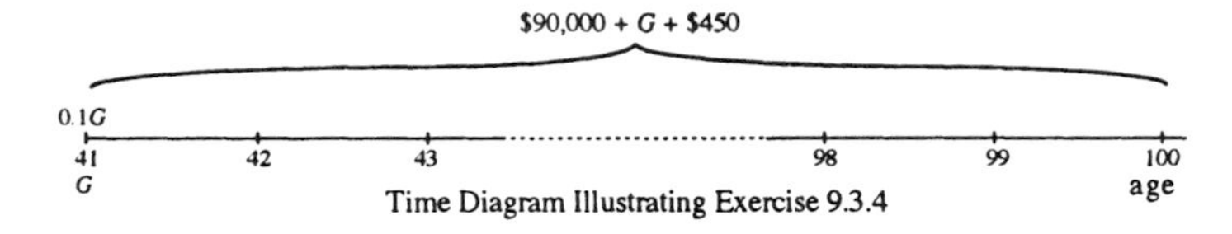

Time Diagram Illustrating Exercise 9.3.4

From this diagram we write the equation of value:

$$G = (\$90{,}450 + G)\frac{M_{41}}{D_{41}} + 0.1G.$$

When we solve this for G, we get

$$G = \frac{\$90,450M_{41}}{0.9D_{41} - M_{41}} = \$89,294.21.$$

Remark: This is a very high premium, but, remember: It is returned as part of the benefit.

9.3.5 The time diagram below should illuminate the situation:

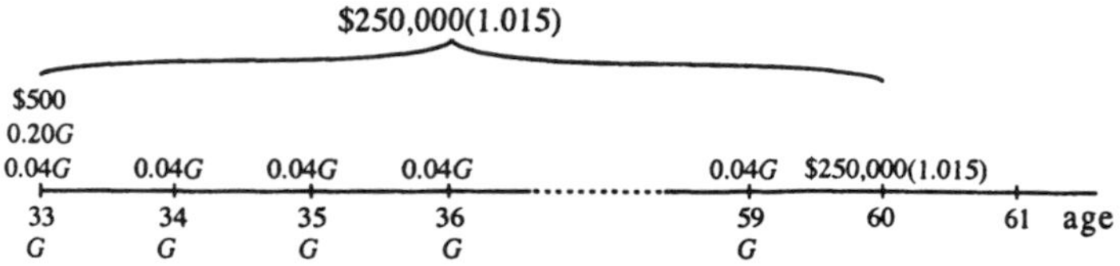

$250,000(1.015)

\$500 0.20G 0.04G	0.04G	0.04G	0.04G	....	0.04G	$250,000(1.015)		
33	34	35	36		59	60	61	age
G	G	G	G		G			

Time Diagram Illustrating Exercise 9.3.5

The equation of value can be written

$$G\frac{N_{33} - N_{60}}{D_{33}} = \$253{,}750\frac{M_{33} - M_{60} + D_{60}}{D_{33}} + \$500 +$$

$$0.20G + 0.04G\frac{N_{33} - N_{60}}{D_{33}},$$

which we solve for G to get

$$G = \frac{\$500D_{33} + \$253,750(M_{33} - M_{60} + D_{60})}{0.96(N_{33} - N_{60}) - 0.20D_{33}} = \$6,179.92.$$

9.3.6 Below is a time diagram for Mauricio's annuity:

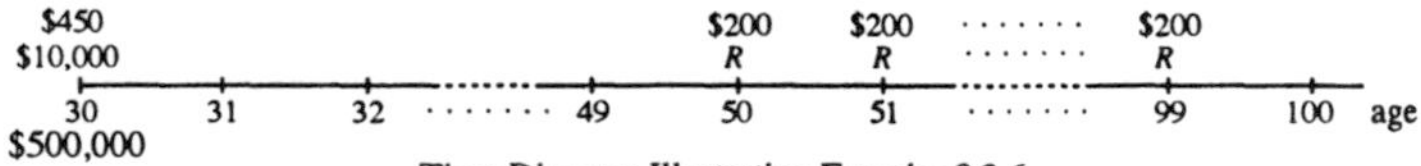

\$450 \$10,000				\$200 R	\$200 R	....	\$200 R		
30 \$500,000	31	32	49	50	51	....	99	100	age

Time Diagram Illustrating Exercise 9.3.6

Mauricio's gross premium is \$500,000, so the 2% commission is \$10,000. We let R stand for the amount of Mauricio's annuity. An equation of value for the problem is

$$\$500{,}000 = \$10{,}450 + (R + \$200)\frac{N_{50}}{D_{30}}.$$

The solution for R is

$$R = \$489,550\frac{D_{30}}{N_{50}} - \$200 = \$97,738.14.$$

9.3.7 First, let's look at a time diagram:

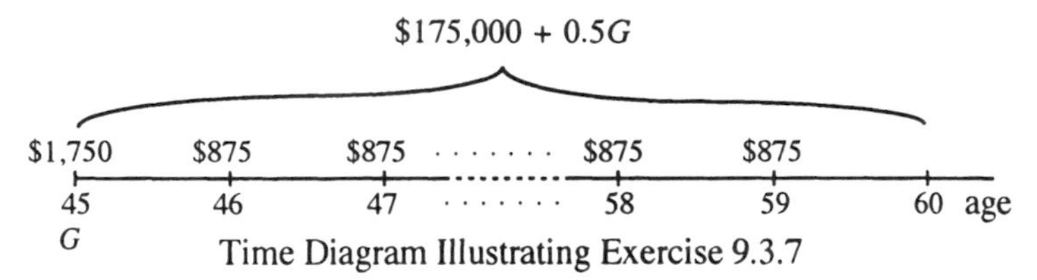

Time Diagram Illustrating Exercise 9.3.7

We can write down an equation directly:

$$G = (\$175{,}000 + 0.5G)\frac{M_{45} - M_{60}}{D_{45}} + \$875 + \$875\frac{N_{45} - N_{60}}{D_{45}},$$

and solve for G to get

$$G = \frac{\$875D_{45} + \$875(N_{45} - N_{60}) + \$175{,}000(M_{45} - M_{60})}{D_{45} - 0.5(M_{45} - M_{60})}$$

$$= \$21{,}978.01.$$

9.3.8 Consider the time diagram for Liam's policy:

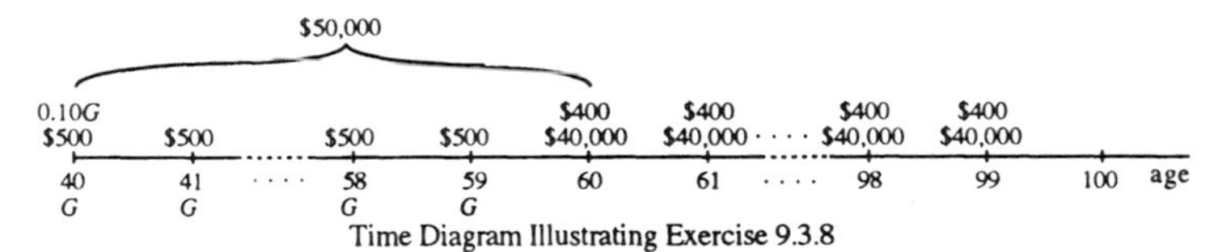

Time Diagram Illustrating Exercise 9.3.8

The equation of value may be written

$$G\frac{N_{40} - N_{60}}{D_{40}} = \$50{,}000\frac{M_{40} - M_{60}}{D_{40}} + 0.10G +$$

$$\$500\frac{N_{40} - N_{60}}{D_{40}} + \$40{,}400\frac{N_{60}}{D_{40}}.$$

We solve this equation for G and find

$$G = \frac{\$50{,}000(M_{40} - M_{60}) + \$500(N_{40} - N_{60}) + \$40{,}400N_{60}}{(N_{40} - N_{60}) - 0.1D_{40}}$$

$$= \$11{,}746.81.$$

9.3.9 Here is a time diagram for this annuity:

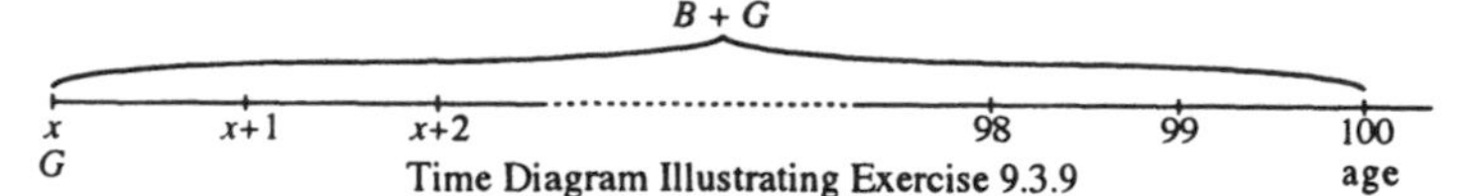

Time Diagram Illustrating Exercise 9.3.9

On the one hand, the net single premium, P, must satisfy

$$P = (B+G)\frac{M_x}{D_x}.$$

On the other hand, Formula 9.5 dictates that

$$G = (P + c)(1 + k).$$

We have two equations in two unknowns, from which we get

$$G = \frac{(BM_x + cD_x)(1+k)}{D_x - M_x(1+k)}$$

9.3.10 The equation of value for this annuity is

$$G = (R+c)\frac{N_x}{D_x} + pG\ ,$$

so

$$G = \frac{(R+c)N_x}{(1-p)D_x}.$$

9.3.11 We will use a time diagram here. In the diagram, we let j = 0.02926829268, and k = 0.03178484108, giving rise to the two values i_1^* = (0.055 - 0.02926829268)/1.02926829268 = 0.025, and i_2^* = (0.055 - 0.03178484108)/1.03178484108 = 0.0225 needed as inputs to Formula 7.7.

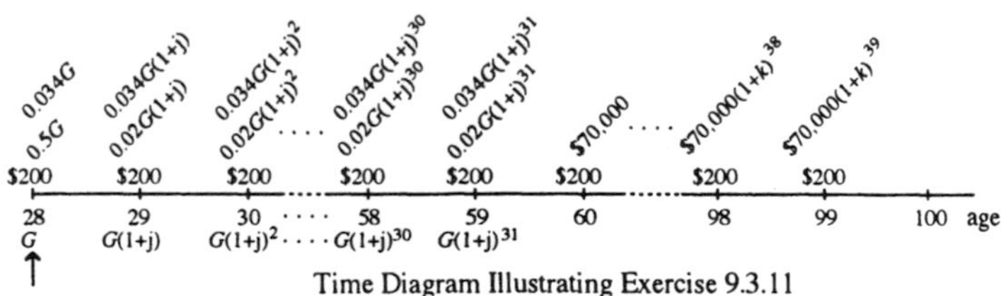

Time Diagram Illustrating Exercise 9.3.11

From the time diagram, and with the help of Formulas 7.6 and 7.7, we can deduce the equation of value:

$$G\frac{N_{28}(0.025)-N_{60}(0.025)}{D_{28}(0.025)}=\$200\frac{N_{28}(0.055)}{D_{28}(0.055)}+0.48G+$$

$$0.054G\frac{N_{28}(0.025)-N_{60}(0.025)}{D_{28}(0.025)}$$

$$+\ \$70,000\frac{N_{60}(0.0225)}{D_{60}(0.0225)}\frac{D_{60}(0.055)}{D_{28}(0.055)}.$$

Solved for G, we get

$$G=\frac{\$200\dfrac{N_{28}(0.055)}{D_{28}(0.055)}+\$70,000\dfrac{N_{60}(0.0225)D_{60}(0.055)}{D_{60}(0.0225)D_{28}(0.055)}}{0.946\dfrac{N_{28}(0.025)-N_{60}(0.025)}{D_{28}(0.025)}-0.48}$$

$$=\$6,361.38.$$

In a similar manner, we find Edsger's net annual premium to be

$$G=\$70,000\frac{\dfrac{N_{60}(0.0225)D_{60}(0.055)}{D_{60}(0.0225)D_{28}(0.055)}}{\dfrac{N_{28}(0.025)-N_{60}(0.025)}{D_{28}(0.025)}}=\$5,775.01.$$

Bibliography

Ayres, F. Jr., *Mathematics of Finance*, New York: McGraw-Hill, 1963.

Bowers, N. L., Gerber, H. U., Hickman, J. C., Jones, D. A., and Nesbitt, C. J., *Actuarial Mathematics*, 2nd ed., Schaumburg, Illinois: Society of Actuaries, 1997.

Cissell, R., Cissell, H. and Flaspohler, D. C., *Mathematics of Finance*, 8th ed., Boston: Houghton Mifflin, 1990.

Davids, L. E., *Dictionary of Insurance*, 6th ed., New Jersey: Rowan & Allanheld, 1983.

Francis, J. C., and Taylor, R. W., *Investments*, 2nd ed., New York: McGraw-Hill, 2000.

Jordan, C. W. Jr., *Life Contingencies*, 2d ed., Chicago: Society of Actuaries, 1975.

Kellison, S. G., *The Theory of Interest*, 3d ed., New York: McGraw-Hill/Irwin, 2008.

Menge, W. O., and Fischer, C. H., *The Mathematics of Life Insurance*, 2d ed., Ann Arbor: Ulrich's Books, 1965.

Neill, A., *Life Contingencies*, London: Heinemann, 1977.

Parmenter, M. M., *Theory of Interest and Life Contingencies, with Pension Applications*, 3d ed., Winsted, Connecticut: ACTEX, 1999.

Report of the Special Committee to Recommend New Mortality Tables for Valuation, *Transactions of the Society of Actuaries*, Vol. 33, (1981), pp. 617-669.

TI-83 Plus Graphing Calculator Guidebook, Texas Instruments Incorporated, 1993.

TI-84 Plus Silver Edition Guidebook, Texas Instruments Incorporated, 2005.

Zima, P. and Brown, R. L., *Mathematics of Finance*, 2d ed., New York: McGraw-Hill, 1996.

Index

Q

R

S

T

W

X

Y

About the Author

Since receiving his Ph. D. in Applied and Mathematical Statistics at Rutgers University, Kenneth Kaminsky has taught at universities in Sweden and the United States, where he is currently a professor in the Department of Mathematics at Augsburg College. Some of his mathematical and other cartoons can be seen at *www.mathcartoons.com*.

Breinigsville, PA USA
24 September 2010

245984BV00003B/1/P

9 780761 853091